MANAGING WITH STATISTICS

MANAGING
with
STATISTICS

T. W. LEWIS
R. A. FOX

OLIVER & BOYD
EDINBURGH

OLIVER & BOYD LTD
Tweeddale Court
Edinburgh

05 001976 7
First published 1969
© 1969 The Authors
All rights reserved

Set in Times Roman 327 and printed in
Great Britain by T. & A. Constable Ltd, Edinburgh

PREFACE

IT is now generally agreed that no holder of a position in the ranks of middle management is properly trained for his function as a link-man between the department and the policy-maker unless he has acquired a knowledge of statistical presentation and elementary statistical methods. His superior is expected to be at ease with both.

This is not to say that either need be a specialist. It would be helpful if in each company of any size some management staff were trained to a level which could be described as 'a good grounding' and a few others to a 'middle grade' level, which to be of any real value should include a practical knowledge of sample survey work and forecasting methods. The impediment to this aim is mainly in the minds of present and potential managers. Both are over-conscious of their slender mathematical background, and do not believe that they can understand new ideas based on abstractions.

This fear is unfounded. The most that is required is a knowledge of very elementary algebra and the type of mind that can make a satisfactory stab at a not-too-simple crossword puzzle. Fortunately some professional bodies are now including in their examinations a paper or part-paper on Statistics, and are thus dissipating the fear in the best possible way, by compelling their students to face it. Equally, older men who were secretly measuring their good fortune at having qualified before statistical pass-words like 'non-response' and 'exponential smoothing' were heard in management corridors, find themselves faced with the subject on management courses organised either by their own firm or by their industry in fulfilment of the requirements of the Industrial Training Act.

This book is intended to help all such people, whose inhibitions the authors have been forced to analyse while lecturing over a reasonably long period to many of their kind. The plan is to ask the reader to approach the learning of statistics in the same way that he might approach a study of Italian if he already knew a little painfully acquired Latin. That is, we ask him to treat the subject as an entirely new one, but to accept that such rusty mathematics as he may cull from his recollections may have a helpful bearing on his comprehension. Readers will find that the introduction to most topics is fairly lengthy, that there is some measure of repetition, and that arithmetical examples take precedence over symbolism. Algebra, when it is employed, is always simple, and we have tried to introduce it in the spirit of 'you never thought that you would find a use for that apparently arid piece of nearly forgotten information'.

In those few instances where it might present a somewhat forbidding appearance, it has been tucked away in appendices, for the attention of the more adventurous reader. Fortunately these appendices constitute extensions to the text and not an essential part of it.

It is our joint experience that many students, once they have built and then asserted their confidence, and irrespective of their age and experience, see a relevance for the subject in their own sphere of interest, and, through that, develop their knowledge to a degree unthought of at the beginning of their studies. We would be happy to think that this book will help many to do just that.

The aim is thus seen to be a very limited one. There may be some statistical sages who will blanch at the way in which we have disregarded rigour where it has stood in the way of simplicity of presentation. We would like to affirm our respect for their viewpoint and to say that we do not suggest that the purist approach should be abandoned, but only deferred to a stage which is certainly beyond the bounds of this book. It is generally more efficient to breed a race of managers who can both call for and use material which is statistically germane to their problems, than to encourage the idea that these jobs are for the expert only.

We are anxious to express our gratitude to those who have permitted us to use examples, tables and quotations from other works. In particular, we are indebted to the Literary Executor of the late Sir Ronald A. Fisher, F.R.S., to Dr Frank Yates, F.R.S., and to Oliver and Boyd Ltd., Edinburgh, for permission to reprint Table XXXIII from their book *Statistical Tables for Biological, Agricultural and Medical Research*, sixth edition, 1963; the Biometrika Trustees for permission to reproduce Extracts from Tables 1, 8, 12, 18 and 41, *Biometrika Tables for Statisticians*, Volume I, third edition, 1966; Mr C. St J. O'Herlihy and the Royal Statistical Society for permission to reproduce extracts from *Demand for Cars in Great Britain*, published in *Applied Statistics*, 1965, Volume XIV, Numbers 2 and 3; Methuen and Co. Ltd. for permission to reproduce Table 7.4 from *Analysing Qualitative Data* by A. E. Maxwell; to Mr R. E. Beard and the Institute of Actuaries for permission to reproduce extracts from *Some Statistical Problems Arising from the Transaction of Motor Insurance Business* published in the *Journal of the Institute of Actuaries Students' Society*, Volume 17, Part 4, July 1964; and *The Economist* for permission to use the pictograph in Chapter I. Our thanks are also extended to the following examining bodies for permission to reproduce some of their past examination questions: the Association of Certified and Corporate Accountants (A.C.C.A.), the Institute of Cost and Works Accountants (I.C.W.A.), the Institute of Marketing (I.O.M.), and the Institute of Statisticians (I.O.S.). Where we have suggested answers to these questions, they are, of course, our own solutions.

Our thanks are also due to Mr James Elmslie, who has given valuable help and advice with the proofs.

Finally, it would be churlish not to mention the indefatigable efforts of our wives in undertaking most of the typing and much of the proof reading, as well as their patience in accepting, for a long period, households which have been disrupted if not disorderly.

CONTENTS

Introduction. *Reliability of data:* Rounding of figures ... Errors. Frequency distributions. Charts and diagrams. *Graphs:* Discrete and continuous variables ... Plotting a graph ... Logarithmic or ratio scale. Cumulative frequency curves ... The Lorenz curve.

Introduction. Elementary symbolism. *Measures of central location:* The arithmetic mean ... The advantages and disadvantages of the use of the arithmetic mean ... The properties of the arithmetic mean ... The median ... Calculating the median of a grouped frequency distribution ... The mode ... The geometric mean ... The harmonic mean. *Dispersion:* Mean deviation ... The inter-quartile range ... Variance and standard deviation ... The pooled variance of two or more distributions.

Basic ideas. *Problems in the construction of index numbers:* The purpose of the index ... The availability and comparability of data ... Selection of the items to be included in the index ... Choice of the base period ... Choice of the weights. *The construction of index numbers:* Base weighted index ... Current weighted index ... Other fixed weights ... Quantity indices ... Value indices ... Relation between price, quantity and value indices ... Fixed and moving weight indices. *The continuity of index numbers:* Splicing ... Chain index numbers ... Changing the base. *Some uses of index numbers:* Adjusting for price increases ... Standardization ... Problems in interpreting index numbers.

APPENDIX. The algebra of index numbers.

Introduction. Secular or long-term trend. Cyclical movements. Seasonal variations. Random movements. *Determining the trend line:* Least squares ... Moving average. Seasonal variation.

APPENDIX. Derivation of the Normal equations.

LIST OF IMPORTANT TABLES

ABBREVIATIONS FOR EXAMINING BODIES
WHOSE QUESTIONS APPEAR IN THE EXERCISES

I.C.W.A.   Institute of Cost and Works Accountants.
A.C.C.A.   Association of Certified and Corporate Accountants.
I.O.M.   Institute of Marketing.
I.O.S.   Institute of Statisticians.

1: GETTING THE FACTS STRAIGHT

The Tabulation and Presentation of Numerical Information

INTRODUCTION

THE debate over the extent to which management can be taught is subsiding without either side losing face. Management, it is now said, presents two aspects, one human and the other numerical. Within the boundaries of the former come subjects such as management structure, the relations with trade unions, and industrial psychology. This book deals with the numerical information which an efficient management should demand from its subordinate officers and functional specialists, and advises management how to make the best use of the information when it has been produced.

'Statistics' is a contraction of the words 'state arithmetic'. It has its origin in the economic data supplied to the British administration during the Napoleonic Wars and the Continental Blockade. There is little point in formulating a definition of the subject because its boundaries are no longer clear, but broadly, statistics is two things. First, it is any set of numerical facts which has been systematically collected on a subject. Bowley says, 'Statistics are numerical statements of facts in any department of enquiry . . .', while Yule refers to statistics as 'Quantative data affected to a marked extent by a multiplicity of causes'.

But if the subject consisted of nothing more than this it would indeed be a boring business, and the second meaning applied to the word is that body of techniques and devices, mainly derived from probability theory, which helps management to take decisions.

It is not difficult to think of examples which fit the first of the two concepts. The Census of Population Reports, the National Income and Expenditure Blue Books and other such publications spring to mind, but it applies equally as well to a firm's sales analysis or production schedules. The better known of the statistical techniques which fall within the second concept are quality control through

1

inspection, market research and the selection of the best industrial process from several under test. Since the Second World War (and partly because of it) much attention has been given to a number of topics generically known as Operational Research. Many of these are especially relevant to the science of management such as the theory of queues, linear programming and network analysis.

Too often managers shy away from any serious study of statistics because they do not believe themselves capable of thinking in the abstract. The top-level executive insists that he is entitled to be fed with statistical results expressed in everyday language, the junior manager convinces himself that this is a subject he can, and ought, to leave to the specialist. Both are wrong. Managers are more worthy of their salt if they can develop, through some training, a sense for the accuracy of figures and a feeling for the plausibility of results. Professor Sir Roy Allen writes, 'The Administrator . . . requires statistics to support his arguments and to illuminate the problems he handles. . . . [He] must follow the discriminating middle course between the cynic who thinks statistics can prove nothing, and the uncritical believer in the veracity of every figure' (*Statistics for Economists*). And according to Marjorie Deane in a paper on 'The Training of Statisticians for Economics and Business', '. . . statisticians may consider themselves artists, but they cannot be abstract ones; it is silly to try to imagine that they can evolve their own language and outsiders can go hang if they do not understand. In business they are answerable to management.' This is sound sense, but nowadays the manager is expected to go part of the way to meet them in the translation. He is not expected to pursue the subject in depth, but is the better for understanding the statistician's approach to the common problems of his trade. In short, a knowledge of the statistical approach is now needed by those who receive statistical reports so that they may at least question them intelligently and have faith in their findings.

RELIABILITY OF DATA

It is the statistician's duty to question the authenticity of figures presented to him before analysing them, because it is clearly a waste of time to apply sophisticated statistical methods to figures whose accuracy is doubted.

There is a story that a population census was conducted in the early part of the war in a country whose administration was untrained in the appropriate techniques. The result was much lower than expected and the operation was repeated about six months later, the second result being much larger than the first. The explanation, it was suggested, lay in the fact that on the first occasion the populace thought the census was related to military service, and on the second to the issue of ration books. Whatever may be the degree of truth in the story, it points to the moral that all original data must be examined critically. Analysed nonsense is still nonsense.

A statistician must always be deciding whether data is worth collecting. Sometimes it is possible to obtain a less accurate but nevertheless acceptable result by using material which is less costly to collect. On these occasions he is forced to work with figures which can only provide him with an approximate solution. For example, it is very difficult to calculate the volume of production from some industries. This is particularly true of machinery, because clearly the number of machines manufactured is no measure of the volume of machinery output: ten newspaper printing-machines are not, in this context, equal to ten domestic lathes. The problem is solved by employing indicators of the volume of production, such as the value of the output, the quantity of materials used, or even the numbers of persons employed.

There are a number of pitfalls in using collected information. The ton has different values in North American, Spanish and British statistics, and there are other differences in international weights and measures to mislead the unwary. One must always approach data involving ages with reticence, particularly women's ages. Some British government statistics excluded Northern Ireland for some years and included it for others. The Census of Population Report has given the population of towns on the basis of 'civilian population' one year and 'home population' the next, the latter figure including the armed forces stationed in the town. This would be significant knowledge for a brewery undertaking a pilot sales scheme for a new brand of lager in Aldershot.

In general, it is always most important to know the definition of the terms and accuracy of the figures in collected information. A trained statistician will, if the option is open to him, dictate the form of the information he requires; the novice is recommended to be as suspicious of his original data as a lap-cat sniffing around a new dish.

ROUNDING OF FIGURES

Very often one is presented with numerical information which, although from a reliable source, contains approximations which are either explained in the text or understood from the way in which the data is presented. In a General Election it has been known for a ballot to be counted seven times because the result was so close, and for a different result to be obtained on each occasion. The human error is always with us: life would be a sombre affair if it were not so. In all probability no declared election result for a single constituency is in fact ever correct, but if it were possible to know the truth of the matter, most of the declared majorities are probably correct to within 50 votes.

This means that we may often read figures which present an unwarranted appearance of great accuracy. The result of the Oxbridge by-election may be posted on the boards as:

M. BOARD	(Conservative)	33 462
H. GOWN	(Labour)	28 639
	Majority	4 823

but it would in fact be *more* conservative to say that the majority was 4823 ± 50.

Statisticians are the first to recognise the human-induced shortcomings of their material, and a great deal of data is therefore rounded. A few examples will explain what is meant by this.

27 863 467 rounded to the nearest million is   28 000 000
27 863 467 rounded to the nearest thousand is 27 863 000
27 863 467 rounded to the nearest hundred is 27 863 500

42·6374 rounded to the nearest unit is   43
42·6374 rounded to the first decimal place is   42·6
42·6374 rounded to the second decimal place is 42·64

In all the above examples it is obvious that we are merely selecting the number which is nearer the truth. 27,863,467 is nearer 28,000,000 than 27,000,000. Clearly there is a problem with 27,500,000 because it is as close to 28,000,000 as it is to 27,000,000. It is not very satisfactory to follow the convention 'always round to the higher value' or its opposite, because if future use of the data includes the addition of many rounded figures, there is a constantly increasing bias upwards or downwards. The most acceptable convention is to round to the even integer preceding the 5. On the assumption that it will always be a matter of chance whether the figure preceding the 5 is odd or even, this convention will tend to diminish the errors when a large number of rounded figures are added together.

In this context

> 27·5 rounded to the nearest whole number is 28
> 28·5 rounded to the nearest whole number is 28

Sometimes a different terminology is used: one speaks of rounding to so many significant figures. Further examples will explain this.

> 27 863 467 rounded to 2 significant figures is 28 000 000
> 27 863 467 rounded to 4 significant figures is 27 860 000

The number of significant figures is counted from the left provided there are units to the left of the decimal point.

> Thus 27·863 rounded to 4 significant figures is 27·86
> 1·007 contains   4 significant figures
> and   0·107 contains   3 significant figures

because the zero before the decimal point could be continued indefinitely, but 0·007 contains 1 significant figure because the unit in which figures are measured is arbitrary: in this case it *could* be measured in thousandths.

ERRORS

The errors contained in rounded figures will affect the accuracy of any addition, subtraction, multiplication or division in which these figures are employed.

ADDITION AND SUBTRACTION

38·1 and 40·7 are two numbers rounded to one decimal place. They can therefore be more accurately written as $(38·1 \pm 0·05)$ and $(40·7 \pm 0·05)$.

The highest and lowest results obtained from their *addition* are

Lowest	Highest
38·05	38·15
40·65	40·75
78·70	78·90

The result can be expressed as $38·1 + 40·7 = 78·8 \pm 0·1$.

The highest and lowest results obtained from their *subtraction* are:

Lowest	Highest
40·65	40·75
38·15	38·05
2·50	2·70

The result can be expressed as $40·7 - 38·1 = 2·6 \pm 0·1$.

The general rule is: *The total error contained in the sum or difference of any number of rounded figures is the sum of the errors in the separate figures.*

It is confusing and untidy to add together figures rounded to varying degrees of accuracy, because if all the constituent numbers of an aggregate are rounded to the same unit, the total error in the sum is a multiple of the common error.

PRODUCTS AND QUOTIENTS

The derivation of the degree of error in the product or quotient of rounded figures is a little more difficult.

Consider $8·2 \times 0·24$. Both numbers are written to two significant figures, but the proportional errors within each are different.

The proportional error within 8·2 is $\dfrac{0·05}{8·2} = 0·006$ or $0·6\%$.

The proportional error within 0·24 is $\dfrac{0·005}{0·24} = 0·021$ or $2·1\%$.

Now, $8·2 \times 0·24 = 1·968$.

The lowest possible result is $8·15 \times 0·235 = 1·915\ 25$.

The largest possible result is $8·25 \times 0·245 = 2·021\ 25$.

These limits can be expressed as $1·968 \pm 0·053$.

The error is $\dfrac{0·053 \times 100}{1·968} = 2·7\%$.

In other words, the proportional error of the product is the sum of the proportional errors of the multiplied figures. The student can check for himself that this is also true for a quotient.

The general rule is: *The proportional error in a product or quotient of rounded numbers is the sum of the proportional errors in the numbers themselves.*

The reader may be forgiven for thinking that this section about errors is a lot of fuss about nothing. He must however be advised very firmly that a cautious eyeing of derived figures to gauge their accuracy is a necessary trait of the statistician. An appreciation of the properties of rounded figures will also prevent the manager from wrongfully returning a list of rounded percentages because they do not add to 100. Statistical clerks have been known to 'cook' such totals to deprive old 'so-and-so' of the unfounded pleasure of returning them for correction!

FREQUENCY DISTRIBUTIONS

If the General Manager of a firm employing one thousand salesmen were to be confronted by the trade union with a demand for improved terms, he would call for a list of the salesmen's earnings over the past year. He might first be given an alphabetical list of his salesmen with the related earnings beside each name. This would tell him nothing except possibly the exact number of salesmen employed. The information would be more useful if arranged in descending order of total earnings. Such a table would reveal the range of values, i.e. the difference between the highest and lowest earnings, and would also give an impression of the average from the amounts which appear half-way down the list. But beyond yielding these two additional pieces of knowledge, a graduated listing by earnings would not be very helpful.

Much more could be learned from the arrangement depicted in Table 1.1 and known as a *grouped frequency table* or *grouped*

TABLE 1.1. *Grouped Frequency Distribution.*
Salesmen's Salaries 1966 [Vito Food Co. Ltd]

Salary class (1)	Class Midpoint x (2)	Frequency f (3)	Col. (3) Cumulative (4)
£	£		
800 less than 1000	900	50	50
1000 less than 1200	1100	200	250
1200 less than 1400	1300	350	600
1400 less than 1600	1500	150	750
1600 less than 1800	1700	100	850
1800 less than 2000	1900	75	925
2000 less than 2200	2100	50	975
2200 less than 2400	2300	25	1000

frequency distribution. The construction of such a table is not as easy as would at first appear. This is because in utilising the data for certain purposes at a later stage, an assumption is made that the salaries of the salesmen included in each specified salary range are equally spread throughout that range. For example, it is assumed that the salaries of the 350 salesmen with salaries ranging from £1200 to just less than £1400 per annum are distributed throughout that range in equal intervals of £200/350.

The first three would therefore be

$$£1200, \left(£1200 + \frac{£200}{350}\right), \left(£1200 + \frac{£200}{350} + \frac{£200}{350}\right) \text{ and so on.}$$

In practical terms this equal spread is never possible to achieve, but one must get as near to it as possible. Most certainly it would be wrong to arrange a distribution in which the salaries for a large group of salesmen were at one end of the salary class to which they belonged. The task becomes easier when the number of items involved—in this case salesmen—becomes very large.

In order to avoid looseness in terminology, the information contained in the columns of a grouped frequency distribution is specially defined. Referring again to Table 1.1, it will be seen in column (1) that the salaries of one thousand salesmen have been broken down into eight *classes*. The highest and lowest quoted salaries in each class are called the *upper class boundary* and *lower class boundary*, and the difference between them is known as the *class interval*. In Table 1.1 there is a common class interval of £200.

Column (2) lists the *class midpoints* which are each defined as the sum of their upper and lower class boundaries divided by two. It can be seen that the class midpoint is the centre of gravity of its class.

The number of salesmen whose salaries fall within a salary class is known as the *class frequency*.

One can therefore see from Table 1.1 that the salary class whose class boundaries are £1400 and £1600 (or to be more precise, the largest amount which is less than £1600, i.e. £1599·99½), has a class midpoint of £1500 and a class frequency of 150.

It is not always easy to determine the best number of classes in a distribution. There must not be so many that the class frequencies become small and the form of the distribution becomes so attenuated that its essential 'shape' is lost, nor must there be so few that the class frequencies are very large, and important information about the distribution is crushed in a concertina fashion into a single class frequency. The ideal is between these two, but the judgement called for lies little beyond common sense.

It might well have been that the salaries of the sales force were recorded in the firm's ledgers to the nearest pound, £0·50 being rounded to the higher figure. The classes would then read:

$$£800 \quad to \quad £999$$
$$£1000 \quad to \quad £1199$$
$$£1200 \quad to \quad £1399$$

and so on. This means that the class boundaries are in reality:

$$£799·50 \quad to \quad £999·49\tfrac{1}{2}$$
$$£999·50 \quad to \quad £1199·49\tfrac{1}{2}$$
$$£1199·50 \quad to \quad £1399·49\tfrac{1}{2}$$

.   .   .   .   .

In dealing with rounded units of this sort it is usual (and eminently sensible) to approximate the upper boundaries to £999·50, £1199·50, £1399·50 . . . so that the class midpoint is easy to calculate. The loss of accuracy is negligible. The reader will however notice that the class midpoints now have shifted to:

$$(£799·50 + £999·50) \div 2 = £899·50$$
$$(£999·50 + £1199·50) \div 2 = £1099·50$$
$$(£1199·50 + £1399·50) \div 2 = £1299·50$$

This feature, which at this stage seems to be an irritating pressure for accuracy unwarranted by the nature of the original data, has an important bearing on later work and should not be dismissed.

For the sake of simplicity the class intervals in Table 1.1 are all £200. In fact, this situation is somewhat unreal, because the highest earners are unlikely to fall into a tidy final class interval of £200 with a frequency as large as 25—there are more likely to be (say) five geniuses who have large earnings reaching to £3000. The compiler of the distribution might in those circumstances prefer to define the final class as '£2200 and upwards'. This is called an *open ended class*.

There are some important frequency distributions which do not lend themselves to grouping. If, for example, a member of the inspection department of a factory has at regular intervals to test samples of five products which appear consecutively on the conveyor belt, the distribution table of his results from one thousand samples, would probably appear like this:

Nil	imperfect parts out of five	952
One	imperfect part out of five	22
Two	imperfect parts out of five	14
Three	imperfect parts out of five	7
Four	imperfect parts out of five	4
Five	imperfect parts out of five	1
		1000

Both the above frequency distributions have a total frequency of one thousand items. This of course is because they have been contrived. Practical distributions are less obligingly rounded but a simple treatment of the information can convert the actual frequencies into a proportion of the total frequency. To illustrate this point the salary classes from Table 1.1 are below given a more irregular frequency pattern, and a total frequency of 2104.

Salary class £	Frequency	Relative frequency
800 less than 1000	5	2
1000 less than 1200	86	41
1200 less than 1400	243	116
1400 less than 1600	722	344
1600 less than 1800	590	280
1800 less than 2000	317	151
2000 less than 2200	100	47
2200 less than 2400	41	19
	2104	1000

The figures in the third column of this table are obtained by multiplying the separate frequencies in the second column by the factor 1000/2104 (a slide rule or calculating machine is suitable for this process).

They are known as the *relative frequencies*.

It is easier to imagine the proportion of the total frequency which each frequency bears, when the distribution is converted in this way: indeed the figures in the third column are readily changed into percentages by dividing each of them by ten, e.g. 0·2, 4·1 and so on.

However, one should be chary of the information furnished by a distribution for which the relative frequencies are given but not the total frequency, because the value of such data is considerably reduced when the total number of items involved is fairly small. It is not very helpful to be told that the production of product 'A' has leaped by 50% when in fact six have been manufactured instead of four!

In any frequency distribution the item which is being measured, whether it be salary, the hardening time for an industrial adhesive or the length of cuckoo's eggs in nests of the meadow-pippit, is called the *variate* or *variable* because it is the thing whose value is varying. One therefore speaks of the frequency distribution of the variate or variable. The variable in Table 1.1 is the salaries of salesmen.

CHARTS AND DIAGRAMS

Some people have great difficulty in interpreting a table of numerical information: their minds involuntarily shy at an array of figures. Fortunately it is frequently possible to represent such information in diagrammatic form so that the essential features of the table are highlighted.

FIG. 1.1(a). Incomes liable to surtax, 1959-1965. Bar chart.

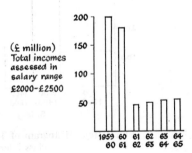

(£ million) Total incomes assessed in salary range £2000-£2500

In Fig. 1.1(a) the incomes liable to surtax in the salary range £2000 to £2500 for fiscal years 1959/1960 to 1964/1965 are presented in the form of a chart. It is a series of rectangles whose heights are proportional to the total incomes assessed. A diagram of this sort is called a *bar chart* or *bar diagram*. Usually the bars are not contiguous, and, although we have drawn them vertically it is sometimes advantageous to draw them horizontally. The sole object of a bar chart is to represent the data visually through the height of the rectangles. It therefore loses its usefulness unless the origin or zero point is clearly marked.

A simple illustration will exemplify the dangers. The Slimo Bread Co. engaged a new sales director whose promise to increase their sales by 50 per cent. was unfulfilled. At the end of his campaign he presen-

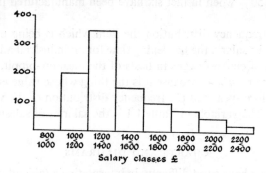

FIG. 1.1(b). Histogram of Table 1.1 data; equal class intervals.

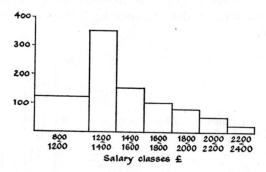

FIG. 1.1(c). Histogram of Table 1.1 data; unequal class intervals.

ted two charts to the Board (Fig. 1.2). It will be seen that the origin of Chart II is not shown, and one hopes that this attempt to hide his failure behind a dishonest chart was as unsuccessful as his sales drive.

Let us again refer to Table 1.1. If we represent the class intervals on the x-axis as in Fig. 1.1(b), and erect rectangles with the class intervals as their bases, so that the *areas* of the rectangles represent the *class frequencies*, we shall have drawn a *histogram*. A histogram

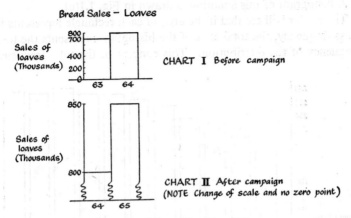

FIG. 1.2. Sales—Slimo Bread Co.

of the data in Table 1.1 would have equal class intervals. Had the first salary class read '£800 but less than £1200' and the other classes read '£1200 but less than £1400', '£1400 but less than £1600' and so on, the height of the rectangle erected on the lowest salary class would have been ½ (200+50) units according to the scale that was being used.

TABLE 1.2 *Historical or Time Series.* Consumers' Expenditure on Entertainment 1950–1960. *At* 1960 *prices.*

			£ million		
	Cinema	%	Other	%	Total
1950	108	53	96	47	204
1951	111	52	103	48	214
1952	112	50	113	50	225
1953	111	49	118	51	229
1954	112	47	127	53	239
1955	108	44	136	56	244
1956	106	43	142	57	248
1957	95	38	154	62	249
1958	85	34	168	66	253
1959	71	28	179	72	250
1960	65	26	186	74	251

Source: National Income and Expenditure 1961 C.S.O. Blue Book.

A histogram of this situation is drawn in Fig. 1.1(c).

The reader will see that if the area of each rectangle represents its class frequency, the total area of the histogram represents the total frequency of the distribution. This concept of the total frequency

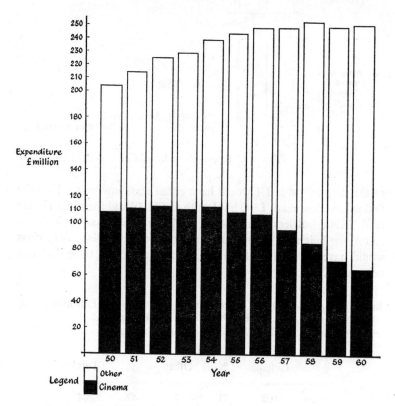

Fig. 1.3. Component bar chart. Consumers expenditure on entertainment, 1950-1960. Data from Table 1.2. (Source: *National Income and Expenditure* 1961, C.S.O. Blue Book.)

of a distribution being depicted by an area will be taken up again in Chapter 7.

Great care must be taken to define fully the variate, and to head the diagram adequately.

Table 1.2 gives an example of an historical or time series. The information relates to expenditure on entertainment and points to a significant social change which took place within the decade

1950-1960. Buried beneath the aridity of such figures lie the secrets of human motivation and emotional fickleness, of the multitude of independent private decisions which in aggregate create a social force strong enough to make a fashion or elect a government.

It would clearly be helpful if a form of bar chart could be constructed which not only traced the growth of total expenditure on

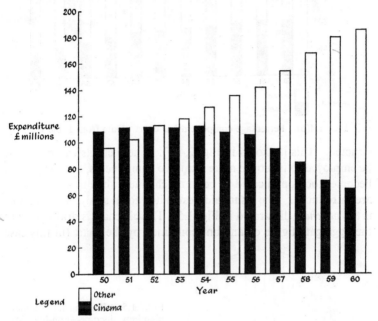

FIG. 1.4(a). Component bar chart. Consumers expenditure on entertainment, 1950-1960. Data from Table 1.2. (Source: *National Income and Expenditure* 1961, C.S.O. Blue Book.)

entertainment, but also recorded the changing fortunes of its component forms. This is done in Fig. 1.3, where a *component bar chart* is illustrated. Although this is a common and convenient way to draw a visual aid for information of this sort, it suffers from the drawback that the expenditure on 'other' forms of entertainment is drawn to a different base level each year; it is therefore difficult to compare its size from year to year. This can be remedied by drawing the expenditure on 'cinema' and 'other' forms side by side as is done in Fig. 1.4(a). However, this chart does not record the total expenditure. It seems one cannot have everything, so the purpose of the drawing has to be known before one can decide on the best type of chart. It could be, for example, that the proportion which the components bear to the total expenditure is the information required. If that is so, then a *percentage component bar chart* has to be con-

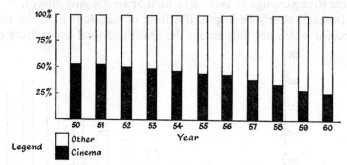

FIG. 1.4(b). Percentage component bar chart. Data from Table 1.2.

structed. This is shown in Fig. 1.4(b). These charts can of course be employed to illustrate more than two components, but the picture becomes too complicated to be of much value when more than four are introduced. The visual aid best suited to an analysis of this sort is the *pie chart* illustrated in Fig. 1.5. The pie chart is drawn so that the aggregate figure or amount contained by the chart (in this case

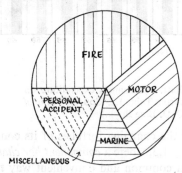

FIG. 1.5. Pie chart. Annual premium income—non-life insurance (1966). XYZ Insurance Co.

annual non-life insurance premium income) is represented by the 360° of a circle: each of the subdivisions of this aggregate is converted into the appropriate proportion of 360°, and radii drawn accordingly. The popularised pamphlet often prepared by the government to explain the country's economy in simple terms, invariably uses the pie chart to demonstrate how the national 'cake' is shared among the different classes of citizens. Perhaps this mixture of culinary metaphors is permissible!

Yet another representation of approximate statistical information is the *pictogram* or *pictograph*, shown in Fig. 1.6. It is an admirable

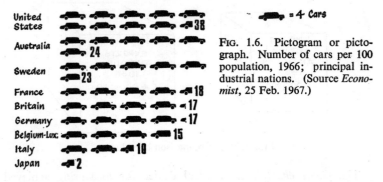

FIG. 1.6. Pictogram or pictograph. Number of cars per 100 population, 1966; principal industrial nations. (Source *Economist*, 25 Feb. 1967.)

device for making rough comparisons, but it is unsuitable for representing anything below a single pictorial unit.

The use of a pictogram points to the human difficulty experienced in comparing areas, and to the still greater difficulty in comparing

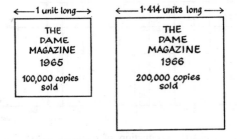

FIG. 1.7(a). Comparison by area.

volumes. Efforts made to simplify numerical information by converting it into pictures frequently fail because of this. If one wishes to illustrate pictorially that the circulation of a magazine has doubled, by drawing two periodicals one of which is twice the area of the other, it would appear as in Fig. 1.7(a). (To double the area of a square, the length of its side is increased from one to approximately 1·414 units.) Similarly, if one represents a cube which is twice the volume of another, the length of the side of the larger cube is approximately 1·26 units of the smaller. This comparison can be seen in Fig. 1.7(b). The reader will readily accept that this method of portraying numerical data is unsatisfactory. In short, it is not visually helpful.

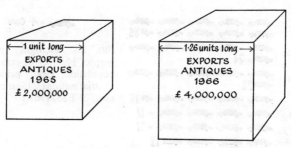

FIG. 1.7(b).   Comparison by volume.

The *Gantt chart* is a practical device for measuring achieved performance against scheduled performance.   Fig. 1.8 shows a Gantt Chart and the table of values from which it is drawn.   The figures recorded against 'schedule' are those which the production engineer forecasts as being necessary to maintain the assembly lines.

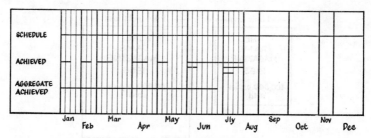

FIG. 1.8.   Gantt chart.   Production of component 'A', Jan.-July 1966. Scale: One vertical width represents one hundred components.

TABLE OF FIGURES FOR GANTT CHART

	Jan	Feb	Mar	Apr	May	Jun	Jly	Aug	Sep	Oct	Nov	Dec
Schedule	400	300	700	500	600	700	400	300	600	600	300	600
Achieved	200	200	300	300	200	900	1000					
Aggregate Achieved	200	400	700	1000	1200	2100	3100					

They take into account such factors as the number of working days in the month, the market's seasonal demand for the product, staff holidays and the need for an adequate stockpile.   The 'achieved' line indicates the actual number of components manufactured in each of the months.   At the end of each month the total production for the year to date is written into the 'aggregate achieved' line. The horizontal axis of the chart is drawn to an appropriate scale, and the tabulated data are drawn in as they are known.

A somewhat bizarre situation has been devised to demonstrate fully the purpose of the chart. Production for the months of January to May inclusive was severely affected by poor labour relations, and because of a work-to-rule and overtime ban, production was much below that scheduled. In June, however, the dispute was settled, and the mutual good will then re-established was reflected in considerable overtime working and a works holiday shutdown postponed from July to August.

The 'monthly achieved' figure is drawn, according to scale, entirely within the month's vertical boundaries. The 'aggregate achieved' figure is represented by a continuous line from the vertical start-line. For example, at the end of March the 'aggregate achieved' line would terminate at a point represented on the scale by seven hundred components from the start-line. It so happens that this coincides with the end-February boundary. The Gantt Chart would thus indicate to the production manager at the end of March that he was exactly March-month's scheduled production in arrears. Indeed, the main purpose of the chart is to provide a quick visual idea of a developing situation of under- or overproduction, so that remedial action may be taken. It thus lends itself to sales-project measurement, education curriculum attainment and other such subjects.

GRAPHS

It is logical that a discussion on the meaning and construction of graphs should follow the section on pictorial or visual aids because a graph is itself a form of visual aid. It has been described as a pictorial presentation of the relationship between variables. There are so many pitfalls in graph construction and interpretation, however, that a study of some of the fundamental concepts involved is called for.

DISCRETE AND CONTINUOUS VARIABLES

The variables involved in some departments of enquiry can take only integral values. Any frequency distribution of people has such a variable. There cannot be a fraction of a person. A variable falling within this definition is a *discrete variable*.

B

There are other variables which can take any value, integral or fractional. The heights of people can be any value between those for a newborn child and a giant. Only the limitations of the measuring instrument defines the accuracy of the recordings. Presumably if there were a micrometer accurate enough, the heights could be taken to several places of decimals. Variables such as these are called *continuous variables*.

Data which can be described by a discrete or continuous variable are respectively called *discrete data* or *continuous data*. If a variable can take only one value, it is called a *constant*, although the reader could be forgiven for thinking that this was a contradiction of terms!

PLOTTING A GRAPH

Most readers will be familiar with the practice of graph plotting on Cartesian co-ordinates. The following few lines are meant to be nothing more than a rapid revision of the simple rules.

Fig. 1.9 shows four points plotted, one in each quadrant, with the co-ordinates appropriate to each. The convention regarding the

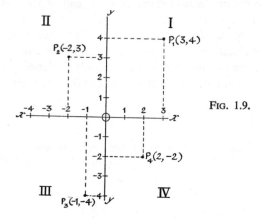

FIG. 1.9.

signs of the co-ordinates can be seen from the plots. In quadrant 1 the sign of both x and y is positive.

Normally a statistician will be dealing with positive quantities, so that in the ordinary course of events he will be concerned only with the first quadrant.

In Fig. 1.9 the x- and y-axes have been plotted with quantitative values, but it is frequently necessary to record historical periods such

as years on one of them. The *x*-axis is usually used for this purpose.
An example of this is given in Table 1.3 and Fig. 1.10. One has to
be most circumspect in reading such a graph. The information
from which the graph has been constructed tells us only that a certain
number of passengers were carried by U.K. airlines in each of seven
years. Presumably the figures relate to calendar years, but it could
be that the statistical source from which the data is derived used the
tax year because it was easier and cheaper to do so. However,
accepting the first assumption as the correct one, the *x* point for 1959
means 'the calendar year 1st Jan. to 31st Dec. 1959', and so on for

TABLE 1.3. *United Kingdom Airlines
Passengers Carried*

	Thousands
1959	392·1
1960	490·0
1961	570·9
1962	641·6
1963	720·6
1964	812·9
1965	905·9

Source: *Monthly Digest of Statistics*, Board of Trade.

the other years. What then would be the value of *y* applicable to a
point on the *x*-axis exactly half-way between 1959 and 1960? The
definition of this *x* point would be 'the period 1st July 1959 to 30th
June 1960', and the basic data does not supply the number of passengers carried in this period.

In the circumstances, the only thing to do is to make the most use
of the relevant information that is available, namely, the number of
passengers carried in 1959 and 1960. In the absence of information
to the contrary one can only guess that the increased traffic recorded
for 1960 was acquired gradually and regularly throughout the period.
The best course, then, is to draw a straight line between the co-
ordinates for 1959 and 1960, and to *interpolate* on this straight line
for the period defined by the midpoint between 1959 and 1960. This
is done in Fig. 1.10. On the basis of this reasoning the remaining
co-ordinates may now be joined, to form a *line graph*. Actually the

most accurate assessment would be obtainable arithmetically from the table, but the source data does not always accompany a graph, and one is frequently forced into the approximation of a graphical interpolation. At the same time it would be sensible to make any other enquiries germane to the problem: within the present context, information concerning fare changes, strikes and unusual weather conditions in the period suggest themselves. As was mentioned in the section on the reliability of data, it is most important for a

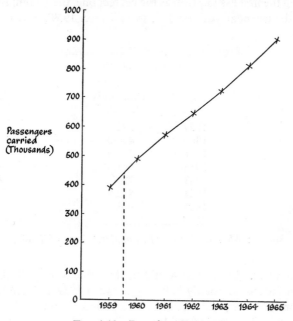

Fig. 1.10.   Data from Table 1.3.

statistician to be aware of the limitations of his material. Intelligent guess-work based on reliable figures is preferable to a hide-bound technical approach.

Let us now consider some further graphs. There is a production process in the construction of electrical relays that requires an operator to place a metal strip splayed like the veins of a fan on top of a circular piece of mica covered with a vulcanising material. The two components are placed on the metal bed of a machine, a handle is pulled, heat is passed, and the metal strip and mica are fused together.

The mechanics of the process are important because they have a bearing on the speed at which the operator can work. If it is assumed (somewhat naïvely) that a worker can produce sixty pieces in every hour throughout his 8-hour shift, his production table would be:

TABLE 1.4(a)

x (Hours)	1	2	3	4	5	6	7	8
Hourly output	60	60	60	60	60	60	60	60
y (Cumulative output)	60	120	180	240	300	360	420	480

If, on the other hand, a more practical outlook suggests that his output would begin higher but diminish through fatigue by one-fifth every hour of his shift, his schedule would be as below. (The fraction one-fifth is of course excessive, and has been selected to illustrate a graphing point.)

TABLE 1.4(b)

x (Hours)	1	2	3	4	5	6	7	8
Hourly output	100	80	64	51	41	33	26	21
y (Cumulative output)	100	180	244	295	336	369	395	416

It could be said as a third possibility, that the operator begins his shift somewhat apprehensively, has difficulty in centring the components on his machine, misjudges the vulcanising time so there are a number of rejects; but he improves with time, and in fact does one fifth more every hour of his shift. The figures would then be:

TABLE 1.4(c)

x (Hours)	1	2	3	4	5	6	7	8
Hourly output	40	48	58	70	84	101	121	145
y (Cumulative output)	40	88	146	216	300	401	522	667

The graphs for Tables 1.4(a), (b) and (c) are drawn in Figs. 1.11(a), (b) and (c). It will be noticed that the hourly output has been aggregated throughout the duration of the shift, and this accumulated value plotted as the y value. The hours of the shift have been plotted on the x-axis. It is in fact conventional to use the x-axis for the *independent* variable and the y-axis for the *dependent* variable.

These terms are self-explanatory. The cumulative production figure recorded on the y-axis is dependent upon the number of hours worked. One would not think of counting the number of units manufactured in order to discover the duration of the shift! Some people prefer to invert the situation by saying that the *causal* variable is recorded on the x-axis.

Let us now consider Fig. 1.11(a). The graph is a straight line, and one can see, if only intuitively at this stage, that any graph drawn from cumulative figures of equal increments would be a straight line. Consider now the equation $y = 60x$, where x is defined as the number of hours of the shift worked and y as the number of components produced.

When $x = 1,\quad y = 60$
$x = 2,\quad y = 120$
$x = 3,\quad y = 180$ and so on.

Clearly, the equation $y = 60x$ is the algebraic shorthand for the situation recorded in arithmetic terms in Table 1.4(a). It is the mathematical 'model' for it. When the worker first sat down at his machine in the morning, i.e. when he had worked for zero hours, his production was nil. The equation produces this situation and it is illustrated on the graph by the line passing through the origin.

But suppose the previous evening the operator had thirty components which he had not handed in: he would begin the new day with these, and would say that by the end of the first hour he had produced 90 components $(30+60)$ and then 60 for every succeeding hour of the shift. This would be represented in algebraic shorthand as

$y = 60x+30$, because

when $x = 0,\quad y = 30$
$x = 1,\quad y = 90$
$x = 2,\quad y = 150$ and so on.

The reader will readily see that the cumulative production from this equation is the same as that from the equation $y=60x$, plus the 30 with which the day's shift began. The graph for the equation $y = 60x+30$ is drawn in Fig. 1.11(a) as a broken (not dotted) line. It is parallel with the straight line for the equation $y = 60x$ but thirty units from it—these units being measured vertically on the y-axis. The intercept on the y-axis (that is, the value of y when $x = 0$) is 30, which is what one would expect from the known facts.

Now the equation $y = ax + c$ where:

y is the dependent variable (i.e. cumulative number of components in the example)

x is the independent variable (i.e. number of hours in the example)

a is a constant (i.e. 60 in the example)

c is a constant (i.e. 30 in the example)

is the general equation for a straight line. This means that a graph of the resultant value of y against the values given to x will always give a straight line whatever values are ascribed to the constants. The reader would do well to convince himself of this by working a few examples, because it is important.

Before leaving this general equation let us consider the meaning of the constant a. For ease of explanation consider constant c as zero, i.e. $y = ax$. Now it is helpful to write

$$a = \frac{ax}{x} = \frac{y}{x}.$$

In Fig. 1.11(a), $\frac{ax}{x}$ or $\frac{y}{x}$ is the same as $\frac{P_1A}{OA}, \frac{P_2B}{OB}, \frac{P_3C}{OC}$, and so on to

$$\frac{P_8H}{OH}.$$

Each of these ratios is by definition, the trigonometrical tangent of the angle P_1OA, or the *slope* of the straight line.

It has therefore been established that in the general equation $y = ax + c$, 'a' is a constant which is the slope of the line, and 'c' is a constant which is the intercept of the line on the y-axis.

Let us now consider the data in Table 1.4(b). When graphed (as in Fig. 1.11(b)) the curve is seen to be 'concave downwards'. This result might have been foreseen because the differences between successive values of y are diminishing. Moreover, they are diminishing regularly and predictably since the number of products manufactured from consecutive hours of work during the shift are in geometrical progression with a common ratio of four-fifths. (The terms of a geometrical progression can be written as $a, ar, ar^2, ar^3, \ldots,$ where a is a constant and r is the common ratio.)

Assuming that mealbreaks were taken at the end of a complete hour and are written out of the tables, the values of y between

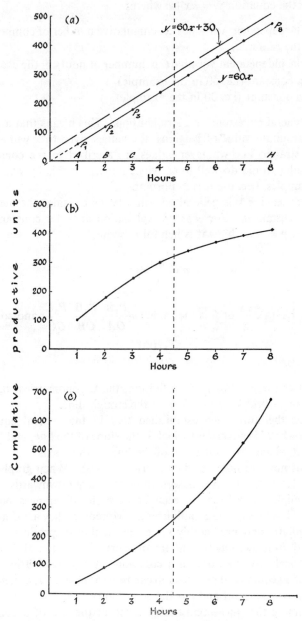

FIG. 1.11. Cumulative output per worker, data from Table 1.4.

integral values of x are not subject to possible disturbing influences as they were in Table 1.3.

A similar approach can be made to the figures in Table 1.4(c) and to the graph in Fig. 1.11(c), except that the curve is 'concave upwards' and the common ratio is six-fifths. Again the curve is regular. Recapitulating, the graph of data in Table 1.4(a) is a straight line, and the graphs of the data in Tables 1.4(b) and 1.4(c) are regular curves. This means that one can interpolate in all three graphs for values of x which are not integral. For example, the values of y for $x = 4.5$ in Tables 1.4(a), 1.4(b) and 1.4(c) are respectively 270 units, 320 units and 260 units from the graphs. This is to say that if the worker were told to stop the machine at the end of $4\frac{1}{2}$ hours' work on the shift, and the total number of products manufactured during that shift were counted, they would amount to the figures quoted above, given the three production situations which the tables specify.

LOGARITHMIC OR RATIO SCALE

So far all graphs have been drawn to arithmetic or absolute-value scales. That is to say, the values marked on the axes are directly proportional to their distances measured from the point of origin. Thus, an x- or y-value registering 6000 units would be twice as far from the origin as one recording 3000 units.

Very often, however, the manager is not as interested in the arithmetic difference between figures as in the proportional increase or decrease which they represent. He is interested in the percentage change which this year's figures represent over last year's, or the annual change over a long period. Also, he wants to compare the performance of his own firm with that of a large competitor, or even with the industry as a whole. It is not easy to achieve these objects on graphs of reasonable size drawn to an absolute scale. For such purposes the *logarithmic* or *ratio scale* graph must be used, and we will now consider its construction.

As has been seen, the graph of an historical series will be a straight line if the values of the variate increase by the same amount for successive time intervals. It would be helpful if a scale could be devised which would give a straight-line graph when y-values increase by the same *percentage* for each successive time interval. Such a straight line could serve as a datum for all other graph shapes drawn to this scale.

Let us first of all consider the figures from Table 1.4(c) which increase regularly by 20%; they have already been recognised as forming a geometrical progression. In Fig. 1.12(a) the hourly output figures of this table are drawn to such a scale on the *y*-axis as will make the resulting graph a straight line. The reader will notice that the increase for each successive hour in the shift may not be *exactly* 20% because it is not possible for an operator to complete a portion of a product, but the slight theoretical inaccuracy that this practical limitation introduces is of no consequence.

The new kind of scale which is required is made by:

(*a*) Marking off seven equal intervals on the *y*-axis.

(*b*) Repeating (*a*) for the *x*-axis.

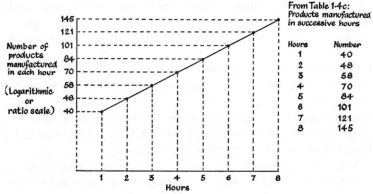

FIG. 1.12(a). Ratio scale, the straight line.

(*c*) Drawing perpendiculars from all the marked points on both axes until each related pair meet (see Fig. 1.12(a)).

(*d*) Joining all points of intersection to form a straight line.

(*e*) Recording against the markings on the *y*-axis delineating the equal intervals, the number of products produced *within* each hour, i.e. against that intersecting on the graph with hour 1, the figure 40; against that intersecting on the graph with hour 2, the figure 48; and so on.

The scale thus drawn on the *y*-axis is one which yields a straight-line graph for figures which increase by a constant proportion for each successive time-interval, i.e. for figures which increase in geometrical progression. The properties of this scale, which is the same as that on a slide rule, must now be investigated.

The first thing to notice is that only a portion of the full scale of values has been drawn. Now it would be possible to continue the scale down the axis by multiplying 40 units by $\frac{5}{6}$, and so on for each successive calculated value. But this would be a tedious business, besides which the process would introduce approximations which would be unacceptable. The best approach is in fact a theoretical one.

If C = the number of products produced *in* the first hour,

M = the number of products produced *in* the nth hour,

p = the constant proportion by which the production in each successive hour is increased,

then, the situation in Table 1.4(c) can be expressed in the following algebraic terms:

During hour	Number of products manufactured
1	C
2	$C + Cp = C(1+p)$
3	$C(1+p) + C(1+p)p = C(1+p)(1+p) = C(1+p)^2$
4	$C(1+p)^2 + C(1+p)^2 p = C(1+p)^2(1+p) = C(1+p)^3$
n	$C(1+p)^{n-2} + C(1+p)^{n-2}p = C(1+p)^{n-2}(1+p) = C(1+p)^{n-1}$

Then $M = C(1+p)^{n-1}$.

Taking logarithms of both sides:

$$\log M = (n-1)\log(1+p) + \log C \qquad (1)$$

Remembering that p is a constant,

C is a constant,

n is the variable,

$\log M$ is dependent on n,

equation (1) is of the general form $y = ax + c$, i.e. a straight line.

It follows that a series of numbers forming a geometrical progression, when plotted on paper ruled to the logarithmic scale, will form a straight line. This then, is the scale which was drawn empirically in Fig. 1.12(a) from the material of Table 1.4(c).

To convince the reader that this is so, the hourly output figures from Table 1.4(c) have been plotted on the logarithmic scale in Fig. 1.12(b). Paper which is printed with a logarithmic scale on one axis and arithmetic scale on the other is called *semi-logarithmic* paper.

Fig. 1.12(c) gives further details about the logarithmic scale. It will be seen that the distances between 1 and 2, 2 and 4, 4 and 8, are all the same, as are the distances between 1 and 3, 3 and 9. Now $\frac{2}{1} = \frac{4}{2} = \frac{8}{4} = 2$ and $\frac{3}{1} = \frac{9}{3} = 3$.

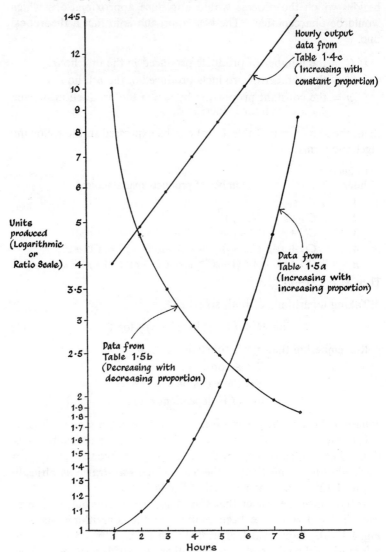

FIG. 1.12(b).   Ratio scales, graph shapes.

In other words, distances on a logarithmic scale between numbers which bear the same ratio to one another are equal. For this reason, a logarithmic scale is sometimes called a *ratio scale*. This fact enables us to complete a ratio scale. For example, the scale in Fig. 1.12(c) is not marked for 2·7 units, and as an exercise in the use of the scale, its position will now be determined. Since the

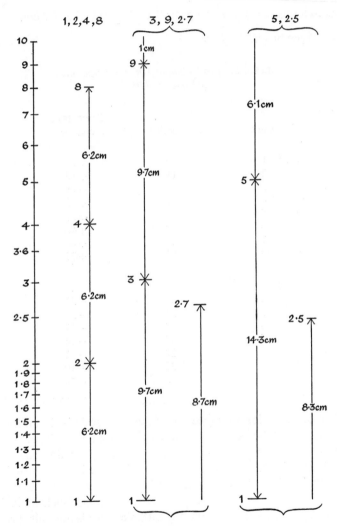

Fig. 1.12(c). Ratio scale, equal ratios cover equal intervals.

distance between 1 and 3, 3 and 9, 9 and 27 is the same, and the
interval for 1 to 3 and 3 and 9 is already defined on the scale, the
interval 1 to 2·7 can be discovered by:

(a) Measuring '9' to '10' on the scale, i.e. 1 cm.

(b) Deducting (a) from the measurement (1 to 3) or (3 to 9), i.e.
    9·7 cm − 1 cm = 8·7 cm.

(c) Measuring (b) from 1 upwards on the scale, i.e. 8·7 cm.

This is done in Fig. 1.12(c).

TABLE 1.5. *Production of Units per Operative
in 8-Hour Shift*

Hours	Production	Percentage change
(a) 1	10	...
2	11	+10
3	13	+18
4	16	+23
5	21	+31
6	30	+43
7	48	+60
8	85	+77
(b) 1	100	...
2	49	−51
3	35	−29
4	29	−17
5	25	−14
6	22	−12
7	20	−10
8	18	−10

The second thing to notice about the ratio scale is that it has no
origin. This follows from the above, because any multiple of zero
is zero, and it cannot therefore have a place on a ratio scale. It is
usual to begin the scale with unity, in which case the scale will finish
with 10, but this value can itself be the beginning of a new scale which
will end in 100, and so on. One complete scale (such as in Fig.
1.12(c)) is called a cycle. Semi-logarithmic paper is normally printed
in one, two or three cycles.

In practice, the value given to the beginning of the scale will be decided by the material with which one is working.

It has been said that a series of values increasing by a constant proportion will yield a straight line on a ratio-scale graph, and that this line can be used as a datum for the other curves. What will these other curves signify? Table 1.5 gives two further series. In the

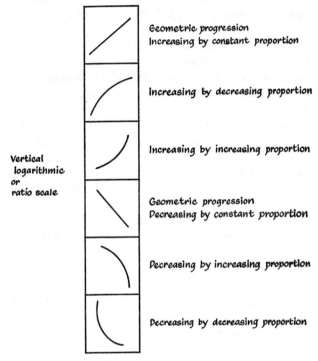

FIG. 1.13. Meaning of graph-shapes drawn to logarithmic or ratio-scale.

first of these the successive values increase by an increasing proportion: in the second they decrease by a decreasing proportion. These two series are plotted on Fig. 1.12(b). The series which increases at an increasing proportional rate is a rising curve showing a concave face to the y-axis: that decreasing at a decreasing proportional rate is a falling curve presenting a convex face to the y-axis. It would take far too much space to plot all the possible graph shapes, but the more important ones are illustrated in Fig. 1.13.

The two important advantages to be gained from the use of a ratio scale are:

(*a*) Since it is graph *shapes* which are being read rather than absolute values, and since also the scale can be made to represent any size of units, the performance of two enterprises, one of which is considerably larger than the other, can be brought into juxtaposition on one graph. For example, a firm may want to compare its results with those of the whole industry, or with those of its group. This it can easily do on a ratio scale graph.

(*b*) It enables the performances of industries manufacturing different but competing products to be compared. The consumption of coal and electricity is a simple illustration; cinema attendances and television sets purchased or hired is another.

One final point should be remembered: a set of values which show an absolute increase each year greater than the previous year appear on an arithmetic scale as a curve rising to the right and convex to the *x*-axis. This could, however, be a series whose *proportional* increase is decreasing, and this fact is cloaked by the curve, which presents too flattering an appearance. It is pulling the wool over the eyes of those members of the management who are statistically uninformed, and who, in more colloquial terms, think that all is well with a curve 'provided it goes upwards and to the right!'

CUMULATIVE FREQUENCY CURVES

At the beginning of this chapter the salaries of one thousand salesmen were analysed. They were not rounded to the nearest pound, so the number of new pence could be any figure from $\frac{1}{2}$p to $99\frac{1}{2}$p Table 1.1 depicted them in the form of a frequency distribution, and in column (4) aggregated the class frequencies from the lowest class to the highest. It is with this column that this section deals.

The first frequency in column (4) is 50, and it should read as 'there are 50 salesmen whose salaries are below £1000 per annum'. The second figure is 250, i.e. (50+200). This aggregate frequency embraces all the salesmen whose salaries fell within the first two salary classes. It therefore should be read as 'there are 250 salesmen whose salaries are below £1200 per annum'. This reasoning is repeated throughout the distribution to the final class, which states the obvious fact that 'there are 1000 salesmen whose salaries are below £2400 per annum'.

The reader can be forgiven for feeling somewhat restless at this apparent pedantry: there is a reason for it. The information in the above paragraph is plotted to arithmetic scale in Fig. 1.14. The cumulative frequency is plotted against *the upper class boundary of*

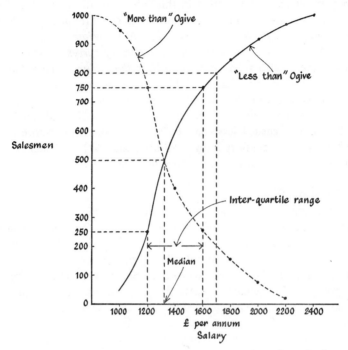

FIG. 1.14. Cumulative frequency curves or ogives, 'less than' and 'more than' ogives. Data from Table 1.1.

the related class. It is *not* plotted against the midpoint. Drawn in this way, one can read from the graph

(*a*) the number of salesmen who earn less than a stated salary,

(*b*) the maximum salary for a specified number of salesmen.

For example, the line horizontal to the axis from the cumulative frequency figure of 800 salesmen cuts the curve at a point vertically above £1700. One can therefore say that there are 800 salesmen who earn salaries less than £1700 per annum—information not immediately available from Table 1.1.

The aggregation which has been made from the lowest class could just as well have been done from the highest class, but in that case the frequency for the highest class would have read 'there are 25 salesmen whose salaries were £2200 per annum or more'. Accumulating, the next figure would read 'there are 75 (25 + 50) salesmen whose salaries were £2000 per annum or more', and so on. These accumulated frequencies are not recorded in this table, but the graph for them has been drawn. In this instance, the aggregated frequency has been plotted against *the lower class boundary of the related class.*

Both curves are called *cumulative frequency curves* or *ogives*: the first curve is also referred to as a *'less than' ogive*. They have a particular relevance to matters which will be dealt with in Chapter 2.

However, before leaving this introduction to the topic, one Parthian shot should be fired. If the 'less than' ogive indicated that 800 salesmen earned less than £1700 per annum, the 'more than' ogive ought to support it by recording that 200 salesmen earned £1700 per annum or more. It does just that.

THE LORENZ CURVE

Most management is done by exception. This means that the ordinary performance is allowed to pass by, but the unusually good (or bad) one is subjected to scrutiny. The business behaviour of most employees of a firm is controlled by policy, but a few exceptions can be made for the geniuses who do not easily fit in to the disciplines of policy. Management reviews the performance of its employees by exception, in order to reward the best and encourage (or dismiss) the poorest. Most annual reports on staff cater for this.

The Lorenz Curve is a simple visual device for determining the limits of what is good or poor, so that management can then proceed to identify the people whose performances call for more detailed examination.

Let us consider the problem of a toy-manufacturing firm giving out work for 'making up' in the home. To each of its domestic workers (whom it tries to select for potential service so that they become skilled) it loans a set of somewhat expensive tools. It is worried by the rejection-rate of the returned work, especially from those who return least, and it is considering cutting down its domestic work and taking on more permanent staff. Table 1.6 and Fig. 1.15 depict the results for March quarter 1966.

The first and third columns of the table are a frequency distribution. Column (4) calculates the production from each class, and the figures are aggregated in column (5). Column (6) represents the figures in column (5) as percentages of the total production. Column

TABLE 1.6. *Lorenz Curve. Number of Articles made up by Domestic Workers during March Quarter 1966* [Woomera Toy Co. Ltd.]

Number of articles (1)	Midpoint (2)	Number of workers (3)	Total production (3)×(2) (hundreds) (4)	(4) Cumulative (5)	(5) Rounded percentages (6)	(3) Cumulative (7)	(7) Rounded percentages (8)
200–399	300	300	900	900	12	300	30
400–599	500	200	1000	1900	26	500	50
600–799	700	150	1050	2950	40	650	65
800–999	900	100	900	3850	53	750	75
1000–1199	1100	80	880	4730	65	830	83
1200–1399	1300	70	910	5640	77	900	90
1400–1599	1500	50	750	6390	88	950	95
1600–1799	1700	30	510	6900	95	980	98
1800–2000	1900	20	380	7280	100	1000	100
		1000	7280				

(7) aggregates the frequencies in column (3) and column (8) depicts the figures in column (7) as a percentage of the total domestic staff.

The Lorenz Curve plots column (6) against column (8). Now, if every worker produced exactly the same amount, 20% of the workers would produce 20% of the total production, and so on. The extent to which the Lorenz Curve bends away from the straight line (diagonal) that this theoretical situation would yield, measures

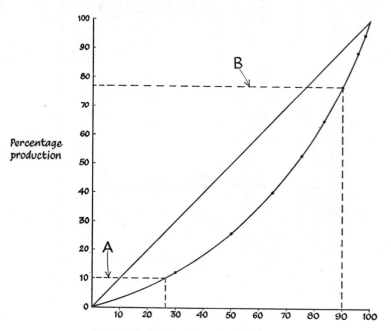

FIG. 1.15. Lorenz curve, data from Table 1.6.

PERCENTAGE WORKERS

Situation A means: The worst 27% of the workers produce only 10% of the total production.

Situation B means: The best 10% of the workers produce 22% of the total production.

the extent to which the production varies as between groups of workers.

What would the management of the Woomera Toy Company read from the curve? It would probably want to know a good deal more about the rejection rates for those workers within the bottom 27% shown in the curve and particulars about their past performance.

If this further investigation showed a history of poor returns from the majority of those in this bottom bracket for the March quarter, with associated unsatisfactory work acceptance, the management would probably decide to take on additional full-time labour and contract its domestic worker staff, retaining only those who were represented on the curve as the top 75% (say) of the total.

The reader will have noticed the weight of frequencies in the lower part of the distribution. This situation is typical of other well-known frequency distributions such as those for incomes and for savings to which the Lorenz Curve also applies.

Prove your prowess [1]

1.1.* The following figures were taken from the *Monthly Digest of Statistics*:

British Railways

Year	Total number of passenger journeys (millions)
1950	981·7
1951	1001·3
1952	989·0
1953	985·3
1954	1020·1
1955	993·9
1956	1028·5
1957	1101·2
1958	1089·8
1959	1068·8
1960	1036·7
1961	1025·0
1962	965·0
1963	938·0
1964	927·6
1965	865·1

Source: Ministry of Transport.

(*a*) Round the figures of the number of passenger journeys to the nearest 10 million.

(*b*) Plot the figures on a graph.   (I.C.W.A., Part I)

[1] An asterisk against the exercise number indicates that the solution is given in detail.

1.2.* Round the commodity export values on the following table to the nearest £1000 and indicate the errors of approximation in the rounded figures.

Commodity	Original values £
Spirits	11 361 617
Beer	1 142 896
Fruit juice and table waters	223 490
Cocoa preparations	
With sugar	547 527
Without sugar	264 771
All other items	283 561
Total	13 823 862

(A.C.C.A., Part I)

1.3.* The following are the ages (to the nearest year) of the 80 employees of a firm, 40 being men and 40 women:

Men				Women			
35	43	45	33	59	45	29	24
20	23	30	24	26	34	25	38
52	20	57	45	41	58	26	47
48	47	27	25	33	51	43	50
25	26	21	40	22	49	25	27
27	44	42	40	31	44	42	29
25	46	34	44	31	30	50	34
41	31	30	35	35	31	35	42
31	42	42	24	48	38	29	28
37	28	39	37	35	23	31	41

(i) Construct frequency distribution tables for

    (a) Men's ages and
    (b) Women's ages.

(ii) Represent the distributions by histograms.

(I.O.M., Part II)

1.4. The following figures represent the weekly expenditure on fruit and vegetables by each of 60 families in a sample:

£	£	£	£	£	£
0·97	0·71	1·05	0·79	1·52	1·37
1·24	1·76	0·77	0·89	0·91	1·02
0·78	0·84	0·91	0·93	1·11	1·24
1·26	0·86	0·97	0·74	0·51	0·37
0·31	1·19	1·47	1·62	0·41	0·47
0·69	0·88	1·08	1·02	1·77	0·69
0·51	0·57	1·27	1·51	1·32	0·66
0·74	0·51	1·57	1·16	1·79	0·64
1·61	0·47	1·16	1·26	1·44	1·82
0·46	0·36	0·21	1·96	2·12	0·86

You are required to:

(a) Arrange the figures in the form of a grouped frequency distribution.

(b) Plot the figures in the form of a histogram.

(I.C.W.A., Part I)

1.5.* Draw a circular or 'pie' chart to illustrate the data in the following table:

Advertising Expenditure by Media

Nature of advertising (1)	Percentage of total expenditure (2)
Press (newspaper, magazine, etc.)	51·4
Catalogues, leaflets, etc.	11·3
Posters and signs	8·4
Window displays, etc.	7·3
Television and radio	5·5
Miscellaneous advertising	16·1
	100·0

1.6. Draw a circular or 'pie' chart to illustrate the data in the following table:

Average Weekly Expenditure of Householders

	£
Housing	1·24
Fuel, light and power	0·86
Food	4·77
Alcoholic drink	0·45
Tobacco	0·87
Clothing and footwear	1·45
Durable household goods	1·09
Other goods	1·04
Transport and vehicles	1·16
Services	1·25
Total	14·18

1.7. From the following data prepare a percentage bar chart:

Shares in the Gross National Product

	Persons £m	Companies £m
1953	13 694	2274
1954	14 482	2600
1955	15 773	2736
1956	16 988	2844
1957	17 951	3036

Source: National Income and Expenditure, 1958.

1.8. Illustrate the following data by means of a histogram and a frequency curve.

Monthly salary £	Number of employees
31–35	2
36–40	3
41–45	4
46–50	9
51–55	12
56–60	7
61–65	4
66–70	1

How does the frequency curve for this distribution differ from a curve showing skewness? (A.C.C.A., Part I)

1.9.*

Weekly commission paid to salesmen in ABC Coy.

Amount of commission (£) (1)	Number of salesmen (2)
2·05 and under 2·50	4
2·50 and under 3·00	7
3·00 and under 3·50	11
3·50 and under 4·00	15
4·00 and under 4·50	25
4·50 and under 5·00	42
5·00 and under 5·50	53
5·50 and under 6·00	38
6·00 and under 6·50	24
6·50 and under 7·00	12
7·00 and under 7·50	5
7·50 and under 8·00	1
	237

(a) Compile a cumulative frequency table of the above data.
(b) Draw an ogive to represent the data in the table.

(I.O.M., Part II)

1.10.

Value of variable	Frequency
20–30	3
30–40	14
40–50	46
50–60	58
60–70	49
70–80	37
80–90	26
90–100	10
Total	243

(a) Plot the ogive using these data.

(b) Estimate by interpolation in the ogive, the number of items with the values of the variable between:

(i) 45–55; (ii) 55–65; (iii) 65–75. (I.O.S., Part I)

1.11.* The sales figures given below relate to the same commodity:

	Total U.K. sales (thousands)	Sales by company A
1954	91	646
1955	95	672
1956	99	704
1957	105	747
1958	111	823
1959	115	902
1960	122	1084
1961	127	1305
1962	133	1561
1963	140	2010
1964	147	2628
1965	154	3478

Plot both sets of figures on the same logarithmic-graph and comment on your findings. (I.C.W.A., Part I)

1.12. Make a semi-logarithmic graph of the two series given below. Comment on the comparison and on the use of ratio scales in general.

Year	Sales (£000)	Selling costs (£000)
1959	153	13·9
1960	130	11·0
1961	168	16·4
1962	197	20·4
1963	214	24·6
1964	191	23·0
1965	208	26·1

(I.O.M., Part II)

1.13. Using ordinary natural scale paper (not semi-logarithmic paper), plot semi-logarithmic graphs of the following two series on the same diagram:

U.K. Passenger Movement by Sea or Air. To and from European Continent and Mediterranean Sea Area

Year	Number of passengers—thousands	
	By sea	By air
1958	4357	2975
1959	4638	3314
1960	4772	4159
1961	5030	4991
1962	5278	5471
1963	5838	6090
1964	6044	6846
1965	6594	7761

(I.O.M., Part II)

Hint. The alternative to plotting natural numbers on a logarithmic scale is to plot logarithms on a natural scale.

1.14.* Construct a Lorenz Curve in respect of the following data:

Weekly Expenditure on Savings Economy Unit Trust Shares

Expenditure	Number of savers	Total expenditure £
under £0·50	620	210
£0·50 and under £1	480	350
£1 and under £5	150	360
£5 and under £15	60	600
£15 and above	50	840

Explain the purpose of the curve and its interpretation in relation to the statistics in the table.   (A.C.C.A., Part I)

1.15. Construct a Lorenz Curve in respect of the following data:

New business orders	Number of salesmen
£5 000 but less than £10 000	4
£10 000 but less than £15 000	17
£15 000 but less than £20 000	53
£20 000 but less than £25 000	10
£25 000 but less than £30 000	8

For what purposes would the Sales Director use this curve?

1.16.* The monthly turnover (£000) of a business organisation for the years 1965 and 1966 were as follows:

Monthly Sales (£000)

Month	Year 1965	1966	Month	Year 1965	1966	Month	Year 1965	1966
Jan.	27	30	May	55	50	Sep.	35	32
Feb.	38	35	Jun.	50	52	Oct.	42	70
Mar.	40	49	July	65	50	Nov.	51	61
Apr.	42	49	Aug.	55	40	Dec.	38	42

Construct a 'Z' Chart for 1966, i.e. a chart showing:

   (*a*) The monthly turnover.
   (*b*) The cumulative monthly total.
   (*c*) The moving annual total.   (I.O.M., Part II)

1.17. Plot a 'Z' chart from the following data, using a different scale for the moving annual total figures:

	Jan.	Feb.	Mar.	Apr.	May	Jun.
1962	27	12	53	30	35	70
1963	43	39	73	59	51	77

	July	Aug.	Sep.	Oct.	Nov.	Dec.
1962	42	33	65	57	59	102
1963	56	52	68	73	69	141

1.18. The production programme (in units) for a workshop, and the production actually achieved, were as follows:

	Programme	Achieved		Programme	Achieved
Jan.	78	50	July	100	140
Feb.	50	25	Aug.	90	100
Mar.	100	125	Sep.	110	110
Apr.	80	90	Oct.	120	130
May	85	70	Nov.	100	95
Jun.	120	90	Dec.	130	120

Draw a Gantt Chart to represent the situation.
What does the chart show?

REFERENCES

A combined bibliography for Chapters 1 and 2 appears at the end of
   Chapter 2.

2: ABOUT AVERAGE

Measures of Location and Dispersion

INTRODUCTION—THE CONCEPTS OF 'AVERAGE'
AND 'SPREAD'

FEW people are gifted enough to see a page of figures and grasp at a glance the pattern of events which it reflects.

Consider for example the information about salesmen's salaries given in Table 2.1. The first thing which people would want to know in trying to understand the full meaning of data such as this, is the value of a representative salary, or what one might call a 'fulcrum' salary around which the others are balanced. This is usually referred to as the 'average' salary, but as will be shown shortly, this phrase has many meanings, and is best avoided for a while. It is true, however, that the first constructive stage in interpreting this information is to find a salary which is centrally placed and which can be appointed as a salary 'delegate'. Common sense and the pattern of frequencies tells us that the salary we are seeking in Table 2.1 will be found not far from the range £1200 to £1400.

But even if we are able to calculate this central value in some way or other, our condensed picture of the distribution will still be very far from complete. The salaries in the table range from £800 to £2400 per annum, and, as can be seen from the frequency column, the majority lie in the lower part of the table, with the remainder becoming more attenuated towards the higher values.

The frequency table could have looked quite different from this and yet have possessed the same central value. A large proportion of the salaries could have been in the '£1400 less than £1600' class with a few rapidly diminishing frequencies in the higher and lower classes.

This line of reasoning suggests that if our mind's eye picture of the salary figures is to be complete, we must know not only the salary which lies at or near the centre of the distribution but also whether the remaining values are clustered tightly around it, or spread widely over a larger range.

These concepts are best thought of in terms of mental pictures.

TABLE 2.1. *Calculation of Arithmetic Mean, Median, Mode and Standard Deviation. Salesmen's Salaries 1966*
[Vito Food Co. Ltd.]

Salary class (1)	Class midpoint x (2)	Frequency f (3)	Col. (3) Cumulative (4)	$\dfrac{\text{Col. (2)} - 1300}{200}$ d' (5)	fd' (6)	fd'^2 (7)
800 less than 1000	900	50	50	-2	-100	200
1000 less than 1200	1100	200	250	-1	-200	200
1200 less than 1400	1300	350	600	0	-300	0
1400 less than 1600	1500	150	750	$+1$	$+150$	150
1600 less than 1800	1700	100	850	$+2$	$+200$	400
1800 less than 2000	1900	75	925	$+3$	$+225$	675
2000 less than 2200	2100	50	975	$+4$	$+200$	800
2200 less than 2400	2300	25	1000	$+5$	$+125$	625
					$+900$	3050
					-300	
					$+600$	

1. $\bar{x}$ = Arithmetic mean = $\bar{x}_a + \dfrac{c\Sigma fd'}{n} = 1300 + \left(\dfrac{200 \times 600}{1000}\right) = \text{£1420.}$

2. Median (Value 500th term) = $1200 + \left(\dfrac{250}{350} \times 200\right) = \text{£1342}\tfrac{6}{7}.$

3. Lower quartile = £1200: Upper quartile = £1600. Semi-interquartile range is $\dfrac{1600 - 1200}{2} = \text{£200.}$

4. Standard deviation = $s = c\sqrt{\dfrac{\Sigma fd'^2}{n} - \left(\dfrac{\Sigma fd'}{n}\right)^2} = 200\sqrt{\dfrac{3050}{1000} - \left(\dfrac{600}{1000}\right)^2} = \text{£320 (approx.)}$

Staying for a while with the idea of a salary distribution, Table 2.2 depicts two contrived distributions which are manifestly different in structure. One can see from inspection that they are both sym-

TABLE 2.2. *Frequency Distributions with Identical Central Values but Different 'Spreads'*

Salary midpoint £ p.a.	Number of salaries (frequency)	
	A	B
900	50	0
1100	100	0
1300	150	50
1500	200	150
1700	250	850
1900	200	150
2100	150	50
2300	100	0
2500	50	0
	1250	1250

metrical about the midpoint of £1700 per annum, and that salary will be the 'representative' salary for both distributions. Beyond that, however, there is little resemblance. Distribution A has no class

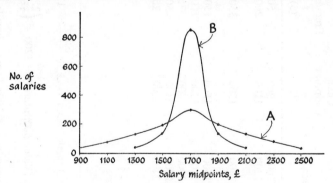

No. of salaries

Salary midpoints, £

FIG. 2.1. Data from Table 2.2.

frequency which is greatly different from its immediate neighbour: it gives the appearance of a slowly changing pattern. Not so distribution B, whose central class frequency is so large that it dwarfs its diminutive colleagues. Moreover, its whole range extends only from £1200 to £2200 per annum. In short, the frequencies surrounding the central class frequency are clustered very tightly around it.

These hypothetical distributions are graphed in Fig. 2.1, which illustrates what is meant by the 'spread' of frequencies around the central value of the variate. Distribution A is widely spread about its central value, and distribution B is closely spread about it. One of the objects of this chapter will be to discover a way of determining this degree of spread. Another will be to examine closely the concept of central value or average and determine ways of measuring that also. In other words we are going to establish measures for *dispersion* and *central location*: but we shall deal with them in the reverse order.

ELEMENTARY SYMBOLISM

Our approach to both these tasks, however, will be simplified if we can first accept the discipline of a little symbolism. Those readers who rise on their haunches at such a suggestion should be assured that the fear of mathematical symbolism disappears when faced four square. It is a gentle giant! After all, mathematical symbols are only a form of shorthand to obviate a great deal of writing and explanation.

Imagine that the salary of every one of the 1000 salesmen from Table 2.1 was known. They could be written: £800, £804, £805, £810, ..., and so on to £2400. If a clerk in the office of the Vito Food Company was asked to calculate the average salary from all the information he had, he would probably acquire an adding machine, add together the 1000 salaries, and finally divide by 1000. In doing so he would be calculating one of the many forms of central location given the definitive name of *arithmetic mean*. In arithmetic symbols this would be expressed as

$$\frac{800+804+805+810+...+2400}{1000}.$$

This method of deriving the arithmetic mean is a general one—it applies to any series of numbers whose mean is required. It can be expressed in general terms by replacing the arithmetic numbers by algebraic symbols.

C

Let x_1 be the first of a series of numbers

x_2 be the second of a series of numbers

x_3 be the third of a series of numbers

....

x_n be the nth and last of a series of numbers.

The arithmetic mean would then be

$$\frac{x_1 + x_2 + x_3 + \ldots + x_n}{n}.$$

Now, it is very tedious and time-consuming to write down the numerator of this quotient every time an expression for the total of a series of numbers is required, so an apt mathematical symbol has been devised. It makes use of the Greek capital letter S, written as Σ and called *sigma*. This letter Σ is shorthand for 'the *sum* of all such terms as'.

Thus

$$\sum_{i=1}^{n} x_i$$

which a few minutes ago would have presented a repelling prospect, we now know to be the shorthand grammalogue for 'the sum of all such terms as x counted from x_1 to x_n inclusive'. In other words

$$\sum_{i=1}^{n} x_i = x_1 + x_2 + x_3 + \ldots + x_n.$$

It does not matter what letter of the alphabet is used as the suffix for x: some books use x_j in which case the whole symbol would be

$$\sum_{j=1}^{n} x_j.$$

The arithmetic mean of a series of numbers x_1, x_2, x_2, ... x_n can now be written

$$\frac{1}{n} \sum_{i=1}^{n} x_i$$

but even this is rather longwinded, and it is usually abbreviated to

$$\frac{\Sigma x}{n}.$$

But in the same way that even shorthand grammalogues have their own contractions, it is conventional to further contract

$$\frac{\Sigma x}{n} \text{ to } \bar{x}, \text{ called '} x \text{ bar'}.$$

Summarising: the arithmetic mean of $x_1, x_2, x_3, \ldots x_n$ is

$$\frac{1}{n} \sum_{i=1}^{n} x_i.$$

This is more commonly written as $\dfrac{\Sigma x}{n}$ and can be further abbreviated to $\bar{x}$. The formulae employed to measure dispersion use these same symbols, but they will not be developed until a later stage in this chapter.

MEASURES OF CENTRAL LOCATION

During the above introduction to symbolism the reader was made familiar with the measure of central location which is most commonly understood, the arithmetic mean. Let us now look at this in a very practical way through the frequency distribution of Table 2.1. If we were asked to deduce the mean of this distribution by inspection, it would not be unreasonable to argue: 'The largest frequency is in the £1200 less than £1400 class and 600 of the salaries fall below £1400. The arithmetic mean is therefore probably at the lower end of the class at about £1250 p.a.'

This would be incorrect. The mean is in fact £1420 (the method of calculating this from a grouped frequency distribution will shortly be demonstrated) and it is important to understand why this is so. Although 600 salaries fall below £1400, the remaining 400 are spread over salary classes which reach as far as £2400, and those falling in the higher classes tend to pull the arithmetic mean away from the lower values where one would intuitively say it lies. This can be seen in pictorial form in Fig. 2.2. The graph is not symmetrical, and that portion of it running away to the right is called the *tail*. It is the variate values forming this tail which influence the arithmetic mean away from those which command the larger frequencies. (It ought to be said in parenthesis that a curve of this shape—called a *bell-shaped curve*—has two tails, one on each side of the peak; but when the curve is lop-sided or *skew*, the trailing part of the curve is frequently referred to as *the* tail.) The curve in Fig. 2.2 is positively skewed.

Now, it so happens that the frequency distribution in Table 2.1 is not grossly out of symmetry, but it would not be difficult to imagine some that are. Consider, for example, amounts paid as motor claims by an insurance company. They represent payments for accidents which caused damage to a car, or personal injuries to a third party, or a combination of both. A frequency distribution of payments for car damage would vary from (say) £15 for a damaged

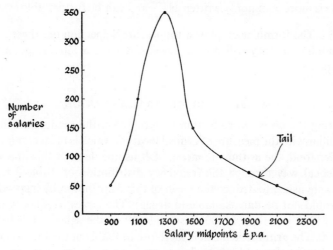

FIG. 2.2. Graph of grouped frequency distribution. Data from Table 2.1.

bumper-bar to (say) £600 for a vehicle write-off, with the maximum frequency at a figure of about £40 for a bent wing and broken head-lamp. Personal injuries to a third party, however, give rise to very large claims amounting to thousands of pounds. In considering motor claim payments as a whole, therefore, one is considering heterogeneous data which would yield a very distorted frequency distribution with an arithmetic mean pulled away from the variate-values carrying the larger frequencies. In such circumstances as these the arithmetic mean loses its quality of 'representativeness' and one is forced into questioning if it has any meaning at all. It needs but one motorist to drop into a brown study whilst he is at the wheel of his car, plough through a waiting bus-queue outside the law courts and maim a bevy of discoursing judges, to lift considerably the arithmetic mean of claim payments! The lack of real meaning to this sort of average has prompted one jaundiced economist to write, 'If a man has one foot on a hot stove and the other in a refrigerator, a statistician would say that on average he was comfortable'. This is an unkind cut, but his point is well taken, and we are forced into searching for a form of average which, in the circumstances mentioned above, would be centrally representative of the distribution.

The solution is the most obvious one: if a central value is required, the most suitable would seem to be the value of the variate positioned exactly in the middle of the distribution. If the thousand salesmen of Table 2.1 were lined up in a single row in ascending order of the salary they earned, lowest on the left and highest on the right, then the new representative value which is being sought would be the salary of the salesman standing in the middle of the long line. The reader is correct to point out that there is no man *exactly* in the middle because there is an even number of salesmen, but the error is negligible if we settle for the man standing 500th from the left of the line.

The name given to this form of average is the *median*, and, again, the procedure for calculating its value and a fuller discussion of its properties will be given at the end of this introductory section.

The limitations to the use of the arithmetic mean as a measure of central location would now seem to be met by the use of the median in appropriate circumstances. However, the phrase 'central location' itself suggests a third measure. It is the value of the variate which occurs most frequently in the distribution. In an ungrouped distribution it is quite clearly (by definition) the variate with the largest frequency, and is called the *mode*. This measure is of value to such enterprises as shoe manufacturers who are interested to know that $8\frac{1}{2}$ is the *modal* size of their range, or to off-the-peg tailors who will wish to know the modal value for height, chest measurement, etc.

The arithmetic mean, the median and the mode are not the only measures of central location employed by the statistician, but they are the most important. Some further measures known sometimes as the *minor means* will be dealt with later on in this chapter. but for the moment we will stop and investigate the properties of the first three and learn how to calculate them.

THE ARITHMETIC MEAN

The arithmetic mean of a set of ungrouped numbers $x_1, x_2, ..., x_n$ has already been defined as

$$\bar{x} = \frac{\Sigma x}{n}.$$

Now, it could be that some of these values of the variate are identical. Let us suppose that there are

$$f_1 \text{ values of } x_1$$
$$f_2 \text{ values of } x_2$$
$$\vdots$$
$$f_k \text{ values of } x_k$$

Now, we know

$$\sum_{i=1}^{k} f_i = n$$

because there were n members of the first series and all we have done is to collect those of the same value and aggregate them. (For example, 1, 1, 2, 3, 3, 3, 4, 4, 5, 5, 5, 5, where $n = 12$, can be restated as $(2 \times 1), (1 \times 2), (3 \times 3), (2 \times 4), (4 \times 5)$.)

Therefore

$$\bar{x} = \frac{\sum_{i=1}^{k} f_i x_i}{\sum_{i=1}^{k} f_i} = \frac{f_1 x_1 + f_2 x_2 + ... + f_k x_k}{f_1 + f_2 + ... + f_k}$$

$$= \frac{\Sigma f x}{n}$$

Weighted arithmetic mean. The last section leads to the consideration of a concept based on the arithmetic mean, called the *weighted arithmetic mean*, which is used extensively in the construction of index numbers (see Chapter 3).

The principle is best illustrated by an example.

In the early days of the Second World War the government removed the tax payable on pleasure yachts—because none were being made. The cartoonist of a famous daily newspaper made capital of the announcement by depicting in his drawing a queue of hard-bitten 'cloth-cap' housewives outside the fashionable showroom of a boatyard, with their purses in their hands. The humour of the cartoon lay in the incongruity of the situation. The housewives would have been more affected by a reduction in the price of meat or bread, less affected by, say, cheaper radio sets. Clearly in gauging the impact on the living standards of lower income groups, of changes in the prices of different commodities, regard must be had for the proportion of total income which is spent on the separate commodities. This can be achieved by *weighting* the separate price changes by this proportion. For example, if the prices of the components of household expenditure for 1965 were the following percentages of the 1964 prices, and the proportions of total spending were as shown against each commodity, the weighted arithmetic mean of price percentage changes would be

$$\frac{(112 \times 25)+(108 \times 20)+(105 \times 8)+(115 \times 15)+(102 \times 10)+(115 \times 22)}{100}$$

$$= \frac{11\,075}{100} = 110 \cdot 75.$$

Commodity (1)	1965 price (Percentage of 1964 price) (2)	Percentage of income spent on commodity (3)
		%
Rent and rates	112	25
Food	108	20
Drink	105	8
Clothing	115	15
Entertainment	102	10
Other expenses	115	22
		100

In the above table, each percentage appearing in column (3) has been used as a *weight*.

The choice of subject to be used as a medium for weighting a series of values is left to the statistician. All that need be said at this stage, is that the subject must have a specific bearing on the mean value which is being calculated.

The advantages and disadvantages of the use of the arithmetic mean. The arithmetic mean has the big advantage of being widely understood and therefore trusted. It also makes use of all the known data, since inconveniently high or low values cannot be excluded at the whim of the person using the material. As has been suggested earlier, however, its main function as a representative value fails if there are some extraordinarily high or low values included. Unfortunately, on occasions the value of the arithmetic mean does not coincide with any of the values in the distribution. Sometimes this latter characteristic gives an absurd result like 4·12 people living in a single household or 28·6 ships unloading at port at the same time.

The properties of the arithmetic mean. The properties of the arithmetic mean are used in the mathematical development of the subject in the following pages. The reader must accept the mathematical treatment of the subject in the next few sections as constituting the grammar of his new management language, and therefore will know that the temporary discipline it demands has its rewards in later comprehension.

Property (a). The algebraic sum of the deviations of a set of numbers from their arithmetic mean is zero (i.e. $\Sigma(x - \bar{x}) = 0$), e.g. the mean of 1, 6, 8, 11, 14 is $\frac{40}{5} = 8$.

Deducting 8 from each of the numbers in the series and adding the result: $-7 -2 +0 +3 +6 = 0$.

Property (b). If in k series of numbers

the first series has a total frequency of f_1 and a mean of m_1
the second series has a total frequency of f_2 and a mean of m_2
the kth series has a total frequency of f_k and a mean of m_k,

then the mean of all series combined is

$$\bar{x} = \frac{f_1 m_1 + f_2 m_2 + \ldots + f_k m_k}{f_1 + f_2 + \ldots + f_k},$$

e.g. if series 1 is $2+8+20$, $f_1 = 3$ $m_1 = 10$

2 is $1+3+4+8$, $f_2 = 4$ $m_2 = 4$

3 is $12+30$, $f_3 = 2$ $m_3 = 21$,

Then

$$\bar{x} = \frac{(3 \times 10) + (4 \times 4) + (2 \times 21)}{9} = \frac{88}{9} = 9\tfrac{7}{9}.$$

It will be appreciated that all that is being done is to calculate the sum of the values of the variate in each of the three series before finally adding them and dividing by the total frequency. The data is being reconstituted so that the formula

$$\frac{\Sigma x}{n}$$

can be employed.

If $\qquad\qquad \bar{x} = \dfrac{\Sigma x}{n}$

then $n\bar{x} = \Sigma x$: this calculation has been made for each of the series, and the arithmetic mean of the combined series obtained by dividing the combined aggregate by the total frequency.

Property (c). If $\bar{x}_a$ is a guessed (i.e. arbitrarily chosen) mean, x_i is any one of the series of values $x_1, x_2, ..., x_n$ and $d_i = x_i - \bar{x}_a$,

then $\bar{x} = \bar{x}_a + \dfrac{\Sigma d}{n}$

Since $\qquad\qquad \bar{x} = \dfrac{x_1 + x_2 + ... x_n}{n}$

$$\bar{x} = \frac{(d_1 + \bar{x}_a) + (d_2 + \bar{x}_a) + ... + (d_n + \bar{x}_a)}{n}$$

$$\therefore \bar{x} = \frac{n\bar{x}_a + \sum_{i=1}^{n} d_i}{n} = \bar{x}_a + \frac{\Sigma d}{n}$$

or, for a grouped frequency distribution and using the class mid-points:

Since $\qquad \bar{x} = \dfrac{f_1 x_1 + f_2 x_2 + ... + f_k x_k}{n}$,

therefore $\qquad \bar{x} = \dfrac{f_1(d_1 + \bar{x}_a) + f_2(d_2 + \bar{x}_a) + ... + f_k(d_k + \bar{x}_a)}{n}$,

and $\qquad\quad \bar{x} = \dfrac{\sum_{i=1}^{k} f_i d_i + \bar{x}_a \sum_{i=1}^{k} f_i}{n} = \dfrac{\Sigma fd}{n} + \bar{x}_a$

$$\text{(since } \Sigma f = n\text{).}$$

As a numerical illustration: if a series of numbers is 8, 13, 15, and x_a the guessed mean is 14,

then
$$d_1 = 8-14 = -6$$
$$d_2 = 13-14 = -1$$
$$d_3 = 15-14 = 1$$

Substituting in the formula

$$\bar{x} = \bar{x}_a + \frac{\Sigma d}{n}$$

$$\bar{x} = 14 + \left(\frac{-6}{3}\right) = 12$$

which can be seen from inspection.

Again: if (2×10), (3×12), (2×14) represent the frequencies and midpoints of a grouped frequency distribution, and $\bar{x}_a$ the guessed mean $= 11$,

then
$$d_1 = 10-11 = -1$$
$$d_2 = 12-11 = 1$$
$$d_3 = 14-11 = 3$$

Then substituting in the formula

$$\bar{x} = \frac{\Sigma fd}{n} + \bar{x}_a = \frac{(2 \times -1) + (3 \times 1) + (2 \times 3)}{7} + 11$$

therefore
$$\bar{x} = \frac{(-2) + (3) + (6)}{7} + 11$$

and
$$\bar{x} = \frac{7}{7} + 11 = 12,$$

which again can be seen from inspection.

Property (*d*). The sum of squares of the deviations of a set of numbers from their arithmetic mean is a minimum.

Example. Consider the series in property (*a*) above

$$1, 6, 8, 11, 14. \quad \bar{x} = 8.$$

$$(1-8)^2 + (6-8)^2 + (8-8)^2 + (11-8)^2 + (14-8)^2$$

$$= 49 + 4 + 0 + 9 + 36 = 98.$$

Repeat for $\bar{x}_a = 7$.

$$(1-7)^2 + (6-7)^2 + (8-7)^2 + (11-7)^2 + (14-7)^2$$

$$= 36 + 1 + 1 + 16 + 49 = 103.$$

And for $\bar{x}_a = 9$

$$(1-9)^2+(6-9)^2+(8-9)^2+(11-9)^2+(14-9)^2$$
$$64\ +\ \ 9\ +\ \ 1\ \ +\ \ 4\ \ \ +\ \ 25\ \ \ = 103.$$

This is, of course, no proof of the proposition but it demonstrates the likelihood of its truth.

The reader will, it is hoped, have noticed in this section on the properties of the arithmetic mean how a few very simple numerical examples can take the steam out of some pretentious looking mathematical formulae.

Calculating the arithmetic mean from the data of Table 2.1. In Table 2.1 the basic material from Table 1.1 has been further developed to include the calculations for the arithmetic mean. Now, the formula for determining the arithmetic mean from a grouped frequency distribution has already been proved to be

$$\bar{x} = \bar{x}_a + \frac{\Sigma fd}{n}.$$

It would be possible to take the values of the class midpoints in column (2) and the frequencies in column (3), to use £1300 as the guessed or arbitrary mean, and to substitute directly in the formula. However, much work can be saved by making use of the fact that the values remaining after £1300 has been subtracted from the midpoints, are all multiples of 200. In column (5), therefore, each of the differences (d_i) has been divided by 200. The results are said to be in *working units*. We have to remember now, in applying the formula, that $\frac{\Sigma fd}{n}$ (using the usual notation) is 200 times too small. For this reason column (5) has been headed not 'd' but 'd''', in which 'd''' is the shorthand for 'd expressed in working units'. We can now write

$$\bar{x} = \bar{x}_a + \frac{\Sigma fd}{n} = \bar{x}_a + c\frac{\Sigma fd'}{n}$$

where $c =$ the constant factor, in this instance 200.

Substituting now in the formula

$$\bar{x} = 1300 + 200\left(\frac{600}{1000}\right) = 1300 + 120 = \text{£}1420 \text{ p.a.}$$

a result which was quoted in the earlier part of this chapter.

The figures in this table have been easy to use and fairly small, but in practice some distributions have very large total frequencies, and it is for these situations that this device for calculating the arithmetic mean is designed. One should also remember that the work is reduced by the use of calculating machines when large and difficult frequencies are involved.

THE MEDIAN

Recapitulating from the earlier section on measures of central location, the median is the value of the variate in the middle of the series, or the arithmetic mean of the two middle values if the series has an even number of terms.

One considerable advantage attaching to the median is that it is very often the value of an actual member of the series. Moreover, it can be calculated even though some of the extreme values are unknown, although it is essential that we should know of their existence. Although the usefulness of the median is limited because it is not mathematically tractable, it clearly has a specific role in the understanding of statistical data.

Calculating the median of a grouped frequency distribution. The median of the distribution in Table 2.1 is, strictly speaking, the arithmetic mean of the 500th and 501st salaries when arranged in order of size. They are so arranged in the distribution. It will suffice, however, to calculate the value of the 500th. The error will be negligible.

Now, column (4) shows the cumulative frequencies from the lower values of the variate. According to the figures in column (4), there are 250 salaries below £1200 p.a. and 600 below £1400. The value of the 500th salary must therefore lie somewhere in the salary class '£1200 to £1400'. It will be an easy matter to determine its exact value if we remember the assumption that the salaries falling within each salary class are equally spread throughout the class. The $(500 - 250)$th or 250th salary along the class £1200 to £1400, which has a frequency of 350, is required. By simple proportion and remembering the class interval of £200, this becomes

$$1200 + \left(\frac{250}{350}\right) \times 200 = £1342\tfrac{6}{7} \text{ p.a.}$$

By concocting appropriate symbols it is possible to create a formula for this calculation, but the reader is recommended to make a common-sense approach. For the sake of completeness, however:

Let n = total frequency
$\quad l$ = lower class boundary of the median class
$\quad n_1$ = sum of the frequencies for all classes lower than the median class
$\quad f$ = frequency of the median class
$\quad c$ = interval of the median class.

$$\text{Then median} = l + \left(\frac{\frac{n}{2} - n_1}{f}\right) c.$$

The median can also be calculated graphically. In Fig. 1.14 the 'more than' and 'less than' ogives cross at the ordinate representing 500 salesmen. This is to be expected, because if there are 500 salesmen with salaries of *less than* £1340 p.a., there must be, in a total sales force of one thousand, 500 with a salary *greater than* that amount. By definition, then, the salary on the graph corresponding to the point of intersection of the two curves, is the median salary. Within the limits of graphical accuracy the result tallies with the arithmetical result.

The methods employed to determine the median can be used to find the value of the salary exactly one-quarter of the way along the distribution from its lower or upper ends: or one-tenth, two-tenths and so on, from the lower end of the distribution. These former values are called *quartiles* and will be mentioned again in the section on dispersion; the latter values are referred to as *deciles*. The idea can be extended still further to *percentiles*; these are used frequently by the teaching profession in grading the results from examinations.

A brief summary of the above in numerical terms would be: In a distribution having a total frequency of 1000, the quartiles would be the values of the 250th and 750th terms when the values were arranged in order of size: the deciles, the values of the 100th, 200th, etc., terms: the percentiles, the values of the 10th, 20th, etc., terms.

THE MODE

The mode of a series of numbers is that which occurs most frequently in the series. There may be one or more modes—indeed there may be no mode at all, or so many that the concept fails. In the *unimodal* series 5.6.6.6.7.7.8.9 the mode is 6. If the series had been 5.6.6.7.7.8.9 it would have been classed as *bi-modal*. In any ungrouped series the mode or modes can be found by arrangement and inspection, but since the identity of individual values is lost in the classes of a grouped distribution, the true mode or modes are difficult to determine accurately.

Consider again Table 2.1, the modal class (i.e. that with the greatest frequency) is £1200 to £1400. One is tempted to quote the class midpoint as the mode, but the more purist approach is to take note of the frequencies of the classes adjoining the modal class, and to incline the modal value towards the greater of them according to the formula:

$$\text{Mode} = l + \left(\frac{D_1}{D_1 + D_2}\right)c,$$

where l = the lower boundary of the modal class
D_1 = difference between frequency of the modal class and the frequency of the preceding class, counted as positive
D_2 = difference between the frequency of the modal class and the frequency of the following class, counted as positive
c = interval of the modal class.

THE GEOMETRIC MEAN

In Chapter 1 the construction and use of ratio-scale graphs was demonstrated. In these graphs the visual representation of a rate of change was considered, and in particular the straight line graph produced by a geometric progression such as 3, 9, 27, 81, or, in algebraic terms, a, ar, ar^2, ar^3, where r is the common ratio. A review of this idea will help the reader to understand the use of the geometric mean.

The geometric mean is defined generally as $\sqrt[n]{(x_1 x_2 x_3 ... x_n)}$. It is used whenever the proportional change of successive values of a series is known. For example, if the cost of stationery rose 7%, 13% and 22% in successive years, the average proportional rise would *not* be the arithmetic mean of 14%, but would be derived from the geometric mean of the values 107, 113, 122.

Now $\sqrt[3]{(107 \times 113 \times 122)} = 113 \cdot 8$

i.e. the average proportional rise is 13·8%.

The reader will appreciate that the geometric mean cannot be calculated from negative values. It is equal to or less than the arithmetic mean.

THE HARMONIC MEAN

The harmonic mean is defined as the reciprocal of the arithmetic mean of the reciprocals of the values,

i.e.
$$H = \frac{1}{\dfrac{\dfrac{1}{x_1} + \dfrac{1}{x_2} + ... + \dfrac{1}{x_3}}{n}} = \frac{1}{\dfrac{1}{n}\sum\dfrac{1}{x}},$$

It is seldom used in statistical reckoning, and is mentioned here only for the sake of completeness. One situation to which it must be applied, however, is as follows.

There are two grades of petrol on the market, one for selling at 5 gallons for £1, the other at 4 gallons for £1. Now if one pound's worth of each petrol is bought, nine gallons of petrol have cost £2, and the average price is 4·5 gallons for £1. If, however, equal quantities of petrol are bought—say 20 gallons, the cost would be £4 + £5. The average cost would then be $\frac{40}{9}$ gallons for £1 = $4\frac{4}{9}$ gallons. This is not the arithmetic mean, but the harmonic mean. Substituting in the formula

$$H = \frac{1}{\dfrac{\frac{1}{5}+\frac{1}{4}}{2}} = \frac{2}{\frac{9}{20}} = \frac{40}{9} = 4\tfrac{4}{9} \text{ gallons for £1.}$$

The distinction is not easy to grasp. If one is dealing with rates expressed as 'x per y' and y remains constant, the arithmetic mean is appropriate. If x remains constant, the harmonic mean must be calculated.

DISPERSION

In the introduction to this chapter it was suggested that, in order to gain a 'mind's-eye' picture of a distribution, we needed first a form of average, and secondly a measure of the 'spread' of all the other values around the average. A more correct term for 'spread' is 'dispersion'.

MEAN DEVIATION

One obvious indicator of the dispersion of values around the arithmetic mean is the sum of the differences of all values from the arithmetic mean (all counted as positive) divided by the total frequency. This would be symbolically expressed as

$$\frac{1}{n} \Sigma \mid x - \bar{x} \mid,$$

the vertical lines being the shorthand for 'the difference counted as positive'.

Numerically: In the series 5, 8, 9, 14, $\bar{x} = 9$

$$\mid 5-9 \mid + \mid 8-9 \mid + \mid 9-9 \mid + \mid 14-9 \mid = 4+1+0+5 = 10.$$

The indicator $\frac{10}{4} = 2\cdot5$ is called the *mean deviation*, a term which explains itself: the average amount by which each term in the series differed from the arithmetic mean of the series was 2·5 units.

For a grouped frequency distribution the expression

$$\text{Mean deviation} = \frac{1}{n} \sum_{i=1}^{n} \left| x_i - \bar{x} \right| = \frac{1}{n} \Sigma \left| x - \bar{x} \right|$$

becomes

$$\frac{1}{n} \sum_{i=1}^{k} f_i \left| x_i - \bar{x} \right| = \frac{1}{n} \Sigma f \left| x - \bar{x} \right|.$$

The mean deviation can also be expressed as a mean difference from the median. When it is, the sum of the differences is less than that derived from any other parameter.

THE INTERQUARTILE RANGE

The quartiles have previously been mentioned in connection with the calculation of the median. The *lower quartile* is the value of the variate exactly one quarter of the way along the distribution: this means that 25% of the total frequency lies *below* the lower quartile value. Similarly the value *above* which lie 25% of the total frequency is the *upper quartile*. The middle 50% of the total frequency therefore lies between the two quartiles. The difference between the values of the lower and upper quartiles is consequently taken as a measure of dispersion and is known as the *interquartile range*. Alternatively, this same difference is halved and called the *semi-interquartile range*. It is calculated in the same way as the median. The graphical method of measuring its value is depicted in Fig. 1.14. The reader will notice that the value of

(lower quartile + semi-interquartile range) or

(upper quartile − semi-interquartile range)

is *not* the value of the median, nor will it be, except in a symmetrical series.

VARIANCE AND STANDARD DEVIATION

Although the mean deviation and interquartile range are simple and effective measures of the dispersion of a distribution about its average, neither are in fact the one that is normally used. This is because they are not mathematically tractable. The principal and most important measure of dispersion is the *standard deviation.*

In a way the standard deviation is a development of the mean deviation. When calculating this, the arithmetic mean was subtracted from each value, and the result treated as positive. The standard deviation uses the same concept but deals with the 'sign' problem by *squaring* each difference.

But first of all a definition.

The standard deviation of a set of numbers x_1, x_2, x_3, ..., x_n is

$$s = \sqrt{\left\{\frac{1}{n}\sum_{i=1}^{n}(x_i-\bar{x})^2\right\}} = \sqrt{\left\{\frac{1}{n}\Sigma(x-\bar{x})^2\right\}}$$

in the usual notation.

The quantity beneath the square root sign is known as the *variance* of the distribution: the qualities and constituents of the variance will be dealt with in Chapter 12 where the analysis of variance technique is discussed.

For a grouped frequency distribution this becomes

$$s = \sqrt{\left\{\frac{1}{n}\sum_{i=1}^{k}f_i(x_i-\bar{x})^2\right\}} = \sqrt{\left\{\frac{1}{n}\Sigma f(x-\bar{x})^2\right\}}.$$

Conventionally, the mathematical symbol for the standard deviation is '*s*' as above or the Greek small letter for '*s*'—again called sigma— and written σ. This matter will be developed in Chapter 8.

The value for $\bar{x}$ is usually likely to be fractional, and if there were no short-cut, the calculation of *s* from the above formula would be very difficult and lengthy.

Fortunately there is a quick method, but the first thing is to derive mathematical authority for its use, on the assumption that we all like to know what we are doing. The reader is given every assurance that, however formidable the following algebra may look, it is in fact of the most elementary kind, and it will be of the greatest benefit to his ego to know that he can easily follow it with a modicum of concentration.

Now, if $s = \sqrt{\left\{\frac{1}{n}\Sigma(x-\bar{x})^2\right\}}$, then $s = \sqrt{\left\{\frac{1}{n}\Sigma(x^2-2x\bar{x}+\bar{x}^2)\right\}}$

$$= \sqrt{\left\{\frac{1}{n}(\Sigma x^2 - 2\bar{x}\Sigma x + n\bar{x}^2)\right\}}.$$

But

$$\frac{1}{n}\Sigma x = \bar{x}.$$

Therefore

$$s = \sqrt{\left\{\frac{\Sigma x^2}{n} - 2\bar{x}^2 + \bar{x}^2\right\}}$$

i.e.

$$s = \sqrt{\left\{\frac{\Sigma x^2}{n} - \bar{x}^2\right\}}.$$

This result is a very important one, which is expressed by the jingle: 'The standard deviation of a distribution is the square root of *the mean of the squares less the square of the mean*'.

(If the reader is uncertain why $\Sigma\bar{x}^2 = n\bar{x}^2$, he should take three values and list the results, e.g. of the three numbers 2, 3, 7, $\bar{x} = 4$.

$$(2-4)^2 = 2^2 - 2.2.4 + 4^2$$
$$(3-4)^2 = 3^2 - 2.3.4 + 4^2$$
$$(7-4)^2 = 7^2 - 2.7.4 + 4^2$$

Therefore $\Sigma(x-4)^2 = \Sigma x^2 - 2.4\Sigma x + 3.4^2$.

Now, in the notation previously used

$$d_i = x_i - \bar{x}_a \quad \text{or} \quad x_i = d_i + \bar{x}_a$$

Therefore, dropping the suffices,

$$\sqrt{\left\{\frac{\Sigma x^2}{n} - \bar{x}^2\right\}} = \sqrt{\left\{\frac{\Sigma x^2}{n} - \left(\frac{\Sigma x}{n}\right)^2\right\}}$$

$$= \sqrt{\left[\frac{\Sigma(d+\bar{x}_a)^2}{n} - \left\{\frac{\Sigma(d+\bar{x}_a)}{n}\right\}^2\right]}$$

$$= \sqrt{\left\{\frac{\Sigma(d^2+2d\bar{x}_a+\bar{x}_a^2)}{n} - \left(\frac{\Sigma d + n\bar{x}_a}{n}\right)^2\right\}}$$

$$= \sqrt{\left[\frac{\Sigma d^2 + 2\bar{x}_a \Sigma d + n\bar{x}_a^2}{n} - \left\{\frac{(\Sigma d)^2 + 2n\bar{x}_a\Sigma d + n^2\bar{x}_a^2}{n^2}\right\}\right]}$$

$$= \sqrt{\left\{\frac{\Sigma d^2}{n} + 2\bar{x}_a\frac{\Sigma d}{n} + \bar{x}_a^2 - \frac{(\Sigma d)^2}{n^2} - 2\bar{x}_a\frac{\Sigma d}{n} - \bar{x}_a^2\right\}}$$

$$= \sqrt{\left\{\frac{\Sigma d^2}{n} - \left(\frac{\Sigma d}{n}\right)^2\right\}}.$$

For a grouped frequency distribution this becomes

$$s = \sqrt{\left\{\frac{\Sigma f d^2}{n} - \left(\frac{\Sigma f d}{n}\right)^2\right\}}.$$

If there is a common factor in the values of the midpoints—say 'c'— and if $\dfrac{d}{c} = d'$ as before,

then
$$s = c\sqrt{\left\{\frac{\Sigma f d'^2}{n} - \left(\frac{\Sigma f d'}{n}\right)^2\right\}}.$$

This formula is employed in the calculation of the standard deviation in Table 2.1.

When it is finally obtained, the standard deviation is in the same units as the distribution, e.g. if the distribution is of rod lengths that have been manufactured on a machine under test, the units will be in (say) millimetres or inches.

This makes the comparison between two or more standard deviations difficult, and the units are reduced to a standard by dividing the standard deviation by the arithmetic mean. The result is expressed as a percentage and called the *coefficient of variation* or *coefficient of dispersion*.

THE POOLED VARIANCE OF TWO OR MORE DISTRIBUTIONS

Earlier in this chapter we illustrated how the collective mean of two or more distributions could be calculated, provided we knew, or could discover, the separate means of the component distributions as well as their total frequencies.

In the same way we can calculate the variance of a combination of two or more distributions. The formula for doing this is:

$$n\sigma^2 = n_1(\sigma_1^2 + d_1^2) + n_2(\sigma_2^2 + d_2^2) + \ldots + n_r(\sigma_r^2 + d_r^2)$$

where $n_1, n_2, ..., n_r$ are the total frequencies of each of the component distributions,

$d_1, d_2, ..., d_r$ are the differences between the mean of the combined distributions and the separate means of the component distributions (i.e. if $\bar{x}$ is the mean of the combined distribution and $\bar{x}_1, \bar{x}_2, ..., \bar{x}_r$ are the means of the component distributions then $d_1 = \bar{x} - \bar{x}_1$; $d_2 = \bar{x} - \bar{x}_2$, etc.,

$\sigma_1^2, \sigma_2^2, ..., \sigma_r^2$ are the variances of the constituent distributions,

n is the combined frequency of all distributions,

σ^2 is the variance of the combination of all the component distributions.

This result can be derived quite simply from the jingle learnt earlier that 'the standard deviation of a distribution is the square root of (the mean of the squares less the square of the mean)' and the reader might want to try to derive this for himself.

Prove your prowess [1]

2.1.* The figures below show the amount spent on entertaining customers by each of a firm's 92 salesmen in a given month.

Calculate (a) the arithmetic mean,
              (b) the standard deviation.   (I.C.W.A., Part I)

Amount spent £ (1)	Number of salesmen (2)
less than 2	5
2 less than 4	6
4 less than 6	8
6 less than 8	14
8 less than 10	21
10 less than 12	12
12 less than 14	9
14 less than 16	9
16 less than 18	6
18 and over	2

---

[1] An asterisk against the exercise number indicates that the solution is given in detail.

2.2. The following distribution shows the turnover (£000s) of the branches of a group of multiple shops in March 1967.

Turnover (£000s)	No. of shops
£5 and under £10	8
£10 and under £15	18
£15 and under £20	42
£20 and under £25	62
£25 and under £30	30
£30 and under £35	10
£35 and over	4
	174

Using an assumed mean of £22 500, calculate:

(a) the arithmetic mean, and

(b) the standard deviation of the distribution.

(I.O.M., Part II).

2.3. Repeat Question 2, using an assumed mean of Zero.

2.4. The yields of grain (in pounds) from small plots are grouped in classes of width 0·2 lb, the values of x being the mid-values of the classes. In the table below, f indicates the frequency within each class:

x	2·8	3·0	3·2	3·4	3·6	3·8	4·0	4·2	4·4	4·6	4·8	5·0	5·2
f	4	15	20	47	63	78	88	69	59	35	10	8	4

Calculate for this distribution (a) the mean, (b) the standard deviation, (c) the coefficient of variation. (I.O.S., Part I)

2.5.* Calculate the median, mode, upper and lower quartiles for the distribution:

Height (inches)	No. of men
60 less than 63	4
63 less than 66	14
66 less than 69	59
69 less than 72	33
72 less than 75	8
75 less than 78	2

2.6. Calculate the median, mode, upper and lower quartiles and the first three deciles for the following data:

Breaking stress (short tons)	Number of cables tested
8·0 less than 8·5	9
8·5 less than 9·0	23
9·0 less than 9·5	44
9·5 less than 10·0	68
10·0 less than 10·5	45
10·5 less than 11·0	11
	200

2.7. You are provided with the following raw sums in a statistical investigation of two variables, x and y:

$$\Sigma x = 235 \quad \Sigma y = 250 \quad \Sigma x^2 = 6750 \quad \Sigma y^2 = 6840$$

Ten pairs of values are included in the survey.

Compute the standard deviations of the x and y variables.

(I.O.S., Part I)

2.8. Male deaths in England and Wales during 1961 for disease classification 720–749 are tabulated below.

Calculate the mean age of death and the standard deviation. For the purpose of calculation you may take the last age group as 85–100.

Age	Number of deaths	Age	Number of deaths
0–	12	50–	87
1–	2	60–	157
5–	4	70–	96
10–	17	75–	80
15–	29	80–	78
20–	26	85–	48
40–	33		
		Total	669

(I.O.S., Part I)

2.9.* The population of a northern town diminished from 546 000 in 1956 to 512 000 in 1959. What was the average annual percentage decrease?

2.10. The bacterial count in a culture increased from 1200 to 3700 in three days. What was the average daily percentage increase?

2.11. The catering manager of a Dining Club is allowed to spend the same amount of cash each year for the purchase of meat.

The average cost per lb of meat purchased rose in four successive years from 30p to 35p to 45p to 55p per lb.

What was the average cost per lb over the whole period to the catering manager?

2.12. In assessing the overall Examination results for an academic prize, subjects are allotted an 'importance index'.

They are: Mathematics 25; English 15; French 20; Chemistry 20; Woodwork 10; Art 10.

Students Swot, Plod and Grind achieve the following results. The final overall result has to be expressed as marks out of 100. Who won the prize? What were the three competing results?

	Swot %	Plod %	Grind %
Mathematics	70	40	50
English	60	50	80
French	50	80	70
Chemistry	80	60	60
Woodwork	20	80	40
Art	10	90	50

2.13.* Two groups of items are taken from each of two factories of the same firm; the statistical quantities obtained are as follows:

Factory	No. of items	Mean length (in inches)	Standard deviation (in inches)
A	100	4·1	0·1
	200	4·0	0·2
B	100	4·2	0·1
	200	4·1	0·2

Calculate the mean length and standard deviation of length of all 600 items. (I.O.S., Part II)

REFERENCES

ALLEN, R. G. D. 1957. *Statistics for Economists.* 3rd Edn. London, Hutchinson.

BROOKES, B. C., and DICK, W. F. L. 1953. *Introduction to Statistical Method.* London, Heinemann.

CONNOR, L. R., and MORRELL, A. J. H. 1957. *Statistics in Theory and Practice.* 4th Edn. London, Pitman.

FREUND, J. E., and WILLIAMS, F. J. 1958. *Modern Business Statistics.* New York, Prentice-Hall.

MORONEY, M. J. 1956. *Facts from Figures.* Harmondsworth, Penguin Books.

3: MAKING COMPARISON EASY
Index Numbers

BASIC IDEAS

THE reader will already be familiar from his own experience with the idea of comparisons based on percentages. If, for example, one's income increases over a period from £2500 p.a. to £3000 p.a. one can express this increase as either £500 p.a. or 20%. This chapter examines in some detail the general question of expressing comparisons by means of percentages and, in particular, the question of making such comparisons when the quantities being compared represent some aggregation of items instead of a single item as in the above example.

The comparison of aggregates is a commonplace in business, in fact comparison is one of the basic activities of both statisticians and businessmen. Comparison adds a dimension to the data. Figures in isolation rarely reveal as much information as when they are compared with other figures. For example, the actual value of last month's salary cheque, though of burning interest to the recipient, provides additional interest when it is compared with values in previous months or when it is compared with someone else's cheque. In business, unfortunately, the aggregates to be compared are rarely as clear-cut as a monthly salary cheque. Industrial output, for example, consists of such diverse items as tons of steel, gallons of acid, yards of cloth, millions of bricks, etc. Changes in the prices and quantities of all these items are most conveniently expressed as percentages; comparison of the aggregate is then based on some form of average of these percentages. Such an average is an example of what is usually meant by the term 'index number'. This expression, however, is commonly used to mean two slightly different things and it will be well to distinguish between them.

First, the term *index number* is frequently applied to any series of values which are expressed as percentages of some base period value. Reverting to the example of an individual's income, we might find that over five recent years the salary of the individual rose as follows:

Year	Salary at 1st Jan. £ p.a.	Salary as percentage of 1963	1967
1963	2000	100	67
1964	2400	120	80
1965	2750	138	92
1966	3000	150	100
1967	3000	150	100

We can say that based on 1963 = 100 the current salary of our individual stands at 150 or that based on 1967 = 100 his salary in 1963 was 67. Similarly we can refer any of the intermediate years to either of our base years.

We note in passing that the arithmetic for the two calculations is consistent. Both calculations show an increase in 1967 compared with 1963 of 50% and both show that the 1963 level was only two-thirds of the 1967 level.

We also note another point; one which is a frequent source of ambiguity in discussing index numbers. A rise in the level of the index can be expressed either as a rise of *x per cent* or as a rise of *x percentage points*. The difference between the two can be considerable. Consider our simple example above: we have said that based on 1963 = 100 the index in 1967 is 50% higher. We could also have said that the index has risen by 50 *percentage points* (or just *points*, for short). In this case, because we are comparing with the base year, the two statements are equivalent, but consider now the increase from 1964 to 1967. Here the percentage increase in the level of the index is 25 but the index has risen by 30 points.

The second meaning of the term *index number* is a quantity which shows by its variations changes in some aggregate which is not susceptible of direct measurement in itself or which comprises a number of items of varying importance, perhaps expressed in different units, which have to be combined, e.g. industrial production. Another simple example will illustrate this second meaning. Again reverting to the individual whose salary we have been studying, let us suppose that in 1963 his total expenditure was as follows, the balance being saved:

Item	Expenditure 1963 £
Food	500
Housing	300
Other	600
	1400

Assume now that in 1967 *he is still buying exactly the same selection of goods and services* but that these are now costing him as follows:

Item	Expenditure 1967 £
Food	700
Housing	600
Other	1000
	2300

We wish to express in a single convenient figure the increase in the individual's *cost of living*. We can do this very easily in this example by simply expressing his expenditure in 1967 as a percentage of the 1963 figure. We obtain a figure of 164, i.e. his 1967 *cost of living* is 64% higher than in 1963.

Compared with our previous example we are now measuring a more complex quantity, namely the *cost of living*. The method we have used to construct the index number is known as the *aggregative* method. What we have done is to establish the individual's consumption pattern in 1963 and price this particular set of goods and services, first at 1963 and then at 1967 prices. We could equally have started with the 1967 pattern of consumption and priced this particular pattern at 1963 and 1967 prices. (For the sake of simplicity we assumed that the two were identical anyway.) Having obtained comparable expenditure for the two years, we can then express either of them as a percentage of the other.

In practice, index numbers are rarely calculated in this way. We do not normally have full details of expenditure on the various items in the index. What we often do have is, first, information about the consumption pattern at a certain date and, secondly, information about the percentage increases in the prices of the various items in the index. Using our same example we might now obtain the following table:

Item (1)	Expenditure 1963 £ (2)	1967 prices as percentage of 1963 prices (3)	Col. (2) × Col. (3) (4)
Food	500	140	70 000
Housing	300	200	60 000
Other	600	167	100 200
	1400		230 200

The percentage figures in column (3) are referred to as *price relatives*. We now calculate a weighted average of these price relatives taking as weights the expenditure in 1963. Since we are also taking 1963 as our base year, we refer to the index as a *base weighted* index.

The weighted price relatives are shown in column (4) of the table and with the usual methods of weighted averages we obtain the index for 1967 as:

$$\text{Index} = \frac{230\ 200}{1400} = 164,$$

which is the value we obtained from the aggregative method.[1]

The second of the two meanings of the term *index number* is clearly a much more complex concept than the first meaning, and gives rise to many problems in the construction and use of index numbers. However, it is this type of index number which is more common and some statisticians reserve the term for this kind of index, referring to the simpler process as *standardising*. This raises its own problems, however, because standardising is also a technique, closely related

[1] The weights used in the calculation were rather cumbersome, adding up to 1400. Since it is the relative and not the absolute size of the weights which matters, it often helps to express the weights as proportions or percentages. This simplifies the final division sum in the calculation.

to index number construction, with a specific meaning of its own, as we shall see later.

We shall try to avoid ambiguity by using the term *index number* with its more complex meaning unless otherwise stated.

We have seen that index numbers are concerned mainly with the problem of measuring changes in *aggregates* or *groups* of prices, quantities, etc. The majority of index numbers are price indices and the majority of comparisons are over time, but index numbers are used to compare quantities and values as well as prices, and to make comparisons between firms, between departments of firms, between regions, between industries, etc., as well as comparisons over time.

The feature of most well-known index numbers is that they attempt to measure changes in rather ill-defined concepts such as the level of retail prices, industrial production, share prices, etc. A more unusual application of an index number is the estimation of intelligence by means of I.Q. scores.

Index numbers are widely used in business and government. They are often written into contractual agreements to guard against price changes during the contract period, e.g. construction contracts, wage agreements, pension scales, etc., and apparently in the United States have even been written into alimony settlements.

We now proceed to a discussion of the general problems which arise in the construction of index numbers.

PROBLEMS IN THE CONSTRUCTION OF INDEX NUMBERS

We saw in the previous section the two main alternative approaches in constructing an index number: the *aggregative approach* and the *weighted average of price relatives approach.* In practice there are usually a number of reasons why neither approach can be applied in the same form as in our simplified example. In the example the index number was clearly defined—in practice the definition is usually rather vague or the intended coverage rather general; in the example we assumed implicitly that all items entering into the budget were priced and included in the calculations—in practice a severely restricted sample of items and their prices has to be selected.

In practice, therefore, given the specification of the purpose and coverage of the index, we have to select from those prices and quantities available, a relatively small set which we hope will reflect changes in the much larger group in which we are interested. We then have the problem of combining this small set of items in the best possible way, i.e. we have to consider what type of average and what weights, if any, we should use. These and other problems are discussed below.

THE PURPOSE OF THE INDEX

When a new index is proposed, or a major revision of an existing one undertaken, the purpose is usually specified in advance. This is especially true where an internal index, i.e. one set up by an organisation for its own use rather than for publication, is concerned. With a number of general purpose indices, however, the purpose is rather vague. Index numbers, like many other published statistics, often emerge as a 'by-product of administration' and are put out because it is thought (very often correctly) that they will be of use to a wide circle of users. Even when an internal index number is specified, however, important and unexpected additional uses often come to light at a later date. The index number constructor can therefore perform a valuable service by (a) trying to find out how other departments and organisations might use his index and (b) giving details of the coverage, collection of data, etc., to enable the would-be user to assess the value of the index.

This is not just altruism; it is possible that offers of assistance, including financial assistance, will be made.

THE AVAILABILITY AND COMPARABILITY OF DATA

It is self-evident that *some* data must be available before an index number can be constructed, but with the notable exception of some well-known published index numbers the necessary data is rarely collected specifically for the purpose of constructing the desired index. This leads to problems of *availability* and *comparability*.

If the lack of data is sufficiently serious, of course, the proposed index has to be abandoned before it is started. Assuming the availability of sufficient data to initiate the index, the problem of availability is mainly a long-term rather than a short-term one. Over a period of time, items on which the index was originally based might disappear and new items which were not in existence when the index was initiated might become more and more important. Obvious examples of the latter are TV sets, frozen foods, electronic products, etc.

Comparability of data is a more common short-term problem. Here the problem is usually one of changing definitions and coverage in the data, or changing quality in the items included in the index. This last factor is one of the most difficult problems in index number construction and interpretation, and is discussed more fully in the section on updating the index.

Basically, then, the need is for goods and services precisely defined which change neither definition nor quality between periods. This is easy to specify but almost impossible to achieve in practice.

SELECTION OF THE ITEMS TO BE INCLUDED IN THE INDEX

In simple single-series indices, the items virtually select themselves, e.g. an index of crude steel production requires only a knowledge of the intricacies of crude steel statistics. In a general purpose index, however, it is impossible to include more than a small sample of items, and even for the selected items it is impossible to take more than a small sample of prices.

The items and prices to be included in the index are usually selected on a non-random subjective basis, the aim being to achieve a representative set, variations in which will reflect changes in the aggregate under study.

Several stages of selection are involved and the process can be illustrated by an example from the Index of Retail Prices. The first stage, after specifying coverage, etc., is to arrange all possible items of expenditure into a relatively small number of homogeneous sub-groups, e.g. food, housing, transport, entertainment. Each homogeneous group is then further subdivided, e.g. food might be subdivided into meat, cereals, beverages. From each of these subdivisions a representative sample is selected, e.g. from the meat subdivision a few types and cuts of meat are selected.

Only the specific items included at the final stage of selection are priced. Representative prices by region and kind of retail outlet are obtained and it is these prices which form the basis of the index.[1]

---

[1] The problem of collecting representative prices is becoming rather more difficult at the moment with the collapse of resale price maintenance over a wide range of products. The index number constructor is saved a certain amount of work when his index contains items such as confectionery, for which (at the time of writing) prices can be obtained direct from the manufacturer without the fear of large deviations from the published prices. Contrast the situation with that for, say, instant coffee or frozen peas, where prices vary from retailer to retailer and day to day.

The question of what items to select for inclusion in the index is closely related to the problem of availability and comparability of data. If the index is to be relevant the items included should be topical, and provision should be made for the inclusion of new items as and when they appear.

CHOICE OF THE BASE PERIOD

The *base year* of an index is the year with which all other years are compared. It is convenient to refer to the year being compared as the *given year*.

There are two generally desirable features of a satisfactory base period: it should be a period of relative stability and it should be fairly recent.

The stability criterion is important because the choice of an exceptionally high or low year as base would distort all subsequent values of the index. For example, an index of wool prices based on 1951 = 100 would exaggerate the general decline in wool prices in the last ten or so years, since wool prices were exceptionally high in 1951 because of stockpiling caused by the Korean War.

A fairly recent base is advantageous simply because it keeps the values of the index fairly close to the 100 mark and thereby helps comprehension of changes in the index. If, for example, the base year is too far in the past, current values might be around the 400, 500 or 600 level. If two consecutive values of the index are, say, 573 and 604, further calculation is necessary to appreciate the size of the increase; furthermore there is a psychological effect. If a price index stands at 604 this very fact tends to give people the impression that prices are high.

A fairly recent base also ensures that comparisons are based on relatively homogeneous quantities, e.g. the 1967 Rolls Royce bears a much closer resemblance to the 1960 model than to the 1914 model; a price relative based on 1960 would therefore be more meaningful.

CHOICE OF THE WEIGHTS

Most real life index numbers are based on a weighted arithmetic mean of price or quantity relatives; the question is what weights should we use?

The problem is basically one of averaging a set of numbers, i.e.

price or quantity relatives, which are not of equal importance. The need for weighting is obvious. If, among the items on which we spend our income, we consider two items, namely food and cuff-links, it is clear that whereas a modest percentage increase in the price of food will have an impact on our total expenditure, a very large percentage increase in the price of cuff-links will have a barely perceptible effect.

If we look back at our first example of an index number (p. 78) where we compared the aggregative method with the weighted average of price relatives method, we see that the appropriate weights to give us the correct average of the price relatives were provided by the relative expenditure on the various items in the index. This is a general rule; price or quantity relatives are usually weighted by corresponding values.

In most indices the weights refer to either the base or the given year, but there is no reason why the weights should not refer to some other year, nor need this year be one of the years between the base and given years. The broad distinction is between *fixed weights* and *current weights*. With *fixed weights*, the same weights are used from one year to the next, with *current weights* a new set of weights is constructed each year. This is merely another way of saying that either we select a fixed set of goods and services and price this set at base year and all subsequent years' prices, or we establish the given year pattern of goods and services and price this set at given year and all previous years' prices.

Collecting data for the estimation of weights is often a mammoth task, e.g. the Ministry of Labour Household Budget Inquiries are based on very large random samples of households. Index constructors, unless they are officially sponsored, rarely have the resources to establish the precise weighting pattern but, fortunately, small variations in the weights of an index number have little effect on the results. This is just an arithmetical phenomenon, and can be tested by the reader. Its relevance to index number construction is that an approximation to the true weights is usually quite good enough.

THE CONSTRUCTION OF INDEX NUMBERS

We have already noted that most index numbers in practice are calculated using weighted arithmetic averages of price or quantity

D

relatives. Other kinds of averaging methods are sometimes used, e.g. the geometric mean, an unweighted arithmetic mean, but we will confine our attention to the most common method and refer the reader to some of the many works in which the other methods are discussed.

We will illustrate the mechanics of index number construction with an example of a price index, and then briefly discuss quantity and value indices. The formulae used are discussed in the Appendix.

PRICE INDICES

Suppose that a building firm wishes to construct an index of raw material prices for 1967 compared with 1960. We will assume that the index is to be based on the following items: bricks, timber, plasterboard, sand, and copper piping. We will further assume that representative prices for each of these items can be obtained without difficulty. The necessary data is set out in the following table:

TABLE 3.1

Item	1960 Prices (in appropriate units) (1)	1960 Quantities (in appropriate units) (2)	1967 Prices (in appropriate units) (3)	1967 Quantities (in appropriate units) (4)
Bricks	10	1000	16	1200
Timber	20	700	21	800
Plasterboard	5	1500	6	1200
Sand	2	850	2	850
Copper piping	7	100	14	80

From Table 3.1 we can calculate the weights and the price relatives. These are set out in Table 3.2.

Base weighted index. The calculation of the base weighted index is shown in Table 3.3.

The base weighted index gives a value for 1967 of 126·45 based on 1960 = 100 (126·4 to one place of decimals).

Current weighted index. The arithmetic is slightly different from that of the base weighted index. The price relatives are inverted,

TABLE 3.2

Item	1960 Weights Actual (1)	1960 Weights Relative (2)	1967 Weights Actual (3)	1967 Weights Relative (4)	Price Relatives (1967 ÷ 1960) (5)
Bricks	10 000	0·30	19 200	0·42	160
Timber	14 000	0·41	16 800	0·36	105
Plasterboard	7 500	0·22	7 200	0·16	120
Sand	1 700	0·05	1 700	0·04	100
Copper piping	700	0·02	1 120	0·02	200
	33 900	1·00	46 020	1·00	

Notes:
1. The weights are obtained by multiplying prices and quantities in Table 3.1, e.g. bricks: 10 × 1000 = 10,000 in column (1) of Table 3.2.
2. The relative weights are simply the actual weights converted to proportions. It is important that they should add up to 1·00.

TABLE 3.3

Item	1960 Weights (1)	Price relatives (1967 ÷ 1960) (2)	Weight × Price relatives (3)
Bricks	0·30	160	48·00
Timber	0·41	105	43·05
Plasterboard	0·22	120	26·40
Sand	0·05	100	5·00
Copper piping	0·02	200	4·00
			126·45

i.e. we take base year divided by given year and we take the reciprocal of the weighted average of these relatives, using 1967 weights. The calculations are shown below:

TABLE 3.4

Item	1967 Weights (1)	Price relatives[1] (1960 ÷ 1967) (2)	Weight × Price relatives (3)
Bricks	0·42	0·625	0·2625
Timber	0·36	0·952	0·3427
Plasterboard	0·16	0·833	0·1333
Sand	0·04	1·000	0·0400
Copper piping	0·02	0·500	0·0100
			0·7885

$$\text{Index} = \frac{1}{0 \cdot 7885} \times 100 = 126 \cdot 8.$$

The current weighted index gives a value for 1967 of 126·8 based on 1960 = 100. The different weighting systems, therefore, give virtually the same answer.

Other fixed weights. It is possible that our building firm, although able to estimate the price relatives fairly accurately, is unable to make any good guess about the weighting pattern in either 1960 or 1967. Let us assume, however, that some information is available about the expenditure pattern in say 1958. This year is outside the range of the index but there is nothing to prevent the firm using the 1958 expenditure pattern to weight the price relatives. This is an example of a fixed weights system where the weights do not refer to the base year.

One final variation of the weights for our example might be mentioned. Bearing in mind that the two weighting systems give two rather different values of the index, it might occur to the reader to average the two sets of weights, e.g. Bricks: $\dfrac{0 \cdot 30 + 0 \cdot 42}{2} = 0 \cdot 36$.

This in fact has been suggested by eminent statisticians and econo- mists and this form of index, using (current ÷ base year) price relatives, is called a *Marshall-Edgeworth-Bowley* index.

The two more common forms of the index also have names. The

[1] Expressed as a fraction in this instance.

base weighted index is often referred to as the *Laspeyre* index and the current weighted index as the *Paasche*, in each case after the person who first proposed it.

For the sake of simplicity we considered only two years in our above example. In practice, of course, an index usually covers a series of years, in which case one calculates the price relatives for each year and weights according to which form of index is being used.

QUANTITY INDICES

In a quantity index the only difference compared with a price index is that instead of averaging price relatives we average quantity relatives. Instead of pricing the same set of goods at two different price levels, we in effect value two different sets of goods at constant prices, i.e. we reduce the heterogeneous units of bricks, timber, etc., to the common unit of money.

Returning to our building materials index, the firm might wish to construct a base weighted index of its material consumption. A simple aggregation of the items used would involve adding together thousands of bricks, standards of timber, yards of sand, etc., so a weighted average of quantity relatives is constructed to overcome this problem.

From Tables 3.1 and 3.2 we can obtain the necessary data for calculating our quantity index. This is set out in Table 3.5 below:

TABLE 3.5

Item	1960 Weights (1)	Quantity relatives (2)	Weight × Quantity relatives (3)
Bricks	0·30	120	36·00
Timber	0·41	114	46·74
Plasterboard	0·22	80	17·60
Sand	0·05	100	5·00
Copper piping	0·02	80	1·60
			Index = 106·94

It is left to the reader as an exercise to calculate the current weighted index.

VALUE INDICES

A value index is simply the ratio of prices × quantities in the given year to prices × quantities in the base year.   From Table 3.2, columns (1) and (3), we calculate this index as

$$\text{Index} = \frac{46\ 020}{33\ 900} \times 100 = 135 \cdot 8.$$

Value indices are calculated from simple aggregates expressed in terms of money and in themselves are of limited use.

RELATION BETWEEN PRICE, QUANTITY AND VALUE INDICES

Although the value index is of little use in itself, it has an interesting relationship with the two other kinds of index.   The value index is obtained if we multiply the corresponding base weighted price index by the current weighted quantity index and the current weighted price index by the base weighted quantity index.   In our example the values of these various index numbers were as follows:

	Price	Quantity
Base weighted	126·5	106·9
Current weighted	126·8	107·5

Value   index = 135·8 = (126·5 × 107·5) = (106·9 × 126·8)   within the range of errors due to rounding.

Thus a change in value can be explained by a combination of a price and a quantity index.

FIXED AND MOVING WEIGHT INDICES

From a practical point of view fixed weight indices, whether the weights refer to the base year or some other year, have the advantage that the weights do not have to be recalculated every year.

From a more theoretical point of view both base weighted (Laspeyre) and current weighted (Paasche) indices are probably biased.   The Laspeyre usually overestimates and the Paasche underestimates changes in prices.   The reason for this is that price changes cause the pattern of consumption to change.   When prices rise, consumption of those items whose prices have risen most goes down, hence the Laspeyre index will give too much weight to those items.

Conversely, when prices fall, consumption of those items whose prices have fallen most goes up, hence the Laspeyre index will give insufficient weight to those items.

The argument concerning the downward bias in the Paasche index follows similar lines.

The difference between the Laspeyre form and the Paasche form of an index is generally small unless major changes have occurred between the base year and the given year, and the Laspeyre form is usually chosen. If neither form appears satisfactory the two indices can be averaged. The geometrical mean of the two indices is known as *Fisher's Ideal Index*, after Irving Fisher who proposed it in 1920. It is ideal in that it fulfils certain desirable requirements for an index number, but is not often used in practice.

THE CONTINUITY OF INDEX NUMBERS

When an index has been running for a number of years a number of problems arise concerning the availability and comparability of data, the introduction of entirely new items and changes in the weighting pattern. If the index is to remain useful it must be revised as and when necessary. Revision can take various forms, three of which we shall proceed to discuss.

SPLICING

It may become necessary to drastically revise an index, to take account of new items and a change in the weighting pattern, without destroying the continuity of the index. *Splicing* is a method of achieving this end, and is illustrated in Table 3.6 (the figures are hypothetical).

We here assume that the new weighting system comes into force with the January 1967 index, and that at the same time the base is changed from Jan. 1956 = 100 to Jan. 1967 = 100. The new index can be spliced on to the old provided the January 1967 index is calculated with both the old and the new weighting pattern.

To splice backwards we calculate the splicing factor $\dfrac{100}{127} = 0.787$ and multiply all values of the 'old' index by this factor, thereby obtaining the values in column (3) of Table 3.6.

To splice forwards we calculate the splicing factor $\dfrac{127}{100} = 1.27$

(which is the reciprocal of the backwards factor) and multiply all values of the new index by this factor, obtaining the values in column (4) of Table 3.6.

TABLE 3.6. *Index of Retail Prices*

	Based on old weights Jan. 1956 = 100 (1)	Based on new weights (2)	Spliced indices Backwards (3)	Forwards (4)
Jan. 1966	120		94·4	
Feb. 1966	121		95·2	
· · ·	· · ·	· · ·	· · ·	· · ·
Dec. 1966	126		99·2	
		Jan. 1967 = 100		
Jan. 1967	127	100		
Feb. 1967		102		129·5
Mar. 1967		104		132·1

CHAIN INDEX NUMBERS

A fundamental feature of the index numbers so far considered is that irrespective of choice of weights, method of averaging, etc., comparison is always with a base year. With a chain index, however, the comparison is always with the immediately preceding year, e.g. if we construct a chain index to cover the years 1965 to 1967 we have two separate calculations, one for 1966 based on 1965 and one for 1967 based on 1966.

We can express these indices as $I_{65, 66}$ and $I_{66, 67}$ respectively. If we now wish to compare 1967 with 1965 we multiply the two chain indices to obtain $I_{65, 67} = I_{65, 66} \times I_{66, 67}$ (for this calculation it is necessary to express the index as a proportion rather than a percentage).

For example, suppose we have calculated $I_{65, 66} = 1\cdot02$ and $I_{66, 67} = 1\cdot04$, our estimate of $I_{65, 67}$ is $1\cdot02 \times 1\cdot04 = 1\cdot06$.

So far we have said nothing about the weights to be used. In a true chain index the weights change every year. In our example, assuming base weighted indices, $I_{65, 66}$ would be based on 1965 weights, $I_{66, 67}$ on 1966 weights. The effect of this is that $I_{65, 67}$

calculated from the chain index will not give the same answer as a direct base weighted index of 1967 on 1965.

Chain index numbers are used when the items in the index and the weighting system need constant revision. The fact that the chain index comparison gives different answers from fixed based indices (except for adjacent years) is irrelevant since in a situation where a chain index was necessary the fixed base index would by definition be even more unreliable, if indeed it could be calculated at all.

A chain index, therefore, is an extreme example of splicing. An alternative form of 'chain index' is sometimes put forward in which the weights remain fixed throughout the life of the index. In terms of our example, the same weights would be used for $I_{65,66}$ and $I_{66,67}$; they could for example be based on the 1965 consumption pattern.

This kind of index has the possibly slight advantage that multiplication of successive values of the index, say $I_{65,66} \times I_{66,67}$ leads to the same answer as a direct calculation of $I_{65,67}$. The only difference between this and the more usual fixed base index, however, is that in this index successive values are always calculated on the preceding year. Unless these year-to-year ratios are of interest it is easier to construct an orthodox fixed base index. The term 'chain index' is better reserved for the 'true' chain index.

CHANGING THE BASE

Splicing and chain indices are two particular methods of changing the base of an index. The base of an index is frequently changed by a much simpler method, namely by setting the new base equal to 100 and recalculating all the values in the index. For example, an index based on 1950 = 100 might have its base changed to 1963 = 100 as follows:

Year	Index number 1950 = 100	Index number 1963 = 100
1960	110·7	100·6
1961	112·3	102·1
1962	112·9	102·6
1963	110·0	100·0
1964	111·5	101·4
1965	113·0	102·7
1966	114·6	104·2

This method of shifting the base is just a matter of simple arithmetic. As with chain indices, the calculations do not in general lead to the same answer as if the index were completely reworked, but the method is simple and frequently used. In any case, it is often impossible to do otherwise.

SOME USES OF INDEX NUMBERS

Index numbers have a variety of uses in business, where decisions for the future are frequently based on comparisons of past data. They are particularly useful in costing, management accounting and forecasting. In this section we shall briefly describe two of their more valuable uses.

ADJUSTING FOR PRICE INCREASES

A frequently heard comment is that 'the £ is not worth what it was'. A similar comment is 'the cost of living is rising much faster than my income'. Index numbers can be used to test the truth of such statements and, if necessary, to correct for a rise in prices. To take a concrete example, we return to our first example (p. 76), the individual whose salary we studied. We can construct the following table:

Year	Salary at 1st Jan. £ p.a. (1)	Index of retail prices (monthly averages) 1963 = 100 (2)	Real salary £ p.a. (3)
1963	2000	100	2000
1964	2400	103·3	2323
1965	2750	108·2	2542
1966	3000	112·5	2667
1967	3000	114·4	2622

Actual increase since 1963 $= \dfrac{3000}{2000} \times 100 = 150.$

'Real' increase since 1963 $= \dfrac{2622}{2000} \times 100 = 131.$

Column (3) of our table is obtained by dividing the figures in column (1) by those in column (2), and multiplying by 100.

The table shows that although actual income increased by 50% between 1963 and 1967, the accompanying rise in prices reduced the value of the increase to only 31%.

This use of index numbers has many applications. The process is usually referred to as *deflating* and the index used for the purpose as a *deflator*.

STANDARDISATION

Consider the problem of comparing labour turnover between two companies. (We shall consider only resignations in our example.)

TABLE 3.7

| Staff category | COMPANY A | | | COMPANY B | | |
	Avge. no. employed (1)	No. of resigna- tions (2)	Turnover rate (%) (3)	Avge. no. employed (4)	No. of resigna- tions (5)	Turnover rate (%) (6)
Adult men	3000	300	10	2000	180	9
Adult women:						
Married	600	360	60	2000	1000	50
Single	300	60	20	1500	225	15
Junior men	700	175	25	500	100	20
Junior women	400	120	30	1000	250	25
All staff	5000	1015	20·3	7000	1755	25·1

Let us assume that the crude rate, i.e. the number of resignations as a percentage of the average total labour force in a given year is as follows:

Company A: 20·3   Company B: 25·1.

We might conclude from these figures that labour relations at Company A are slightly better than at Company B.

Suppose now we investigate the figures behind these crude rates and classify the labour force into broad categories. We might end up with Table 3.7.

The crude rates are unchanged but the picture now is very different. Company B, which has the highest crude rate is now revealed as having lower rates *for every single category of staff*. The higher crude rate is due solely to the fact that Company B employs relatively more of the 'high turnover' categories of staff.

In order to make a fair comparison between turnover rates we need, therefore, to base our comparison on some standard labour force, i.e. we need to apply the two sets of rates to the same distribution of staff. This is basically an index number problem with the rates as relatives and the numbers of staff as weights. We need to apply a common set of weights in order to get comparable weighted averages for the two companies.

A 'standard' labour force can conveniently be obtained by averag-

TABLE 3.8

| | | COMPANY A | | COMPANY B | |
	Standardised labour force (1)	Turnover rate (%) (2)	Standardised no. of resignations (1)×(2) (3)	Turnover rate (%) (4)	Standardised no. of resignations (1)×(4) (5)
Adult men	2500	10	250	9	225
Adult women:					
Married	1300	60	780	50	650
Single	900	20	180	15	135
Junior men	600	25	150	20	120
Junior women	700	30	210	25	175
	6000		1570		1305

ing the two companies (although there is nothing to stop us using one or other of the companies on its own). The standard labour force and the new calculations of turnover rates are set out in Table 3.8. The standardised turnover rates are:

$$\text{Company A: } \frac{1570}{6000} \times 100 = 26 \cdot 2$$

$$\text{Company B: } \frac{1305}{6000} \times 100 = 21 \cdot 8.$$

The standardised turnover rates confirm the conclusion to be drawn from the table, namely that Company B enjoys the better labour relations.

The technique of standardisation can be applied to a wide range of problems in business. Obvious applications are to comparisons of accident rates, and sickness absence rates but the method can be used in studying productivity and other aspects of business.

An example of this kind of application can be found in the steel industry. Steel passes through a number of processes in its manufacture: from blast furnace, through steel furnace, ingot casting, primary rolling, secondary rolling and finishing. At any stage, defects are likely to be found which result in some or all of the batch being scrapped. Between 5 and 10% of output might be scrapped, the cost to the firm running into £millions per year, so obviously great efforts are made to keep the level of defects down to a minimum, and to spot defective steel at as early a stage as possible.

Some measure is required of the output lost through defects, and this measure must take into account the different value of a ton of steel at different stages in the production process. One way of achieving this is to construct an index number of total percentage defective tonnage. In this index number the relatives are the percentages scrapped at various stages in the production process and the weights are the value of a ton of steel at each stage of production. This standardised defect rate can then be directly compared with similar rates for previous years, or between various works or companies.

PROBLEMS IN INTERPRETING INDEX NUMBERS

The interpretation of index numbers, especially general purpose indices, raises a number of problems, the first of which is the *relevance* of the index. If an index of retail prices is constructed, the weights are usually based on the expenditure pattern of some average family. Any given individual might fairly argue that the weights chosen did not represent *his* consumption pattern and that the index was therefore irrelevant. The only answer to this argument is that (*a*) the index does not set out specifically to measure the changes in *his* expenditure and (*b*) even though the weights are not precisely the same the difference probably has little effect on the final value of the index. At the same time, however, it would probably be admitted that separate index numbers for different groups in the population would result in index numbers of greater relevance, and this step has in fact been proposed for the official Index of Retail Prices which at the moment covers about nine-tenths of the population.

A second problem in interpreting an index number is that the index *disguises the absolute level of activity*. For example, if a country decides to manufacture cars and in the first three years produces 600, 900 and 1800 vehicles the index of car production is:

Year 1   100

Year 2   150

Year 3   300

Any comparison with an established car manufacturing country based on the index for these three years would be very favourable to the new manufacturer. This is a major problem in making international comparisons of all kinds.

Another example comes from the latest annual report of a major electrical cable manufacturer. 'Our consumption of copper fell by 5% to some 303 000 tons, whereas our consumption of aluminium increased by 29% to 40 000 tons (the electrical equivalent of 80 000 tons of copper).'

Although the statement gives a superficial impression of an overall increase there is nothing really misleading in it because the company has taken care to relate the percentages to the absolute levels. The final remark in parenthesis is particularly useful and interesting. Without the absolute values, however, the percentages would be misleading.

The third problem is one which has already been mentioned: the problem of the *base year*. This also is a common source of confusion in making international comparisons. If the base year was a particularly good one for country A and a particularly bad one for country B, erroneous conclusions about rates of growth, trends in foreign trade, etc., could easily be drawn.

A fourth problem is again one to which we have already referred: the problem of *quality changes*. Cars provide a good example of the problem. If a particular model appears with a more powerful engine, better seats, etc., but also at a higher price, has the price really gone up? Those who would say that it has, argue that whatever improvements have been made the 'transport' component of the cost of living has gone up because you are no longer able to buy the model at its old price. Those who would say that quality changes should be taken into account when assessing price changes argue that to class an increase in the standard of living, i.e. a better car, as a price increase makes nonsense of the whole concept of measuring price increases. One should therefore try to allow for the change in quality. Unfortunately the only methods for doing this are highly hypothetical, e.g. one might try to estimate how much last year's model would have cost this year *if it had been made.*

In the case of people living on fixed incomes there is little doubt about the effect of price/quality increases. If the old product at the old price is no longer available, the fixed income people will have to reduce consumption of some items in order to consume the same quantity as before of the 'improved' product.

APPENDIX: THE ALGEBRA OF INDEX NUMBERS

NOTATION

(i) Suppose the index is based on n items.

(ii) Let the price and quantity of the ith item in the index be denoted by p_i and q_i respectively, where $i = 1, ..., n$.

(iii) The value of the ith item will, therefore, be $p_i q_i$.

(iv) The base year will be denoted by the subscript 0 and the given years by the subscripts 1, 2, ...

We can now write, for example, base year values as

$$\sum_{i=1}^{n} p_{i0} q_{i0}.$$

This is unnecessarily cumbersome, however, and we will drop the subscript i, simplifying the expression to $\Sigma p_0 q_0$. It must be remembered, however, that this expression always implies summation of the n individual products of price $\times$ quantity.

PRICE INDEX NUMBERS

BASE WEIGHTED (LASPEYRE)

(*a*) *Aggregative Form*

$$I_{01} = \frac{\Sigma p_1 q_0}{\Sigma p_0 q_0} \times 100. \tag{1}$$

This is the familiar 'basket of goods' approach, pricing the base year quantities q_0 at base year and given year prices.

(b) Weighted Average of Price Relative

Multiply the numerator of Equation (1) by $\dfrac{p_0}{p_0}$ $\Big($ actually, by the n

successive sets of $\dfrac{p_0}{p_0}\Big)$ to give:

$$I_{01} = \frac{\Sigma p_1 q_0 \left(\dfrac{p_0}{p_0}\right)}{\Sigma p_0 q_0} = \frac{\Sigma p_0 q_0 \left(\dfrac{p_1}{p_0}\right)}{\Sigma p_0 q_0}. \tag{2}$$

I_{01} is now a weighted average of the price relatives $\left(\dfrac{p_1}{p_0}\right)$, the weights being the $p_0 q_0$, i.e. base year values.

Note. $\left(\dfrac{p_1}{p_0}\right)$ can be expressed as a proportion or a percentage. If it is expressed as a percentage we do not need the '$\times 100$' term in our calculation. $\left(\dfrac{p_1}{p_0}\right)$ is usually given in percentage terms but there are occasions, e.g. in constructing chain indices, where it is necessary to use proportions.

CURRENT WEIGHTED (PAASCHE)

(a) Aggregative Form

$$I_{01} = \frac{\Sigma p_1 q_1}{\Sigma p_0 q_1} \times 100. \tag{3}$$

Here the 'basket of goods' to be priced consists of given year quantities q_1.

(b) Weighted Average of Price Relatives

The procedure is slightly more complicated than with a base weighted index and leads to a slightly unfamiliar looking result.

Multiply the denominator of Equation (3) by $\left(\dfrac{p_1}{p_1}\right)$ to give

$$I_{01} = \frac{\Sigma p_1 q_1}{\Sigma p_0 q_1 \left(\dfrac{p_1}{p_1}\right)} = \frac{\Sigma p_1 q_1}{\Sigma p_1 q_1 \left(\dfrac{p_0}{p_1}\right)}. \tag{4}$$

The current weighted index is, therefore, the reciprocal of a weighted average of price relatives where the weights are given year values and the price relatives are worked 'backwards', i.e. base year ÷ given year.

QUANTITY INDEX NUMBERS

Quantity index numbers are obtained from price index numbers by simply changing the p's and q's in the price index formulae. Thus we have:

BASE WEIGHTED (LASPEYRE)

(a) *Aggregative Form*

$$I_{01} = \frac{\Sigma p_0 q_1}{\Sigma p_0 q_0} \times 100. \tag{5}$$

(b) *Weighted Average of Quantity Relatives*

$$I_{01} = \frac{\Sigma p_0 q_0 \left(\dfrac{q_1}{q_0}\right)}{\Sigma p_0 q_0}. \tag{6}$$

CURRENT WEIGHTED (PAASCHE)

(a) *Aggregative Form*

$$I_{01} = \frac{\Sigma p_1 q_1}{\Sigma p_1 q_0} \times 100. \tag{7}$$

(b) *Weighted Average of Quantity Relatives*

$$I_{01} = \frac{\Sigma p_1 q_1}{\Sigma p_1 q_1 \left(\dfrac{q_0}{q_1}\right)}. \tag{8}$$

VALUE INDEX NUMBERS

This index is meaningful only in the aggregative form:

$$I_{01} = \frac{\Sigma p_1 q_1}{\Sigma p_0 q_0} \times 100. \tag{9}$$

RELATION BETWEEN PRICE, QUANTITY AND VALUE INDICES

Laspeyre Price Index $\times$ Paasche Quantity Index $=$ Value Index

OR Paasche Price Index $\times$ Laspeyre Quantity Index $=$ Value Index,

i.e.

$$\frac{\Sigma p_1 q_0}{\Sigma p_0 q_0} \times \frac{\Sigma p_1 q_1}{\Sigma p_1 q_0} = \frac{\Sigma p_1 q_1}{\Sigma p_0 q_0} = \frac{\Sigma p_1 q_1}{\Sigma p_0 q_1} \times \frac{\Sigma p_0 q_1}{\Sigma p_0 q_0}. \tag{10}$$

CHAIN BASE INDICES

The true chain base index, taking a price index as our example, has one of the two alternative forms below:

BASE WEIGHTED

$$I_{01} = \frac{\Sigma p_1 q_0}{\Sigma p_0 q_0}, \quad I_{12} = \frac{\Sigma p_2 q_1}{\Sigma p_1 q_1}, \quad I_{23} = \frac{\Sigma p_3 q_2}{\Sigma p_2 q_2}, \ldots \tag{11}$$

CURRENT WEIGHTED

$$I_{01} = \frac{\Sigma p_1 q_1}{\Sigma p_0 q_1}, \quad I_{12} = \frac{\Sigma p_2 q_2}{\Sigma p_1 q_2}, \quad I_{23} = \frac{\Sigma p_3 q_3}{\Sigma p_2 q_3}, \ldots \tag{12}$$

In either case, an estimate of I_{03} is given by:

$$I_{03} = I_{01} \times I_{12} \times I_{23}.$$

The fixed weight 'chain' index has the following form:

$$I_{01} = \frac{\Sigma p_1 q_a}{\Sigma p_0 q_a}, \quad I_{12} = \frac{\Sigma p_2 q_a}{\Sigma p_1 q_a}, \quad I_{23} = \frac{\Sigma p_3 q_a}{\Sigma p_2 q_a}, \ldots \tag{13}$$

where the q_a are quantity weights based on some year a which might or might not be one of the years 0, 1, 2 or 3.

An estimate of I_{03} is again given by:

$$I_{03} = I_{01} \times I_{12} \times I_{23}$$

but in this case the chained estimate of I_{03} leads to the same result as a direct comparison. We have

$$I_{03} = \frac{\Sigma p_1 q_a}{\Sigma p_0 q_a} \times \frac{\Sigma p_2 q_a}{\Sigma p_1 q_a} \times \frac{\Sigma p_3 q_a}{\Sigma p_2 q_a} = \frac{\Sigma p_3 q_a}{\Sigma p_0 q_a}. \tag{14}$$

The true chain index does not have this property; the inner terms of the expression do not cancel.

Prove your prowess [1]

3.1.* Five particular steel products are to be used in the construction of index numbers for steel. The table below sets out price per ton and consumption in thousands of tons for the years 1964 to 1967:

	1964		1965		1966		1967	
	Price per ton	Consumption '000 tons	Price per ton	Consumption '000 tons	Price per ton	Consumption '000 tons	Price per ton	Consumption '000 tons
	£		£		£		£	
Sheets	44	3074	39	2830	40	2840	39	2743
Plates	41	2751	32	2584	31	2504	32	2347
Heavy sections	32	2041	33	2152	33	1736	33	1684
Bright bars	33	520	32	509	30	462	28	422
Wire rods	35	1479	31	1497	28	1369	29	1417

Calculate:

(a) A Laspeyre Price Index, based on 1964 = 100.

(b) A Paasche Quantity Index, based on 1964 = 100.

(c) A Price Index, based on 1967 = 100 with 1965 weights.

3.2. From the data of Example 3.1 calculate:

(a) A Paasche Price Index, based on 1965 = 100.

(b) A Laspeyre Quantity Index, based on 1964 = 100.

(c) A Value Index, based on 1964 = 100.

3.3. From the details given in the following table, calculate a price index for 1961 using weighted price relatives:

Commodity	Relative weekly sales (lb) 1951	Price (new pence) per lb	
		1951	1961
A	1	10	13
B	3	20	17
C	4	30	60
D	5	40	70

(I.O.M., Part II)

[1] An asterisk against the exercise number indicates that the solution is given in detail.

3.4.* From the data of Example 3.1 recalculate 3.1 (*a*) omitting Heavy sections.

3.5. The following figures were taken from the *Monthly Digest of Statistics*:

Index of Retail Prices
(16th January 1962 = 100)

	1966 Weights	Index at 17th May 1966
Food	298	118·0
Alcoholic drink	67	119·0
Tobacco	77	120·8
Housing	113	129·2
Fuel and light	64	119·4
Durable household goods	57	106·5
Clothing and footwear	91	109·4
Transport and vehicles	116	109·9
Miscellaneous goods	61	112·3
Services	56	119·1

Calculate:

(*a*) an 'all items' price index;

(*b*) a price index for all items except the last (Services).

How would a rise of 10% in the cost of Services affect the 'all items' price index?   (I.C.W.A., Part I)

3.6. A firm wishes to follow the relative movement of the cost of various raw materials which it uses. It decides to construct a simple index number, starting January 1968, based on average prices for 1967. From the following data, calculate an index for the month of January 1968.

Raw materials	Price (£ per ton) Average, 1967	January 1968	Weight
A	16	19	5
B	24	25	1
C	13	18	3
D	8	9	6
E	12	14	4
F	4	8	3

What are the main points the firm should bear in mind when planning
the construction of this type of index number?   (I.O.M., Part II)

3.7. From the following table compute the index number for
industry other than agriculture, for the years 1953 and 1956:

	Relative importance in base year	Index numbers		
		1944	1953	1956
Agriculture	7	100	113·5	119·6
All industries (including agriculture)	40	100	107·7	136·9

(I.O.S., Part I)

3.8. A firm in a certain industry uses an Index of Material Prices
based on movements in the prices of selected materials, weighted by
the quantities consumed in the base year. The index, based on
1950 = 100, for the period 1960–1965 was as follows:

1960	1961	1962	1963	1964	1965
120·3	122·1	126·4	125·2	127·0	131·6

In 1965 the index is completely revised to take into account a
change in the type of materials used. The new index, based on
1965 = 100, shows the following values:

1965	1966	1967
100	106·3	109·4

(a) Splice the new index on to the old, i.e. splice 'forwards'.

(b) Splice the old index on to the new, i.e. splice 'backwards'.

3.9.* From the data of Example 3.1, construct a base-weighted
chain index of prices to compare 1967 with 1964.

3.10. Repeat Example 3.9 for the corresponding quantity index.

3.11. Using the data given in the following table, calculate an Index of Real Wages for each year of the period 1956–1964:

Year (Monthly average)	Index of wage rates 31st Jan. 1956=100	Index of retail prices 17th Jan. 1956=100
1956	105	102
1957	110	106
1958	114	109
1959	117	110
1960	120	111
1961	125	115
1962	130	119
1963	134	122
1964	141	126

(I.O.M., Part II)

3.12.* Demonstrate the calculation of crude and standardised death-rates using the following data:

Age group	Population (in thousands)	Number of deaths	Age distribution of standard population
0–9	21	350	221
10–24	30	102	298
25–44	37	229	285
45–64	17	354	149
65 and over	5	415	47
Total	110		1000

(I.O.S., Part I)

3.13. Six firms in a certain industry are ranked in a league-table of safety performance, based on the number of 'lost-time injuries' per 100,000 man-hours worked. The league table reads as follows:

Firm	Lost-time injury rate
A	0·49
B	1·54
C	1·66
D	3·13
E	3·32
F	4·36

A more detailed analysis of lost-time injuries reveals the following data:

	Dept. 1		Dept. 2		Dept. 3	
Firm	M/hr worked	Number of lost-time injuries	M/hr worked	Number of lost-time injuries	M/hr worked	Number of lost-time injuries
A	480	1	830	2	750	7
B	2300	37	2270	35	1720	25
C	800	4	4100	85	2020	26
D	240	17	540	23	1360	27
E	550	19	270	36	1800	32
F	340	10	860	42	1000	44

Calculate a standardised lost-time injury rate for each firm and draw up a new league table based on the standardised rates.

3.14. A firm operates two factories, North and South, manufacturing the same three products A, B and C. The target profit margins on A, B and C are $12\frac{1}{2}$, 17 and 20% respectively, and the target rate of overall return is 15%. The year-end results show North with an overall return of 14·1% and South with an overall return of 15·3%. The managing director wishes to know why North has failed to meet the target when South has exceeded it, and calls for more detailed figures. The accountant produces the following figures:

	North		South	
Product	Production ('000)	Profit rate %	Production ('000)	Profit rate %
A	550	12	350	$12\frac{1}{2}$
B	300	15	500	15
C	150	20	400	$17\frac{1}{2}$

The managing director, who is reading a book called *Managing with Statistics*, sends for the company statistician to make a more meaningful comparison of the figures. The statistician calculates the standardised rates of return and reports on their implications to the managing director. What are these rates and what do they show?

REFERENCES

FISHER, I. 1923. *The Making of Index Numbers.*

VON HOFSTEN, E. 1952. *Price Indexes and Quality Changes.* Bokforlaget Forum.

MUDGETT, B. D. 1951. *Index Numbers.* John Wiley, New York.

4: THE ARITHMETIC OF REGULARITY

Time Series

INTRODUCTION

ONE of the principal tasks of management is to plan ahead. It must make investment plans, order materials for production several months before they are likely to be needed, make arrangements for projected changes in its staff, and so on. For example, businesses that are concerned in any way with the use of agricultural produce such as the manufacture of conserves or the packaging of frozen or prepared foods, obviously have to order at least a year ahead if they are to be assured of the delivery of their basic commodities from the farms. Furniture manufacturers have to take decisions concerning the quantity and type of timber they require many years in advance of demand. Educationalists have to contract for schools and equipment years ahead of their use: local authorities and the government have to decide about the future for roads. The list is endless. Whether they like it or not, and however difficult they may think it is, managers have to estimate the level of future demand for their products, in addition to making many other predictions about the future which have a bearing on their business or responsibilities. A senior executive who tries to run his firm or department on the basis of 'wait and see' will witness its destruction.

In making guesses about the future the statistician makes the greatest use of what he knows has happened in the past. This he does by studying *time series* of the data from which he wants to make predictions.

THE COMPONENTS OF TIME SERIES

A time series is numerical information about a particular subject presented in historical form. An economic time series which has been built-up over a long period reflects many influences, the most important of which are demographic and social. In trying to analyse these series, the statistician casts an envious eye at the pure scientist, who can control the variables (e.g. temperature, humidity) which

107

influence his results, and accepts with a shrug his inability to control population and migratory movements, tariff policy, taxation changes and the many other factors with which he often works. The reader might be excused for thinking that the quantitative effects of this ravelled skein of causes are beyond analysis, but in fact there are a number of methods of disentangling certain important information from the data of a time series. This chapter is partly about such methods, and it lays the practical foundation for forecasting methods which will be dealt with in a later chapter. It also indicates ways in which the analysis of time series can be used as an analytical tool for the estimation of the success of past policy, and the comparison of a firm's performance with that of the industry of which it is part. The figures in a time series are compounded of the following elements:

Secular or long-term trend. If the figures of a time series stretching over a long period are plotted on a graph, a line can be drawn freehand through the graph in such a way that the aggregate vertical distances of the plots above the freehand line are approximately the same as those below it. Such a line is known as the *trend line*, and it may be linear or non-linear. It represents the general movement or trend of the series over the period, when the short term fluctuations in the data are ironed-out or *smoothed*. The process, which admittedly is a very crude one, is depicted in Fig. 4.1.

There are of course more sophisticated methods of determining the trend line than drawing it freehand, and these will be dealt with later in this chapter.

Cyclical movements. Very often figures for a series plotted over a very long period exhibit fluctuations around the trend line with a frequency of three or more years. These fluctuations may be regular (a constant three-year fluctuation say) or they may be irregular but nevertheless persistent. The most common cyclical movement is that generated by the business cycle, but this has not been so much in evidence since the establishment of world credit facilities through the International Monetary Fund and the wider practice of Keynesian doctrines.

Slow changes in fashion or habits are also responsible for cyclical movements in the sale of some manufactured products, such as hats.

Seasonal variations. The seasonal variation in a series of figures or measurements is the element most readily understood by the layman. Temperature is highest in the summer; umbrellas and raincoats sell more readily in the winter; departmental stores sell most just before Christmas; and, more facetiously, a young man's fancy, if it were measurable, would be seen to be most sensitive in the Spring. These regular seasonal movements or their immediate effects are easily recognised in a time series.

Random movements. A time series is affected at irregular and unpredictable intervals by such things as strikes, floods, hurricanes, heatwaves, wars and other such-like unusual influences. These normally cause a non-typical value or set of values in a series which thereafter resumes its accustomed pattern around the trend. Sometimes, however, the fluctuation is so violent that it forces a permanent change on the series and there is a sudden change in the trend. Insurance rates after the San Francisco fire of 1906 were permanently higher because of the large number of insurance company bankruptcies; labour statistics after the Second World War were notably different from those before it.

The statistician has the job of unravelling the effects of these several movements from the original data, so that he may critically review past results and make predictions about the future. It is clearly to his advantage to determine first the trend line which forms the foundation of the series.

SECULAR OR LONG-TERM TREND

DETERMINING THE TREND LINE

Least squares method. The reader will recall that the trend line, when drawn freehand, must be drawn so that the algebraic sum of the vertical distances of plotted points from the trend line is zero, or as near zero as is practically possible. A more refined approach, known as curve-fitting by the *method of least squares*, is to draw the trend line in such a way that the sum of the *squares* of the vertical distances of the plotted points to the trend line is a minimum. This is accomplished by calculating the equation of such a line, plotting the co-ordinates of the points on it and then joining them. (The student at this stage may wish to draw a parallel between the above operations and the methods of determining the spread of data around a central value called *mean deviation* and *variance* which he met in Chapter 2.)

The trend line may be linear or non-linear. The equation of a non-linear trend curve is determined in a similar way to that of a linear trend line. The latter is however simpler to calculate, and will therefore be dealt with first. It will be stated without proof, although this will be found at the end of the chapter, for the benefit of those whose mathematics is up to it, as well as for those whose spirit of enterprise has been whetted and senses a challenge.

PROBLEM: Find by the method of least squares, the linear equation which best fits the data of Table 4.1.

TABLE 4.1. *Calculation of Constants for Linear and Non-linear Equations Satisfying Least Squares Criterion*

Year	x	x^2	y^*	xy	x^3	x^4	x^2y
1955	-5	25	260	-1300	-125	625	6 500
1956	-4	16	235	-940	-64	256	3 760
1957	-3	9	222	-666	-27	81	1 998
1958	-2	4	218	-436	-8	16	872
1959	-1	1	148	-148	-1	1	148
1960	0	0	160	0	0	0	0
1961	$+1$	1	183	$+183$	1	1	183
1962	$+2$	4	171	$+342$	8	16	684
1963	$+3$	9	134	$+402$	27	81	1 206
1964	$+4$	16	173	$+692$	64	256	2 768
1965	$+5$	25	147	$+735$	125	625	3 675
	0	110	2051	-1136	0	1958	21 794
	Σx	Σx^2	Σy	Σxy	Σx^3	Σx^4	Σx^2y

* y Disposal of low protein oilcake, 1955-65, in thousands of tons.

[Source: Ministry of Agriculture, Fisheries and Food]

PRELIMINARIES

1. The reader will remember from Chapter 1 that the general equation of the straight line is $y = ax+c$. Now obviously this is not altered by the substitution of other letters for the constants a and c, or by the change in position of the items on the right-hand side of the equation. We can therefore write with equal validity that the general equation of a straight line is $y = a+bx$ where a is the value of y when $x = 0$ and b is the slope of the line. The problem is to

find the values of *a* and *b* (known as the *parameters*) which will satisfy the 'least squares' criterion.

2. Considering the straight line $y = a + bx$ in relation to the data of Table 4.1, *y*, being the dependent variable, is the quantity of oil-cake, and *x*, being the independent variable, is the relevant year. (As a sop to the purists it ought perhaps to be said that the words 'dependent' and 'independent' are here being used within their normal dictionary meanings, and not in their restricted probability sense.)

METHOD

1. In order to simplify the arithmetic of the calculation, let the *x* value of the *central year* of the series be zero, and the remaining years be measured in units plus or minus of this. Thus the central year of the series, 1960, is given the value zero, the *x* value of 1959 will be minus one, that for 1961 plus one, and so on. Since there is an odd number of years, Σx will therefore equal zero. The device can also be used with a series covering an even number of terms by dividing the series into six-monthly periods counting them each as a unit, and reckoning the midpoint between the central years as zero. The following series explains the method:

	x
1966	5
1965	3
1964	1
1963	-1
1962	-3
1961	-5
$\Sigma x \;=\;$	0

2. Copy into the Table the *y* values, i.e. the quantities of oil cake.

3. Calculate in the Table the values of Σy, Σx, Σxy and Σx^2 and substitute them in the equations:

$$\Sigma y = an + b\Sigma x \tag{1}$$

$$\Sigma xy = a\Sigma x + b\Sigma x^2 \tag{2}$$

where *n* is the number of terms in the series. (These equations are known as the *normal equations*.)

In this example:

$$\Sigma x = 0$$
$$\Sigma x^2 = 110$$
$$\Sigma y = 2051$$
$$\Sigma xy = -1136$$

Substituting these values in the normal equations:

$$2051 = 11a + 0 \tag{3}$$

$$-1136 = 0 + 110b \tag{4}$$

Therefore

$$a = \frac{2051}{11} = 186 \cdot 5$$

and

$$b = \frac{-1136}{110} = -10 \cdot 3$$

The linear equation satisfying the least squares criterion for the given data is therefore:

$$y = 186 \cdot 5 - 10 \cdot 3x.$$

The same result can be obtained by noting that an algebraic solution to the simultaneous equations (1) and (2) gives the following values for a and b

$$a = \frac{(\Sigma y)(\Sigma x^2) - (\Sigma x)(\Sigma xy)}{n\Sigma x^2 - (\Sigma x)^2}; \quad b = \frac{n\Sigma xy - (\Sigma x)(\Sigma y)}{n\Sigma x^2 - (\Sigma x)^2}$$

i.e. $a = \dfrac{(2051 \times 110) - 0}{(11 \times 110) - 0} = \dfrac{2051}{11}; \quad b = \dfrac{(11 \times -1136) - 0}{(11 \times 110) - 0} = \dfrac{-1136}{110}.$

The reader will have noticed that when $\Sigma x = 0$, the value of a is the arithmetic mean of y.

4. Graph this equation by giving a succession of values for x:

When $x =$		$y =$	$x =$		$y =$
-5	(1955)	238·0	$+1$	(1961)	176·2
-4	(1956)	227·7	$+2$	(1962)	165·9
-3	(1957)	217·4	$+3$	(1963)	155·6
-2	(1958)	207·1	$+4$	(1964)	145·3
-1	(1959)	196·8	$+5$	(1965)	135·0
0	(1960)	186·5			

This is illustrated in Fig. 4.1.

Strictly speaking, since the equation is that for a straight line, only two points need be plotted and joined to give the required line. All points have been plotted here to illustrate the efficiency of the method.

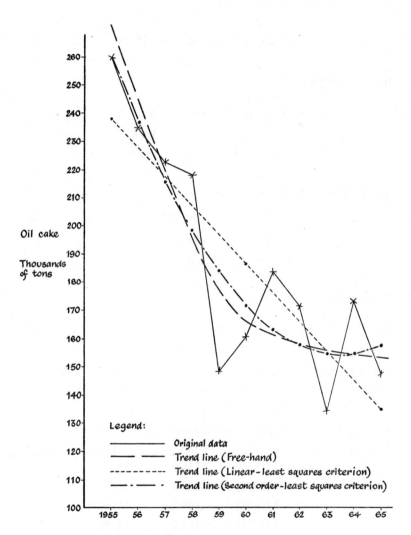

FIG. 4.1. Disposal of low-protein oilcake (thousands of tons). Data from Table 4.1. (Source: Ministry of Agriculture, Fisheries and Food.)

PROBLEM. Find by the method of least squares the equation of the second degree which best fits the data of Table 4.1.

METHOD. This is the same as for the linear equation except that the *normal equations* now become:

$$\Sigma y = an + b\Sigma x + c\Sigma x^2 \tag{5}$$

$$\Sigma xy = a\Sigma x + b\Sigma x^2 + c\Sigma x^3 \tag{6}$$

$$\Sigma x^2 y = a\Sigma x^2 + b\Sigma x^3 + c\Sigma x^4 \tag{7}$$

In this example:

$$n = 11$$

$$\Sigma x = 0 \qquad\qquad \Sigma x^3 = 0$$

$$\Sigma x^2 = 110 \qquad\quad \Sigma x^4 = 1958$$

$$\Sigma y = 2051 \qquad \Sigma x^2 y = 21{,}794$$

$$\Sigma xy = -1136$$

Substituting in the normal equations:

$$2051 = 11a + 0 + 110c \tag{8}$$

$$-1136 = 0 + 110b + 0 \tag{9}$$

$$21{,}794 = 110a + 0 + 1958c \tag{10}$$

From equation (9) $b = \dfrac{-1136}{110} = -10\cdot3.$

We are then left with the simultaneous equations (8) and (10). These give $a = 171\cdot5; c = 1\cdot5.$

The equation of the second degree which best fits the given data according to the least squares criterion is therefore:

$$y = 171\cdot5 - 10\cdot3x + 1\cdot5x^2.$$

The next task is to graph this in the same way as the linear equation.

Now, when $x = -5$ (1955) $y = 260\cdot5$ $x = 1$ (1961) $y = 162\cdot7$

$$-4 \text{ (1956)} \qquad 236\cdot7 \qquad\quad 2 \text{ (1962)} \qquad 156\cdot9$$

$$-3 \text{ (1957)} \qquad 215\cdot9 \qquad\quad 3 \text{ (1963)} \qquad 154\cdot1$$

$$-2 \text{ (1958)} \qquad 198\cdot1 \qquad\quad 4 \text{ (1964)} \qquad 154\cdot3$$

$$-1 \text{ (1959)} \qquad 183\cdot3 \qquad\quad 5 \text{ (1965)} \qquad 157\cdot5$$

$$0 \text{ (1960)} \qquad 171\cdot5$$

This also is drawn in Fig. 4.1.

It can readily be seen that if x is made equal to $+6$, the trend value for 1966 is obtained, i.e. a forecast for one year ahead. This matter will be developed in a later chapter, but the relevance of curve-fitting to forecasting must be appreciated at this stage.

Moving average method. So far we have discussed the fitting of a trend line by freehand and by algebraic methods. Yet another, known as the *moving average* method may be used. The data in Table 4.2 covers a total period of eleven years, which is really rather short for treatment by this process; but it can nevertheless be done, and at the same time a limitation of the method can be illustrated. The essence of the method is to decide, by inspection, if the series exhibits a cyclical movement, i.e. to decide if there is a sustained regular interval between the peaks and troughs of the series: if there is such a movement, it is 'smoothed out'. A graph of the original data does not in fact show a regular and obvious cyclical variation. Because of the short period span of the basic data we might try a three-year and then a five-year period for smoothing and study the appearance of graphs for the derived data. The smoothing process is achieved by taking the successive arithmetic means of three or five years' results and 'centring' them at the second or third year. Table 4.2 shows how this is done. Column (2) records the basic data: Net ton-miles four-weekly averages. In column (3) (upper part) the figures for 1955, 1956 and 1957 are added together and the sum placed against 1956, the central year; then the figures for 1956, 1957 and 1958 are added and the sum placed against 1957, the central year, and so on. In column (4) the sums in column (3) are divided by 3 to give the means, and these when graphed provide the trend line based on a *three-year moving arithmetic average*. This is drawn in Fig. 4.2 and it can be seen that the more violent irregularities of the original data have been ironed out, although this trend curve could clearly be improved with further smoothing. Columns (5) and (6) in Table 4.2 depict the arithmetic necessary for a five-year moving arithmetic average. The basic figures for years 1955–59 are added and the total centred against 1957; those for years 1956–60 are similarly totalled and the sum centred against 1958. Column (6) shows the totals of column (5) divided by 5 to obtain the means, which, when graphed, yield a *five-year moving arithmetic average* trend line. Again this is drawn in Fig. 4.2. On sight it would seem that it is the smoother of the curves and the period nearer the actual

E

TABLE 4.2. *British Railways Freight Traffic—Net ton-miles (millions) Four-weekly averages.* [Source: Ministry of Transport]

Year (1)	Four-weekly average (2)	Sums of threes (3)	Col. (3)÷3 (4)	Sums of fives (5)	Col. (5)÷5 (6)
1955	1640	...	...	...	...
1956	1644	4884	1628	...	...
1957	1600	4656	1552	7654	1531
1958	1412	4370	1457	7439	1488
1959	1358	4195	1398	7143	1429
1960	1425	4131	1377	6778	1356
1961	1348	4008	1336	6547	1309
1962	1235	3764	1255	6417	1283
1963	1181	3644	1215	6176	1235
1964	1228	3593	1198	...	...
1965	1184	...	...	...	...

(1)	(2)	Sums of fours (3)	Sums of eights (4)	Col. (4)÷8 (5)
1955	1640			
		...	...	...
1956	1644			
		← 6296	...	...
1957	1600		← 12 310	← 1539
		← 6014		
1958	1412		11 809	1476
		← 5795		
1959	1358		11 338	1417
		← 5543		
1960	1425		10 909	1364
		← 5366		
1961	1348		10 555	1319
		← 5189		
1962	1235		10 181	1273
		← 4992		
1963	1181		9 820	1228
		← 4828		
1964	1228		...	...
		...		
1965	1184		...	...

cyclical period which a longer original series would have revealed. However, the longer the period chosen for averaging, the fewer are the means left in the series from which the trend line can be drawn, and this can be a disadvantage, as will be shown below.

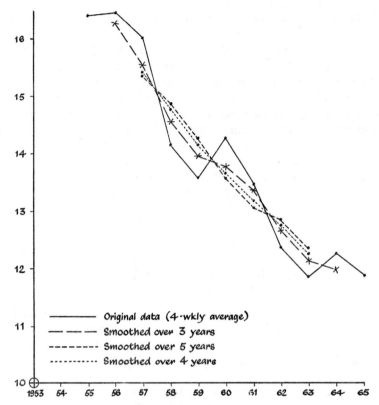

FIG. 4.2. British Railways freight traffic; four-weekly averages, net ton-miles (millions). Data from Table 4.2. [Source: Ministry of Transport.]

For the sake of completeness a moving arithmetic average for four years has been calculated and plotted, and this actually presents a smoother and better fitting curve than either of the other two. The reader will perhaps have noticed that there are four figures between the 'peak' value of 1956 and the trough value of 1959, and the same applies to the peak and trough years of 1960 and 1963 respectively. If the series had been longer and had substantiated this pattern, a four-year moving arithmetic average would have suggested itself from the beginning. The method of centring the moving average for a period of an even number of years will be noted from the calculations in Table 4.2 (lower part), the arrows showing how the value 1539 becomes centred on 1957.

The reader will have noticed that some care has been taken to use the phrase 'moving *arithmetic* average'. It is thus distinguished from the 'moving *geometric* average', which is a form of average some-times employed in deriving trend figures. The process for calculating the moving geometric average is similar to that for calculating the moving arithmetic average, except that the *logarithms* of the original data are used. For example, in Table 4.2 the figures for 1955, 1956 and 1957 are 1640, 1644 and 1600: 'the moving arithmetic average' trend figure based on this data and centred on 1956 is 1628. The moving geometric average—also centred on 1956—would be calculated as follows:

	No.	Logarithm
	1640	3·214 844
	1644	3·215 902
	1600	3·204 120
		9·634 866
Average log (÷3)		3·211 622
Antilogarithm 1628		

Within the accuracy of the tables the derived trend figure appears in this case to be identical with the moving arithmetic average because the original figures are very close to each other, but as a further example, consider the figures for 1957, 1958, 1959. The calculation is as follows:

	No.	Logarithm
	1600	3·204 120
	1412	3·149 835
	1358	3·132 900
		9·486 855
Average log (÷3)		3·162 285
Antilogarithm 1453		

On this occasion the moving arithmetic average gives a trend figure of 1457, whereas the above calculations show that the moving geometric average is 1453. It is in fact a property of unequal num-bers that their geometric mean will always be less than their arithmetic mean, and the student will no doubt wish to use this as a check on his results when necessary.

It has been briefly shown that forecasts of trend values can be made from trend lines fitted by the least squares method by substituting the appropriate value of the independent variable in the equation of the curve. Forecasts can also be made from trend lines drawn from moving average data, but they are less reliable because they must be freehand extrapolations of the curve, and this process becomes liable to considerable error if the final centred mean is positioned too many years from the final year of the original data. In this respect the least squares method is the more advantageous, but it is always wrong to put a blind faith in any extrapolated value. It must be assessed in the light of the current economic (or other) factors which are likely to influence its size.

SEASONAL VARIATION

So far our attention has been centred on the analysis of series of yearly figures. Different considerations arise when quarterly or monthly data are under review. Intuitively we would expect the span of our interest to contract, to become more parochial. Whereas in the last section the averaging process was designed to smooth out cyclical periods covering many years, we are now concerned with fluctuations which are all contained within a calendar year. For example, in the sale of cosmetics a fluctuating demand could be expected which reflected the increased sales for Christmas and Mother's Day. A cosmetic manufacturer with a record of steadily increasing sales would require to know if the upsurge in demand for his product in March and December was large enough to incorporate not only the normal and expected seasonal increase, but also a sustained upward trend. To do this he would require to know what increase in sales would be necessary to accommodate the seasonal change only: any difference between that and his actual sales would represent the contribution towards trend.

There is in fact an amendment to this proposition. Earlier in the chapter, it was pointed out that some fluctuations could be expected from irregular or unpredictable influences. These fluctuations which occur in annual data will clearly be reflected in quarterly or monthly figures too. Moreover, if monthly or quarterly figures are to be averaged over annual periods variations due to cyclical movements will also be present in the derived data. One may therefore say that the figures for any one month or quarter embody the following:

1. a basic trend;
2. a seasonal variation applicable to that month or quarter;
3. the effect of residual influences including cyclical movements.

The following paragraphs and tables will show how figures can be calculated for each of these three elements.

TABLE 4.3(a).   *Home Deliveries—Fertilisers—Phosphates.   Calendar Months (thousands of tons).*   [Source: Board of Trade (Agricultural Depts.)]

	Phosphate '000 tons (1)	Centred sum of twelves (2)	Centred sum of twos (3)	Trend col. (3) ÷24 (4)	Seasonal index. 100 ×col. (1) ÷col. (4) (5)	Adjusted seasonal index (from Table 4.3(b) (6)	Deseasonalised data col. (1)×100 ÷col. (6) (7)	Residual error col. (1)×100 cols. (4)×(6) (8)
1963								
October	36·0							
November	26·5	..	..	..	..	..	..	..
December	28·9	..	..	..	..	..	..	..
1964								
January	42·1	..	..	..	..	..	..	..
February	48·9	..	..	..	..	..	..	..
March	57·9	.. ← 467·3	..	..	..	..	..	..
April	55·9	← 467·6	934·9	39·0	143·3	125·0	44·7	114·7
May	27·8	478·8	946·4	39·4	70·6	76·2	36·5	92·6
June	44·9	480·4	959·2	39·9	112·5	122·1	36·8	92·2
July	38·9	470·1	950·5	39·6	98·2	108·4	35·9	90·6
August	25·7	470·8	940·9	39·2	65·6	70·6	36·4	92·1
September	33·8	472·4	943·2	39·3	86·0	78·4	43·1	109·7
October	36·3	468·9	941·3	39·2	92·6	86·8	41·8	106·6
November	37·7	467·5	936·4	39·0	96·7	85·3	44·2	113·3
December	30·5		939·1	39·1	78·0	73·6	41·4	106·0
1965		471·6						
January	31·8	476·7	948·3	39·3	80·9	90·5	35·1	89·4
February	49·6	478·8	955·5	39·8	125·0	119·1	41·7	104·6
March	59·5	470·8	949·6	39·5	150·6	164·0	36·3	91·8
April	52·4	463·5	934·3	38·9	134·7	125·0	41·9	107·8
May	26·4	451·7	915·2	38·1	69·3	76·2	34·6	90·9
June	49·0	445·7	897·4	37·4	131·0	122·1	40·1	107·3
July	44·0	449·0	894·7	37·3	118·0	108·4	40·6	108·8
August	27·8	438·9	887·9	37·0	75·1	70·6	39·4	106·4
September	25·8	441·7	880·6	36·7	70·3	78·4	32·9	89·9
October	29·0	423·3	865·0	36·0	80·6	86·8	33·4	92·8
November	25·9	428·0	851·3	35·4	73·2	85·3	30·4	85·8
December	24·5		853·6	35·6	68·9	73·6	33·3	93·5
1966		425·6						
January	35·1	421·5	847·1	35·3	99·4	90·5	38·8	109·9
February	39·5	421·6	843·1	35·1	112·5	119·1	33·2	94·5
March	62·3	425·6	847·2	35·3	176·4	164·0	38·0	107·6
April	34·0	422·0	847·6	35·3	96·3	125·0	27·2	77·1
May	31·1	← 425·6	847·6	35·3	88·1	76·2	40·8	115·6
June	46·6	← 425·6	..	..	..	..	..	..
July	39·9	..	..	..	..	..	..	..
August	27·9	..	..	..	..	..	..	..
September	29·8	..	..	..	..	..	..	..
October	25·4	..	..	..	..	..	..	..
November	29·5	..	..	..	..	..	..	..

As the reader will have guessed from the work done in Table 4.2, seasonal changes can be smoothed out by a centred moving average—arithmetic or geometric—taken over the four quarters or twelve months. A moving average calculated in this way will constitute a trend figure. Having derived a trend figure, one may proceed on one of two lines.

1. The *seasonal index* for each month or quarter of each year can be calculated. This is defined as the ratio

$$(actual\ figure)/(trend)$$

expressed as a percentage. The indices for each month or quarter are then separately averaged. The averaging process will reveal the *adjusted seasonal index*, and subsequently, the measure of the residual influences which may be defined as

$$(actual\ figure)/\{(trend) \times (adjusted\ seasonal\ index)\}$$

again expressed as a percentage. This process is fully illustrated with the associated arithmetic in Tables 4.3(a) and (b). Although these

TABLE 4.3(b). *Calculation of Adjusted Seasonal Index Figures*

Year	Jan.	Feb.	Mar.	Apr.	May	Jun.
1964	...	...	...	143·3	70·6	112·5
1965	80·9	125·0	150·6	134·7	69·3	131·0
1966	99·4	112·5	176·4	96·3	88·1	...
Total	180·3	237·5	327·0	374·3	228·0	243·5
Mean	90·2	118·8	163·5	124·7	76·0	121·8
Adjusted seasonal index	90·5	119·1	164·0	125·0	76·2	122·1

Year	Jly.	Aug.	Sep.	Oct.	Nov.	Dec.
1964	98·2	65·6	86·0	92·6	96·7	78·0
1965	118·0	75·1	70·3	80·6	73·2	68·9
1966	...	...	...	...	...	...
Total	216·2	140·7	156·3	173·2	169·9	146·9
Mean	108·1	70·4	78·2	86·6	85·0	73·4
Adjusted seasonal index	108·4	70·6	78·4	86·8	85·3	73·6

Months' Means total 1196·7. Seasonal indices total 1200·0. Obtained from $\dfrac{mean \times 1200}{1196\cdot7}$.

tables derive the trend from a moving arithmetic average the moving geometric average could have been employed with the help of logarithm tables.

2. Alternatively we may express the *seasonal variation* as the arithmetic difference between the actual figure and the trend. As before, when the seasonal variations for each month or quarter are

TABLE 4.4(a). *Home Deliveries—Fertilisers—Phosphates*
Calendar Months (thousands of tons)

	Phosphate '000 tons (1)	Trend from Table 4.3(a) (2)	Seasonal variation col. (1) minus col. (2) (3)	Adjusted seasonal variation from Table 4.4(b) (4)	Deseasonalised data col. (1) minus col. (4) (5)	Residual col. (1) minus col. (2) minus col. (4) (6)
1964						
Apr.	55·9	39·0	+16·9	+9·8	46·1	+7·1
May	27·8	39·4	−11·6	−9·1	36·9	−2·5
Jun.	44·9	39·9	+5.0	+8·4	36·5	−3·4
July	38·9	39·6	−0·7	+3·1	35·8	−3·8
Aug.	25·7	39·2	−13·5	−11·3	37·0	−2·2
Sep.	33·8	39·3	−5·5	−8·1	41·9	+2·6
Oct.	36·3	39·2	−2·9	−4·9	41·2	+2·0
Nov.	37·7	39·0	−1·3	−5·3	43·0	+4·0
Dec.	30·5	39·1	−8·6	−9·7	40·2	+1·1
1965						
Jan.	31·8	39·3	−7·5	−3·7	35·5	−3·8
Feb.	49·6	39·8	+9·8	+7·2	42·4	+2·6
Mar.	59·5	39·5	+20·0	+23·6	35·9	−3·6
Apr.	52·4	38·9	+13·5	+9·8	42·6	+3·7
May	26·4	38·1	−11·7	−9·1	35·5	−2·6
Jun.	49·0	37·4	+11·6	+8·4	40·6	+3·2
July	44·0	37·3	+6·7	+3·1	40·9	+3·6
Aug.	27·8	37·0	−9·2	−11·3	39·1	+2·1
Sep.	25·8	36·7	−10·9	−8·1	33·9	−2·8
Oct.	29·0	36·0	−7·0	−4·9	33·9	−2·1
Nov.	25·9	35·4	−9·5	−5·3	31·2	−4·2
Dec.	24·5	35·6	−11·1	−9·7	34·2	−1·4
1966						
Jan.	35·1	35.3	−0·2	−3·7	38·8	+3·5
Feb.	39·5	35·1	+4·4	+7·2	32·3	−2·8
Mar.	62·3	35·3	+27·0	+23·6	38·7	+3·4
Apr.	34·0	35·3	−1·3	+9·8	24·2	−11·1
May	31·1	35·3	−4·2	−9·1	40·2	+4·9

averaged, the *adjusted seasonal variation* is obtained as well as a figure for the residual influences. This approach is depicted in Tables 4.4(a) and (b), which utilise some of the arithmetic already done in Tables 4.3(a) and (b).

A further word must be said about the construction of Tables 4.3(b) and 4.4(b). In the former the aggregate of the means is 1196·7. Now since we are dealing with percentages over 12 months,

TABLE 4.4(b). *Calculation of Adjusted Seasonal Variation*

	Jan.	Feb.	Mar.	Apr.	May	Jun.
1964	...	...	...	+16·9	−11·6	+5·0
1965	−7·5	+9·8	+20·0	+13·5	−11·7	+11·6
1966	−0·2	+4·4	+27·0	−1·3	−4·2	...
Total	−7·7	+14·2	+47·0	+29·1	−27·5	+16·6
Mean	−3·8	+7·1	+23·5	+9·7	−9·2	+8·3
Adjusted seasonal variation	−3·7	+7·2	+23·6	+9·8	−9·1	+8·4

	Jly.	Aug.	Sep.	Oct.	Nov.	Dec.
1964	−0·7	−13·5	−5·5	−2·9	−1·3	−8·6
1965	+6·7	−9·2	−10·9	−7·0	−9·5	−11·1
1966	...	...	...	...	...	...
Total	+6·0	−22·7	−16·4	−9·9	−10·8	−19·7
Mean	+3·0	−11·4	−8·2	−5·0	−5·4	−9·8
Adjusted seasonal variation	+3·1	−11·3	−8·1	−4·9	−5·3	−9·7

Mean total −1·2. Adjusted seasonal variation total 0.

they must total 12×100 (1200). Since they do not, a correcting factor of 1200/1196·7 must be applied to each of the monthly means to reduce their total to 1200. Corresponding comments apply to Table 4.4(b), but here we are dealing in arithmetic differences which, again because they are being taken over 12 months, should total zero. In fact they add to −1·2. This figure may be reduced to zero by *adding* an appropriate amount to each mean. In this instance, if 0·1 is added to all means the monthly means then sum to zero.

Table 4.3(a) contains a column headed 'Deseasonalised Data'. In this purely seasonal influences have been removed from the monthly recorded figure and it tells the business man, on a short-term basis, whether a higher figure for deliveries is as good as it looks. It is the amount referred to in an earlier paragraph of this section as a 'contribution to trend'. It is left to the reader to plot on the same graph the data for columns (1), (4) and (7) Table 4.3(a) and columns (1), (2) and (5) Table 4.4(a).

Now that the whole of both processes has been illustrated, it can be seen that in one case we have looked at the situation as one in which:

$$(actual\ figure) = (trend) \times (seasonal\ index) \times (residual),$$

and in the other instance we have approached it from the standpoint that

$$(actual\ figure) = (trend) + (seasonal\ variation) + (residual)$$

the term 'residual' meaning an index or amount of residual influences incorporating cyclical and random movement.

The two methods are referred to as *multiplicative* and *additive* respectively.

Both methods are widely used, as well as some variations of them. The multiplicative method is preferred when the series shows a steep upward or downward trend, since, in this situation, the additive model produces an over- or under-estimate of the seasonal effect.

The additive model is particularly useful, however, if several series consisting of a total and its components, e.g. total number of unemployed broken down by regions, are being studied. This is because the additive method ensures that the sum of the seasonally adjusted components is equal to the seasonally adjusted total. This is not the case with the multiplicative method.

As an example of a variation of the two methods, the recorded figure for each month is sometimes expressed as a ratio of the mean monthly figure for the calendar year. But common sense must always be employed. One would expect seasonal indices and variations to change over time, sometimes abruptly. For example an additional week's holiday awarded in a particular industry would affect the seasonal factor for the month in which the holiday was taken. Seasonal indices and variation must therefore take account of such changes either by continual updating or, as in the newer and more sophisticated methods now coming into use, by calculating a moving set of seasonal factors with predictions of next year's factors.

These newer methods involve very heavy computation and are usually carried out by means of a computer 'package'.

Two final points need to be made. First, there is a quite common belief that the sum of a set of seasonally adjusted figures covering a year should equal the sum of the original unadjusted series. This is not true for a series corrected by indices, where, if the two series do agree, it is by the purest chance. As has been seen, seasonal adjustment seeks to estimate the effect which can be expected in a given month as the result of seasonal factors, the figures for some months being reduced, and those for the remaining months increased. The size of these reductions and increases depend not only on the value of the seasonal indices but also on the values of the actual monthly figures.

These monthly figures are influenced by trend, random and cyclical effects and it would be the merest coincidence if the sum of the twelve quantitative adjustments was equal to zero, as would have to be the case for the sum of the adjusted figures to equal the sum of the original data.

The second point is this. More important than the method of seasonal adjustment used is the adjustment of the data to allow for differences in the number of working days in different months, the effect of 4-week and 5-week 'months', changes in the hours worked throughout the year, etc. With 'clean' data most methods of seasonal adjustment will give usable results; with unclean data none of them will.

Summarising, we have endeavoured to demonstrate how the separate quantitative effects of long term and seasonal influences may be disentangled from basic data. Current figures, as they are received, can be 'deseasonalised' by using the appropriate adjusted seasonal index or adjusted seasonal variation. The deseasonalised figures can then be compared with the relevant previous data, and decisions taken about future action.

A knowledge of time series analysis is invaluable in the study of forecasting, and this chapter may be considered an introduction to this subject which is dealt with in chapter 10.

APPENDIX: DERIVATION OF THE NORMAL EQUATIONS

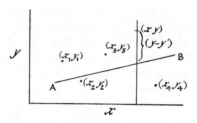

Let the equation of the regression line AB to the series $(x_1 y_1)...(x_n y_n)$ be $y' = a + bx$.

Let y be a representative ordinate from the series.

If the regression line is fitted to the least squares criterion, $\Sigma(y - y')^2$ has to be minimised, i.e. $\Sigma(y - a - bx)^2$ has to be minimised.

Now

$$\Sigma(y - a - bx)^2 = \Sigma(y^2 + a^2 + b^2 x^2 - 2ay - 2bxy + 2abx)$$

$$= \Sigma y^2 + na^2 + b^2 \Sigma x^2 - 2a\Sigma y - 2b\Sigma xy + 2ab\Sigma x. \quad (1)$$

Now the values of a and b in $y' = a + bx$ which will minimise $\Sigma(y - y')^2$ can be derived by partially differentiating equation (1) with respect to a and b and placing the derivatives equal to zero:

$$2na - 2\Sigma y + 2b\Sigma x = 0$$

$$2b\Sigma x^2 - 2\Sigma xy + 2a\Sigma x = 0$$

Rewriting and dividing by 2 we have:

$$\Sigma y = na + b\Sigma x$$

$$\Sigma xy = a\Sigma x + b\Sigma x^2$$

These are the Normal Equations.

Prove your prowess

Note. Because of the extensive examples given in the body of the chapter, no solutions are offered here.

4.1. Product sales within a given country are tabulated below. Estimate the length of the cyclical fluctuation and hence use the method of moving averages to estimate the trend.

Year	Sales	Year	Sales	Year	Sales
1930	458	1937	650	1944	308
1931	672	1938	329	1945	507
1932	611	1939	365	1946	693
1933	266	1940	574	1947	762
1934	207	1941	754	1948	415
1935	410	1942	601	1949	368
1936	720	1943	365	1950	575

(I.O.S., Part I)

4.2. Using the method of moving averages, calculate the trend and seasonal movement of the following series:

Manufacturers' Sales of Gramophone Records, United Kingdom
£'00,000

| Year | Quarters | | | |
	First	Second	Third	Fourth
1962	43	33	37	62
1963	46	38	47	87
1964	60	51	63	82
1965	58	48	56	92
1966	60	50	57	95 (est.)

(I.O.M., Part II)

4.3.

Manufacturers' Sales of Linoleum in the United Kingdom
(Million Square Yards)

| Year | Quarters | | | |
	First	Second	Third	Fourth
1961	12·0	14·2	12·4	12·1
1962	10·1	13·0	11·8	10·6
1963	7·5	10·0	10·1	9·5
1964	9·6	9·4	7·9	...

Using the method of moving averages, calculate:

(a) the trend of the sales, and

(b) the regular seasonal movement of the sales.

(I.O.M., Part II)

4.4. Eliminate the seasonal variation from the following quarterly turnover figures of Mail-order Limited.

Year	Quarter	Turnover £
1956	First	3372
	Second	4106
	Third	4388
	Fourth	3744
1957	First	3214
	Second	4166
	Third	4624
	Fourth	4142
1958	First	3356
	Second	4282
	Third	4470
	Fourth	3696
1959	First	3112
	Second	3990
	Third	4222
	Fourth	3608
1960	First	2980
	Second	3746
	Third	3906
	Fourth	3394
1961	First	2918
	Second	3898

Explain the purpose of removing the seasonal effects from the above data.   (A.C.C.A., Part I)

4.5. The following table gives the numbers of mortgages completed by a Building Society for the months of 1963 and 1964:

	1963	1964
January	58	65
February	32	43
March	44	56
April	76	81
May	85	102
June	99	114
July	70	90
August	61	73
September	85	95
October	86	110
November	77	87
December	64	79

Plot the data and insert the trend line after eliminating seasonal variations.   (I.O.S., Part I)

REFERENCES

CONNOR, L. R., and MORRELL, A. J. H. 1957. *Statistics in Theory and Practice.* 4th Edn. London, Pitman.

CROXTON, F. E., and COWDEN, D. J. 1960. *Practical Business Statistics*, 3rd Edn. New York, Prentice-Hall. (This book has a long section on Time Series.)

FREUND, J. E., and WILLIAMS, F. J. 1958. *Modern Business Statistics.* New York, Prentice-Hall.

HIRSCH, W. Z. 1957. *Introduction to Modern Statistics.* London, Macmillan. (Two chapters on Time Series.)

NEISWANGER, W. A. 1956. *Elementary Statistical Methods.* Revised Edn. London, Macmillan. (Two chapters on Time Series.)

5: WHAT'S IT ALL ABOUT?

The Nature of Statistical Methods

In the four preceding chapters we studied some of the tools of descriptive statistics. These tools, simple though they may be, are a powerful aid in creating order and comprehension out of a mass of figures, and form a very important part of the equipment of the business statistician. More advanced techniques are available, however, which in the appropriate circumstances are even more powerful in helping the statistician to extract the maximum amount of information from a given set of data (or alternatively, obtaining a given amount of information from the minimum quantity of data). These techniques are based fundamentally on the theory of probability, which we shall study in the next chapter. Succeeding chapters consider various applications of probability theory to problems of making valid inferences from data. In explaining and illustrating these applications, a certain amount of mathematics, mostly elementary algebra, is unavoidable, but as far as possible the ideas underlying each technique will be explained in non-mathematical language. Our aim is to give the reader a working knowledge of the techniques available and, more important, an insight into some of the fundamental ideas of statistical theory. In so doing we hope to clear away some of the mystery surrounding the work of the statistician.

After the preparatory work in the chapter on probability we consider the idea of the *theoretical frequency distribution*. A theoretical frequency distribution is represented by a mathematical expression (sometimes called a *frequency function*) which defines the form and properties of the distribution. As an illustration of this idea consider an experiment which consists of tossing a coin 100 times. If we assume that the coin is perfectly made and balanced (we call it an *unbiased* coin) we would expect the chances of throwing Head or Tail to be 50 : 50. Thus in 100 throws of the coin we would 'expect' to throw 50 Heads and 50 Tails. In fact, we would be very surprised if we obtained exactly 50 of each, but we can regard 50 as the theoretical or 'expected' number of Heads and Tails. If we went on to repeat the experiment a number of times, each time throwing the coin 100 times and recording the number of Heads and Tails, we would find that, averaging over all experiments, the number of heads would be very close to 50, but that this number varied from one experiment to another. In other words, we could calculate a mean and variance of the number of Heads. With the aid of a theoretical frequency distribution, however, we do not need to carry out a number of experiments in order to observe the behaviour of our variable. We do not have to construct an empirical frequency distribution. The mathematical representation of the theoretical distribution relieves us of this monumental tedium and enables us to calculate quickly the expected number of Heads and the expected variance for a given number of throws. We say it is a 'model' of the distribution it represents.

The point of a theoretical frequency distribution then is this: if we know that our data conform to, or are drawn from, a particular distribution, we can predict how this data will behave under various conditions. For example, if we know that we are dealing with a population which conforms to another type of distribution called the *Normal* distribution, then we know immediately (as will be explained in Chapter 7) that 95% of all our data will lie within the range $\bar{x} \pm 1\cdot 96\sigma$ in the notation of Chapter 2. Methods are available, as we shall see in Chapter 8, for testing whether our data do in fact conform to a particular frequency distribution.

The idea of a theoretical frequency distribution springs directly from probability theory; so also does our next idea—the idea of a *sample*. We have already briefly mentioned the concepts of population and sample; we shall see in the remaining chapters of the book that the idea of sampling pervades the whole of statistical theory and method. The sample data with which we shall work is, in virtually all cases, only of interest in so far as we can make inferences about the population from which it was drawn. This is the field of *inductive statistics* or *statistical inference*.

In Chapter 8, for example, we shall be considering the problems of estimating population parameters from sample values. We shall also consider the allied questions of whether evidence obtained from a sample supports or contradicts a particular hypothesis we wish to test, or whether, as often happens, the evidence is inconclusive. In tackling such problems the central point of the argument is that results based on a sample cannot be expected to be absolutely accurate—simply because not all units in the population have been considered. Results obtained from samples are subject to *sampling error:* this cannot be eliminated without having 100% coverage of the population, but it can be controlled very closely provided the design of the sample, i.e. the method of selection, the size of the sample and the method of analysis, is given sufficient attention before data collection begins.

Sampling error can be calculated from the sample data, provided the sample is a *random sample.* This is a very remarkable feature of random sampling and enables us to calculate, for example, an estimate of a population parameter together with an estimate of the likely limits between which the 'true' or population value lies.

In a slightly different situation we might wish to compare the results obtained from two samples, i.e. to make an inference about the differences between two populations. In this case an observed difference might be due partly or wholly to sampling error. For example, imagine that we wish to compare the percentage defectives produced by two rival machine tools on the basis of a trial run of each. A straightforward comparison of percentage defectives might be unreliable if the number of observations were small. We seek to test the *statistical significance* of such a comparison, i.e. we try to decide how likely it is that the observed difference has occurred purely by chance. If this likelihood is small, we say that the difference is statistically significant. This idea of statistical significance is one we shall meet repeatedly, especially in Chapter 8 which is devoted to a study of these techniques. At a later stage, in Chapter 11, we shall consider the general question of *sample surveys*, paying particular attention to the sampling of human populations and studying some of the most important types of sample design.

The third major idea we wish to introduce is that of measuring the relationship between two or more variables. This is the idea of *regression* and *correlation* techniques. Regression deals with the problem of relating movements in the value of one variable, the *dependent* variable, to movements in one or more 'explanatory' variables otherwise known as *independent* or *regressor* variables. If height were the independent variable, weight could be the dependent variable. Correlation deals with the more general problem of measuring the *strength of association* between two or more variables. It could, for example, indicate the degree to which movements in the bank rate, level of employment, amount of hire purchase debt, impost of indirect taxes, etc. are accompanied by movements in the volume of sales of a particular product or service. As we shall see in Chapter 9, *correlation coefficients* can be derived very neatly from the results of regression analysis, which, in practice, is used far more widely than correlation.

One of the uses of regression analysis which is currently very fashionable, is for forecasting purposes. The terms *regression models, market models* and *company models* have been coined as synonyms for the *regression equations* fitted to the data under consideration. The simpler models consist of a single equation, the complex models of perhaps twenty equations or more. We shall consider the sense in which an equation or set of equations can be regarded as a model in Chapter 10, which deals with forecasting. Also in Chapter 10 we shall study other well-known techniques of forecasting including the very popular but much overworked method known as *exponential smoothing*. Finally, in Chapter 12, we shall introduce the fourth major idea of statistical theory in the book, the idea of *analysis of variance*. We are already familiar with the idea of the variance of a set of data as a measure of the variability of that data. The analysis of variance is a very powerful method of disentangling data by taking the overall variance and breaking it down into a number of components, each of which can be assigned to a specific factor, plus a 'residual' component, which represents the inherent variabiltiy of the data when the assignable effects have been removed. For example, if a bus company wished to test the effect on fuel consumption of two lubricating oils of different viscosity, it could carry out a service test by selecting a sample of buses and allocating one oil to half the sample and the other one to the remaining half. A direct comparison of results would have little value, however, because the difference due to the different oils would be swamped by the effects of different drivers, different passenger loads, different traffic conditions encountered, and so on. The alternative classical approach to such a problem is to use a laboratory test, holding all variables except the experimental one, i.e. the oil, constant. Thus a number of engines might be run for a certain time under controlled bench-test conditions; alternatively a number of buses might be run round a test track, again under controlled conditions, and the effect of the oils on fuel consumption measured. This approach has the serious shortcoming that laboratory results are rather a poor guide to performance under normal operating conditions.

Analysis of variance techniques enable results to be obtained when the experiment is carried out under normal operating conditions, so that free rein is given to all the factors which influence fuel consumption. The effects of these factors are then estimated separately, and that of immediate interest—the viscosity of the oil—examined.

It follows from the above example that a great deal of attention should be paid to the way in which the data are collected. This leads us to the general area of the *design of experiments*, which is a very large and powerful branch of statistics but one which is beyond the scope of this book. We shall just touch on it at the end of Chapter 12; suffice for now to point out that in many cases a properly designed experiment produces results which virtually 'speak for themselves', requiring no complex analysis. The same comments apply to sampling; time spent on designing an optimum data collection system is amply rewarded by the subsequent ease of analysis. Of course, if the data collection is poor, e.g. biased, irrelevant to the problem, deficient of important items, then no amount of analysis will retrieve the situation. The time to consult a statistician is *before* any data are collected: the common practice of dumping a mass of numerical information in his lap 'for analysis please' is a highly inefficient way of obtaining information, and inevitably leads to a higher proportion of wrong decisions than is necessary.

The problem faced by many statisticians is aptly summed up by a leading teacher of statistics with the words 'Statistics provides solutions to problems the existence of which is not generally recognised'. In the succeeding chapters we shall endeavour to demonstrate both the existence and solution to some of these problems. We begin by considering the elements of probability theory. The reader who finds the chapter heavy going can skim through at the first reading leaving more detailed study to a second reading of the book.

6: WHAT CHANCE DO WE STAND?

Probability

In Chapter 1 the dependence of statistics on the theory of probability was mentioned. We now study the basic grammar of this discipline so that we may better understand the mathematical treatment of some common distributions. Probability is a concept difficult to define rigorously but one which is often, albeit vaguely, used in such statements as: 'I'll probably see you tomorrow'; 'Our profit is unlikely to exceed £500,000'; 'We shall probably have to wait some time for a bus', and so on.

The idea being expressed in such statements is that, based on our own and other people's past experience, and our background knowledge, we expect the event referred to in the statement to occur. This expectation is based largely on the frequency with which such events have happened in the past and the frequency with which we expect them to happen in the long run.

The idea of frequency is a very familiar one in statistics and one of the main weapons in the armoury of descriptive statistics is, of course, the frequency distribution. We can use the frequency distribution to clarify the idea of probability.

Consider the data of Table 1.1 (p. 7). An applicant for a job as salesman with the Vito Food Company might want a rough idea of his financial prospects if he joins the company. The Sales Manager, after pointing out that this depends entirely on the applicant's own efforts, could show him the frequency distribution of salaries. From this, the applicant could see that, other things being equal, his chance of exceeding £2000 was only 75 in 1000, or 0·075, since only 75 of the 1000 salesmen enjoyed salaries of £2000 and above. On the other hand he could see that his chance of receiving less than £1000 was only 0·05, i.e. 50 in 1000. He could perform similar calculations for other salary ranges. In estimating his chances in this way, the applicant is assuming that the relative frequency with which the event of interest occurs at present is a good guide to the probability of the same event occurring in the future.

The ideas illustrated in the above example can be formalised in a definition of probability which is regarded as acceptable by most statisticians. This definition is known as the *relative frequency definition* and states that the probability of an event is equal to the relative frequency with which the event occurs in a large number of trials. More rigorously, the probability of an event is the value to which the relative frequency of successes tends in an infinite sequence of trials.

EXAMPLE. If an unbiased coin is tossed a large number of times the relative frequency of 'Heads' will tend to 0·5. Suppose the first 50 throws give the following results:

Number of throws	Number of heads	Relative frequency of heads
10	6	0·60
20	14	0·70
30	17	0·57
40	19	0·48
50	26	0·52

It can be seen that the relative frequency of heads fluctuates about the value 0·5 and that as the number of throws increases the fluctuations diminish. As more and more throws were made, the relative frequency of heads would fluctuate less and less and in the long run we could imagine it 'settling down' at 0·5.

To take another example, if we said that the probability of gaining Honours in a particular examination was equal to 0·15, we would imply a long run proportion of 0·15 of candidates gaining Honours. We ought not to be surprised if the proportion in any given year is 0·10 or 0·20; this is a random fluctuation.

The relative frequency definition of probability is not the only one and it is a more recent one than the classical definition. The *classical definition* of probability is expressed in terms of equally likely outcomes of a trial. It states that if a particular trial (also called an experiment) has n possible outcomes, and if k of these outcomes are favourable to a particular event then the probability of the occurrence of this event is simply $\frac{k}{n}$.

EXAMPLES. (1) If the experiment consists of tossing a coin, and we require the probability of Heads occurring, we have two possible outcomes, i.e. Heads or Tails, one of which is favourable to the event 'Heads', therefore the probability of Heads is $\frac{1}{2}$.

(2) If the experiment consists of throwing a die and the event in which we are interested is throwing a number greater than 4, there are 6 possible outcomes 2 of which are favourable to the event. The probability of throwing a number greater than 4 is therefore $\frac{2}{6} = \frac{1}{3}$.

A criticism of the classical definition is that it assumes all outcomes are equally likely thereby begging the question of what is meant by probability.

We shall not enter into a discussion of this particular criticism or of rival definitions in general. We merely note that the two alternative definitions given above do not exhaust the list of possible definitions; that both of the above definitions are objective ones; and that in practice the numerical values of probabilities calculated on these two definitions are the same. We can illustrate the application of the classical definition with a die-throwing example.[1]

EXAMPLE. Two dice are thrown.

(a) What is the probability of throwing two sixes?

The total number of outcomes is 36 since each of the 6 outcomes of the first die can be associated with any one of the 6 outcomes of the second. The 36 outcomes can be listed as follows:

1 1	1 2	1 3	1 4	1 5	1 6
2 1	2 2	2 3	2 4	2 5	2 6
3 1	3 2	3 3	3 4	3 5	3 6
4 1	4 2	4 3	4 4	4 5	4 6
5 1	5 2	5 3	5 4	5 5	5 6
6 1	6 2	6 3	6 4	6 5	6 6

The event 6 6 can only occur in one way, therefore the probability of two sixes is $\frac{1}{36}$.

(b) What is the probability of throwing a total of 9?

9 can only occur through combinations of 4 and 5 or 3 and 6. There are 4 such outcomes 4 5, 5 4, 3 6 and 6 3, therefore the probability of throwing 9 is $\frac{4}{36} = \frac{1}{9}$.

[1] We apologise for our pre-occupation with coins, dice, playing cards, urns of coloured balls and other unlikely devices, but they are all particularly useful for illustrating the laws of probability.

(c) What is the probability of *not* throwing a 9?

Since there are 4 outcomes which yield 9, there must be $36-4 = 32$ outcomes which do not yield 9, therefore the probability of not throwing 9 is $\frac{32}{36} = \frac{8}{9}$.

From (b) and (c) we have the important result that the probability of an event occurring plus the probability of it *not* occurring is equal to 1. In symbols, if $P(A)$ stands for the probability of event A occurring and $P(\bar{A})$ stands for the probability of A not occurring:

$$P(A)+P(\bar{A}) = 1.$$

In many situations the complementary probability is easier to calculate than the required probability. In such cases the required probability is obtained by subtracting its complement from 1.

EXAMPLE. Two dice are thrown. What is the probability of not throwing a 10?

The probability of throwing a 10 is $\frac{3}{36} = \frac{1}{12}$ therefore the probability of not throwing a 10 is $1-\frac{1}{12} = \frac{11}{12}$.

Most elementary problems in probability are solved by counting the number of outcomes of the experiment, and in the very simple problems so far considered this has been accomplished by literally counting and listing the outcomes. As soon as problems become a little more complicated, counting in the literal sense becomes impossible, or at least impossibly tedious, so short-cut methods must be sought. Fortunately the use of *permutations* and *combinations* provides us with the required short cuts so we digress briefly to derive the basic formulae of this branch of algebra.

PERMUTATIONS AND COMBINATIONS

PERMUTATIONS

A *permutation* is an arrangement of items in a particular sequence.

EXAMPLE. The letters in the word SUM can be arranged in 6 different sequences:

SUM	MUS	UMS
SMU	MSU	USM

Each sequence, although containing the same three letters, is a different permutation.

Since the purpose of this section is to provide an alternative to literally counting the outcomes of an experiment, we must now seek a formula to give us our answer. The derivation of a formula is simple and proceeds as follows: The first letter can be chosen in 3 ways, i.e. it can be S, U or M. When the first letter has been chosen, two letters remain. The second letter can therefore be chosen in 2 ways. Hence the first two letters can be chosen in $3 \times 2 = 6$ ways. Finally the third letter can be chosen in only one way since only one letter is left. Hence the sequence of three letters can be chosen in $3 \times 2 \times 1$ ways, i.e. 6 ways as we saw above.

This expression can be generalised. If we have n different letters (it is essential for this formula that they should be different) the number of permutations possible is equal to

$$n(n-1)(n-2)...3.2.1.$$

This expression is usually represented by the symbol $n!$ called n *factorial* and the phrase 'the number of permutations of n letters' by the symbol nPn. We therefore have:

$$nPn = n! \tag{1}$$

EXAMPLE. In how many ways can the letters of the word MEDIAN be arranged?

There are six different letters in the word. We therefore have $6P6 = 6! = 6.5.4.3.2.1 = 720$ ways.

We may not wish to take all n items at once. We may, for example, be interested to know how many different sequences of two letters are possible from the word MEDIAN. The list would include such pairs as ME, EM, ED, DE, etc. The formula is obtained by a straightforward application of the previous argument. The first letter can be selected in 6 ways and when this has been done the second letter can be selected in 5 ways. Thus 2 letters can be selected in $6 \times 5 = 30$ different ways i.e. $\dfrac{6!}{4!}$ or $\dfrac{6!}{(6-2)!}$. Generalising this result, we have: The number of permutations of r objects selected from n objects is

$$nPr = n(n-1)(n-2)...(n-r+1) = \frac{n!}{(n-r)!} \tag{2}$$

EXAMPLE. How many permutations of four letters can be obtained from the word MEDIAN?

With $n = 6$, $r = 4$,

$$nPr = \frac{n!}{(n-r)!} = \frac{6!}{2!} = \frac{6.5.4.3.2.1}{2.1} = 360.$$

Notes.   (*i*) 0! is always taken to equal 1.

(*ii*) Formula (2) is quite general since when all *n* objects are taken we have $r = n$ and

$$nPr = \frac{n!}{(n-n)!} = \frac{n!}{0!} = n!$$

(*iii*) A useful relationship is $n! = n(n-1)! = n(n-1)(n-2)!$ etc.

Permutations when some items are alike.   Formulae (1) and (2) do not apply when some items are alike. For example, the number of permutations of the word AVERAGE is not 7! since some of the 7! permutations would be indistinguishable from others because there are 2 A's and 2 E's. The adjustment for similar items is derived from the following argument. Suppose we have *n* items, n_1 of which are alike, and suppose that because of this there are *k* different arrangements instead of *n*!, i.e. *k* is less than *n*! From any one of the *k* arrangements we could, if the n_1 were all different instead of being alike obtain n_1! permutations, since in the given arrangement the n_1 could be arranged among themselves in n_1! different ways. Hence if the n_1 were all different we would have kn_1! permutations instead of *k*. But if the n_1 were all different we would be back to our original expression of *n*!. Therefore

$$kn_1! = n!$$

hence

$$k = \frac{n!}{n_1!}$$

Hence, the number of permutations of *n* objects when n_1 are alike is

$$\frac{n!}{n_1!}$$

This result can be immediately extended to cases where n_1, n_2, etc. objects are alike.

The number of permutations of *n* objects when n_1, n_2, ... n_k objects are alike is given by

$$\frac{n!}{n_1!\, n_2!...n_k!}.$$

EXAMPLE. In how many ways can the letters in the word AVERAGE
be arranged?

We have $n = 7$, $n_1 = 2$, $n_2 = 2$ (for the 2 A's and 2 E's).

Therefore the number of arrangements is

$$\frac{7!}{2!\,2!} = \frac{7.6.5.4.3.2.1}{2.1.2.1} = 1260.$$

COMBINATIONS

In dealing with permutations, the order in which the objects are
arranged is very important. Two different sequences of the same
n objects are counted as two permutations. In calculating probabilities, however, we are frequently interested in counting the
number of sets of r objects selected from n objects where the order
within the set is of no interest. For example, in problems involving
hands at cards, the order in which the cards are dealt is irrelevant.
The number of different selections of r objects out of n where the
order of the objects is irrelevant is called a *combination* and is

denoted by the symbol $\binom{n}{r}$.

The formula for $\binom{n}{r}$ can be derived from that for nPr. Every
combination of r different objects gives rise to $r!$ permutations,
therefore

$$\binom{n}{r} r! = nPr$$

therefore

$$\binom{n}{r} = \frac{nPr}{r!}$$

$$= \frac{n!}{(n-r)!\,r!}.$$

EXAMPLES. (1) How many combinations of two letters can be selected from the word MEDIAN?

We have $n = 6$ and $r = 2$, therefore there are

$$\binom{6}{2} = \frac{6!}{4!\,2!} = \frac{6.5.4.3.2.1}{4.3.2.1.2.1} = 15 \text{ combinations.}$$

(2) How many different hands of 13 cards can be dealt from an ordinary pack of 52? The required number is

$$\binom{52}{13} = 635\ 000\ 000\ 000 \text{ approx.}$$

This is in itself a formidable piece of arithmetic, but tables of factorials are available to ease the burden in practical work.

Notes: (i) $\binom{n}{n} = \dfrac{n!}{n!\,0!} = 1$

(ii) $\binom{n}{r} = \binom{n}{n-r}$

(iii) $\binom{n}{r} + \binom{n}{r+1} = \binom{n+1}{r+1}$

Both of which can be established by expanding the expressions and simplifying.

APPLICATIONS TO PROBABILITY—EXAMPLES

1. What is the probability of holding 4 aces in a bridge hand of 13 cards selected from an ordinary pack of 52?

We have already seen that the total number of bridge hands, i.e. the total number of possible outcomes, is $\binom{52}{13}$. Of these, the number of hands containing 4 aces is equal to the number of ways of selecting 4 aces from 4 aces, multiplied by the number of ways of selecting the remaining 9 cards from 48 cards, i.e.

$$\binom{4}{4} \times \binom{48}{9} = \binom{48}{9}.$$

The required probability is therefore:

$$\frac{\binom{48}{9}}{\binom{52}{13}} = \frac{48!\ 13!\ 39!}{9!\ 39!\ 52!} = 0{\cdot}003 \text{ approx.}$$

2. An urn contains 100 white balls and 50 black balls. Ten balls are drawn together at random. What is the probability of drawing 5 white and 5 black balls?

Five white balls can be drawn in $\binom{100}{5}$ ways and 5 black balls in

$\binom{50}{5}$ ways. Ten balls can be drawn from 150 balls in $\binom{150}{10}$ ways. The required probability is therefore:

$$\frac{\binom{100}{5} \times \binom{50}{5}}{\binom{150}{10}} = \frac{100! \, 50! \, 10! \, 140!}{5! \, 95! \, 5! \, 45! \, 150!} = 0 \cdot 14 \text{ approx.}$$

3. A box contains 100 transistors 20 of which are faulty. Ten are selected for inspection. What is the probability

(a) that all 10 are good?
(b) that all 10 are faulty?
(c) that at least 1 is faulty?

(a) The number of ways in which 10 good transistors can be selected is $\binom{80}{10}$ while the total number of ways in which 10 can be selected is $\binom{100}{10}$. Therefore the required probability is

$$\frac{\binom{80}{10}}{\binom{100}{10}} = \frac{80! \, 10! \, 90!}{10! \, 70! \, 100!} = 0 \cdot 1 \text{ approx.}$$

(b) Ten faulty transistors can be selected in $\binom{20}{10}$ ways. Therefore the required probability is

$$\frac{\binom{20}{10}}{\binom{100}{10}} = \frac{20! \, 10! \, 90!}{10! \, 10! \, 100!} = 0 \cdot 000 \, 000 \, 01 \text{ approx.}$$

(c) We find this probability by noting that the complementary event to event (c) is that no transistors are faulty, i.e. that all 10 are good. We have already obtained this probability in (a) above. The probability that at least 1 transistor is faulty is therefore

$$1 - P \text{ (all 10 are good)}.$$

$$= 1 - 0 \cdot 1 = 0 \cdot 9.$$

MARGINAL AND CONDITIONAL PROBABILITY

We introduced the ideas of probability with the help of a frequency distribution. We again use a frequency distribution, this time a bivariate one, to introduce the ideas of *marginal* and *conditional probability*. Suppose that from a census of firms we obtain the following data:

Size of firm	Type of firm Manufacturing	Distribution	Other	Total
Small	1650	850	1500	4 000
Medium	2050	2500	450	5 000
Large	800	150	50	1 000
Total	4500	3500	2000	10 000

If we were not interested in the *type* of firm we would have a simple frequency distribution of firms by size as in the 'Total' *column* of the table. If we were not interested in the *size* of firm we would have a simple distribution of firms by type, as in the 'Total' *row* of the table. These two simple distributions are called the *marginal distributions* appropriately enough since they occur in the margins of the table.

The idea of marginal probability follows readily from the definition of marginal distributions. For example, if a firm is selected at random from the 10 000, the probability that it is *small* is

$$\frac{4000}{10\ 000} = 0{\cdot}4$$

the probability that it is *medium* is

$$\frac{5000}{10\ 000} = 0{\cdot}5$$

and the probability that it is *large* is

$$\frac{1000}{10\ 000} = 0{\cdot}1.$$

Similarly, the probabilities of being in the *Manufacturing, Distribution* or *Other* sectors are 0·45, 0·35 and 0·2 respectively.

If now we are interested in, say, the distribution by *size*, given that the firms are in the manufacturing section, we are interested only in the *Manufacturing* column of the table. Similarly, if we are interested in the distribution by *type* of large firms, we are interested only in the *Large* row of the table. The probability that a firm is *large*, given that it is a manufacturing firm is

$$\frac{800}{4500} = 0{\cdot}18.$$

Similarly, the probability that a firm is in the *Distribution* sector, given that it is small is

$$\frac{850}{4000} = 0{\cdot}21,$$

assuming random selection within the given category.

Conditional probability is thus the probability of one event occurring given that a second event has already occurred.

EXAMPLE. 'Picture' cards are defined as Jack, Queen, King and Ace. A card is selected at random from an ordinary pack. What is the probability that the card is an Ace given that it is a picture card? There are 16 picture cards in a pack, four of which are Aces. The required probability is therefore $\frac{4}{16} = \frac{1}{4}$.

As a further example of marginal and conditional probabilities the reader might like to consider the distribution of cars by make and size, deciding on his own size categories. He can then consider such questions as 'Given that the next car to pass is a Ford, what is the probability that it is a large car?' and 'Given that the next car to pass is large, what is the probability that it is a Ford?'

SAMPLE SPACE

A useful concept in the understanding of the basic rules of probability is the *sample space*, sometimes called the *event space*. The sample space is defined as the set of points representing the possible outcomes of an experiment.

EXAMPLES. (1) The sample space for the experiment of throwing two dice consists of the set of 36 outcomes.

(2) The sample space for the experiment of tossing a coin consists of 2 points—Heads and Tails.

The concept of the sample space facilitates the understanding of more complex probabilities and enables some of the rules of probability to be represented diagrammatically.

Every point in a sample space has a probability attached to it. The sum of these probabilities must be 1, and all probabilities must be non-negative. In the simple examples so far considered we have automatically attached equal probabilities to the points in our sample spaces, under the assumption of unbiased coins, dice, etc. We can now define probability in terms of the points in the sample space: The probability that an event A will occur is the sum of the probabilities of the sample points that are associated with the occurrence of A.

EXAMPLE. The probability of throwing a 9 with two dice, which we have already found to be $\frac{1}{9}$, can be expressed as the sum of the probabilities in the sample space associated with throwing a 9. There are four sample points favourable to the event. 4 5, 5 4, 3 6 and 6 3. These are four of the 36 points in the sample space and each has a probability of $\frac{1}{36}$. The required probability is therefore $\frac{1}{36}+\frac{1}{36}+\frac{1}{36}+\frac{1}{36} = \frac{4}{36} = \frac{1}{9}$.

THE RULES OF PROBABILITY

We have only considered very simple events up to now, but most applications of probability involve *related* or *compound events*. We therefore require rules for calculating the probabilities of such events.

We also need a form of notation in which to express our rules. We shall denote the probability of event A by $P(A)$, defining other notation as we require it.

RULE 1. $$0 \leqslant P(A) \leqslant 1$$

In words, the probability of an event A has extreme values of 0 (representing impossibility) and 1 (representing certainty).

RULE 2. $$P(A)+P(\bar{A}) = 1$$

$$\text{or} \quad P(A) = 1-P(\bar{A})$$

Where $\bar{A}$ denotes the event *not A*.

MUTUALLY EXCLUSIVE EVENTS

Two events are said to be *mutually exclusive* if the occurrence of either one precludes the occurrence of the other, e.g. Heads and Tails, Hit and Miss are mutually exclusive events.

RULE 3. If two events A_1 and A_2 are mutually exclusive, the probability of either A_1 *or* A_2 occurring, denoted by $P(A_1+A_2)$ is

$$P(A_1+A_2) = P(A_1)+P(A_2).$$

This can immediately be generalised to the case of n mutually exclusive events $A_1, A_2, A_3, ..., A_n$:

$$P(A_1+A_2+...+A_n) = P(A_1)+P(A_2)+...+P(A_n).$$

EXAMPLE. What is the probability of throwing 3 or 4 with one throw of a die? Let event A_1 be the occurrence of a 3 and A_2 be the occurrence of a 4. Then $P(A_1) = \frac{1}{6} = P(A_2)$. Therefore

$$P(A_1+A_2) = \tfrac{1}{6}+\tfrac{1}{6} = \tfrac{1}{3}.$$

Many problems in probability involve events which are not mutually exclusive. The situation which then arises can be illustrated by means of a sample space diagram.

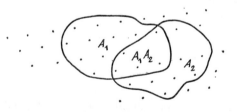

FIG. 6.1.

The set of points in Fig. 6.1 represent the sample space for some unspecified experiment. Consider two events A_1 and A_2 which are not mutually exclusive. We can enclose the sample points which correspond to events A_1 and A_2 and label the areas accordingly. We find that since the events are not mutually exclusive the two areas overlap. In other words, certain outcomes of the experiment are common to both events, which is what we mean when we say that two events are not mutually exclusive. We label the area of overlap

F

A_1A_2 to show that sample points in this area represent the occurrence of events A_1 and A_2 simultaneously.

We can now derive a formula for $P(A_1+A_2)$. From the sample space definition of probability, $P(A_1+A_2)$ is equal to the sum of the probabilities associated with all the sample points included in areas A_1 and A_2 in Fig. 6.1. If, however, we simply add together all the points in A_1 and all the points in A_2 we shall count the points in A_1A_2 twice. We therefore subtract A_1A_2 from A_1+A_2 to correct for the double counting. Thus we have:

RULE 4. $P(A_1+A_2) = P(A_1)+P(A_2)-P(A_1A_2).$

This rule is also fairly easily generalised. For three events $A_1A_2A_3$ we can again draw a suitable diagram. This is shown in Fig. 6.2; the sample points are omitted for the sake of clarity.

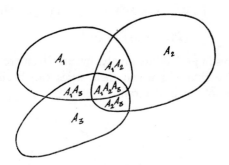

FIG. 6.2.

Note. Area $A_1A_2A_3$ is part of areas A_1A_2, A_1A_3, and A_2A_3.

In Fig. 6.2, simple addition of A_1, A_2 and A_3 would result in double counting of areas A_1A_2, A_1A_3 and A_2A_3, and treble counting of area $A_1A_2A_3$. If, however, we subtract A_1A_2, A_1A_3, and A_2A_3 from $A_1+A_2+A_3$ we shall remove the double counting of A_1A_2, A_1A_3 and A_2A_3, but in doing so we shall remove region $A_1A_2A_3$ altogether. We must, therefore, add $A_1A_2A_3$ back again. We thus obtain:

$$P(A_1+A_2+A_3) = P(A_1)+P(A_2)+P(A_3)-P(A_1A_2)-P(A_1A_3)$$
$$-P(A_2A_3)+P(A_1A_2A_3)$$

The more adventurous reader might like to consider the general case $A_1 + A_2 + \ldots + A_n$ where we have:

$$P(A_1 + A_2 + \ldots + A_n) = \sum_i P(A_i) - \sum_{i<j} P(A_i A_j)$$
$$+ \sum_{i<j<l} P(A_i A_j A_l) - \ldots + \ldots$$

A few minutes spent in expanding this expression for $n = 2$ and $n = 3$ will clarify the subscripts to the summation signs after which the expression for $n = 4, 5$ etc. will present little difficulty.

Rule 4 is known as the *addition rule of probability*. The addition sign has a rather special meaning however: $P(A_1 + A_2)$ is read as 'the probability that either A_1 or A_2 or both will occur'.

Before we can use Rule 4 we need to know how to evaluate expressions such as $P(A_1 A_2)$. $P(A_1 A_2)$ is the probability that events A_1 and A_2 occur simultaneously. A formula for this probability can be derived from a consideration of conditional probabilities, which we have already met in the previous section. A diagram will once more help clarify the argument. Fig. 6.3 below represents two events in a sample space.

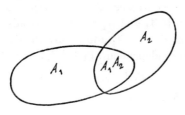

FIG. 6.3.

We are interested in the probability that A_2 occurs, given that A_1 has already occurred. We express this in symbols as $P(A_2 \mid A_1)$.

Since A_1 has already occurred, the sample space is reduced to the area A_1. Within this area, A_2 occurs only in the smaller region $A_1 A_2$, therefore the required probability is given by the number of sample points in $A_1 A_2$, divided by the number in A_1, i.e. $\dfrac{A_1 A_2}{A_1}$.

But the number of sample points in $A_1 A_2$ can be expressed as $P(A_1 A_2)$ and the number in A_1 as $P(A_1)$. We therefore have

$$P(A_2 \mid A_1) = \frac{P(A_1 A_2)}{P(A_1)}$$

which can be rewritten as

RULE 5. $P(A_1A_2) = P(A_1)P(A_2 \mid A_1)$.

EXAMPLE. From our frequency distribution of firms by size and type
(p. 144) we can obtain the probability that a firm is both *small*
and *manufacturing* by inspection of the table. The required
probability is $\dfrac{1650}{10\,000} = 0\cdot165$. We now calculate this probability
from Rule 5. We have $A_1 = $ 'firm is small', $A_2 = $ 'firm is
manufacturing'. Then $P(A_1) = \dfrac{4000}{10\,000}$ and $P(A_2 \mid A_1) = \dfrac{1650}{4000}$.
Therefore

$$P(A_1A_2) = \frac{4000}{10\,000} \times \frac{1650}{4000} = \frac{1650}{10\,000} = 0\cdot165$$

as before.

INDEPENDENT EVENTS

If the occurrence or non-occurrence of A_2 is not affected by the
occurrence or non-occurrence of A_1, A_2 is said to be *independent* of
A_1. For example, if A_1 and A_2 represent 'Heads' in two successive
tosses of an unbiased coin, A_1 and A_2 are clearly independent.

If A_1 and A_2 are independent, $P(A_2 \mid A_1) = P(A_2)$ and

$$P(A_1 \mid A_2) = P(A_1).$$

In this case Rule 5 reduces to:

RULE 6. For independent events $P(A_1A_2) = P(A_1)P(A_2)$.

Independence is often defined in terms of Rule 6, i.e. two events are
said to be independent if $P(A_1A_2) = P(A_1)P(A_2)$.

Rule 6 can immediately be generalised for n events. Given n
independent events $A_1A_2...A_n$ we have

$$P(A_1A_2...A_n) = P(A_1)\,P(A_2)...P(A_n).$$

EXAMPLES. (1) Two coins are tossed. What is the probability of
throwing 2 Heads or 2 Tails?

The events are mutually exclusive, therefore, by Rule 3
P (2 Heads $+$ 2 Tails) $= P$ (2 Heads) $+ P$ (2 Tails). By Rule 6,
P (2 Heads) $= \frac{1}{2}.\frac{1}{2} = P$ (2 Tails), therefore

$$P \text{ (2 Heads or 2 Tails)} = \tfrac{1}{4} + \tfrac{1}{4} = \tfrac{1}{2}.$$

(2) Two cards are drawn from a pack. What is the probability that both are Aces?

Drawing two cards is equivalent to drawing one card and then another without replacing the first. Thus the second event is not independent of the first. The probability of drawing an Ace with the first card is $\frac{4}{52}$. Therefore

$$P \text{ (2 Aces)} = \frac{4}{52} \cdot \frac{3}{51} = \frac{1}{221}.$$

Note. It is important to be certain whether selection is with or without replacement. In the above example if the first card was replaced before drawing the second card, the events would be independent and the probability, by Rule 6, would be $\frac{4}{52} \cdot \frac{4}{52} = \frac{1}{169}$.

(3) Assume that the probability of a child being male is $\frac{1}{2}$. What is the probability that in a family of 4 children they are all of the same sex?

The sex of successive children is independent of the sex of the children already born, therefore by Rule 6 we have:

$$P(4 \text{ males}) = \frac{1}{2} \cdot \frac{1}{2} \cdot \frac{1}{2} \cdot \frac{1}{2} = \frac{1}{16}.$$

Similarly P (4 females) $= \frac{1}{16}$. Hence by Rule 3 we have

$$P \text{ (4 children of same sex)} = \frac{1}{16} + \frac{1}{16} = \frac{1}{8}.$$

(4) A die is tossed. What is the probability of throwing a 2 or an even number? Since 2 is an even number, the events are not mutually exclusive so we have to use Rule 4

$$P(2) = \frac{1}{6}, \ P \text{ (Even)} = \frac{3}{6},$$

therefore

$$P \text{ (2 + Even)} = \frac{1}{6} + \frac{3}{6} - \frac{1}{6} \cdot 1 = \frac{3}{6}.$$

MATHEMATICAL EXPECTATION

At the beginning of this chapter we used the idea of expectation in a general sense in introducing probability. We now discuss a highly specialised meaning of the word, one which is widely used in statistics.

The *expectation* of an experiment is the sum of the values of all possible outcomes, multiplied by the corresponding probabilities. In symbols we have

$$E = \Sigma A_i P(A_i)$$

where E stands for expectation, A_i is the value of the ith outcome and $P(A_i)$ is the probability of that outcome.

EXAMPLE. If two players toss a coin and player A pays 5p to player B
whenever 'Heads' is thrown while player B pays 5p to player A
whenever 'Tails' is thrown, A's expectation is

$$E = (\tfrac{1}{2} \cdot -5) + (\tfrac{1}{2} \cdot 5) \text{ new pence}$$
$$= (-2\tfrac{1}{2} + 2\tfrac{1}{2}) \text{ new pence} = 0.$$

Player B also has zero expectation.

The concept of 'expectation' is thus rather like that of 'average',
but there are important differences. Expectation and *expected values*
will be seen again in the chapters dealing with estimation and hypothesis testing and sampling; in the meantime we present an example
illustrating the use of expectations in making business decisions.

Consider a theatre company faced with a forthcoming one-day
performance. The performance will be given in the middle of summer and the company is faced with two alternatives; it can give an
open-air performance or it can hire a small theatre. How shall it
decide?

If the company was certain that the day would be fine it would
decide on an outdoor performance whereas if it knew that the day
would be a wet one it would hire the theatre. Suppose now the
company estimates the financial results of taking the right and wrong
decision.

Case 1. Give the performance outdoors.
  If the day is fine the profit is £300.
  If the day is wet the loss is £200 (or the 'profit' is − £200).

Case 2. Hire the theatre.
  If the day is fine the 'profit' is − £50 (because people will not
  come into the theatre).
  If the day is wet the profit is £150 (because people are driven
  into the theatre).

Suppose now that the weather records suggest that the probability
of a wet day is $\tfrac{1}{3}$. The expected values of the two alternatives are:

$$\textit{Outdoors} \quad E = £(\tfrac{1}{3} \cdot -200) + (\tfrac{2}{3} \cdot 300) = +£133$$
$$\textit{Indoors} \quad E = £(\tfrac{1}{3} \cdot 150) + (\tfrac{2}{3} \cdot -50) = +£17.$$

Therefore, with a probability of $\tfrac{1}{3}$ for the occurrence of a wet day
the company has a higher expectation of profit from choosing the
outdoor alternative.

Suppose now that the probability of rain is $\frac{2}{3}$ instead of $\frac{1}{3}$. The two expectations are now

$$Outdoors \quad E = -£33$$
$$Indoors \quad E = +£83.$$

In this case the better decision is to hire the theatre.

Finally, suppose that nothing is known from the records about the probability of rain. We could find the level of probability at which the two alternatives gave equal expected values.

Let the required probability be p.

Then the expectations are:

$$Outdoors \quad E = (p.-200)+((1-p).300) = -200p-300p+300$$
$$Indoors \quad E = (p.150)+((1-p).-50) = 150p+50p-50.$$

To find p we equate the two expected values. We have

$$-500p+300 = 200p-50.$$

Therefore $\qquad 300+50 = 200p+500p$

i.e. $\qquad\qquad 350 = 700p.$

Therefore $\qquad\qquad p = \frac{350}{700} = \frac{1}{2}.$

Thus if the probability of rain was $\frac{1}{2}$ the expected values would be equal. For probabilities of rain less than $\frac{1}{2}$ the open air performance has the higher expected value, for p greater than $\frac{1}{2}$ the theatre would have the higher expected value. The decision can now be made according to whether the company feel that p will be greater or smaller than $\frac{1}{2}$, i.e. according to whether they are optimistic or pessimistic.

The above example illustrates the calculation of expectations. We should point out that the expected profit is only one of a number of possible criteria for taking the decision. An alternative criterion is to consider the maximum loss possible with each course of action and to choose the one which minimises the maximum loss. With this criterion, and with the same data as before, the better decision is to hire a theatre since the maximum loss, i.e. the worst that can happen, is £50 compared with a possible loss of £200 if the performance is planned for outdoors and the day is wet. The choice of decision criteria depends on a number of factors. Three of the most important are:

(*i*) The state of the company and of the market. In other words can the company 'afford' the worse possible outcome?

(*ii*) The money involved. Criteria tend to differ between projects costing hundreds of pounds and those costing millions.

(*iii*) The individual concerned with the decision. The 'entre-preneurial' type (or the optimist) is likely to attach more importance to possible gains than to possible losses. The professional manager (or the pessimist) possibly gives more weight to the consequences of being wrong.

RANDOM VARIABLES AND FREQUENCY FUNCTIONS

Consider the sample space associated with the tossing of two dice. There are 36 points in this space but these do not correspond to 36 different scores. In fact, the possible scores range from 2 to 12. We can consider a variable x which assumes the eleven possible scores. Such a variable is called a *random variable*. A more precise definition is that a random variable is a numerical-valued variable defined on a sample space.

We can tabulate the values of the random variable in the above example, associating each value with its probability. We have

x	$f(x)$
2	$\frac{1}{36}$
3	$\frac{2}{36}$
4	$\frac{3}{36}$
5	$\frac{4}{36}$
6	$\frac{5}{36}$
7	$\frac{6}{36}$
8	$\frac{5}{36}$
9	$\frac{4}{36}$
10	$\frac{3}{36}$
11	$\frac{2}{36}$
12	$\frac{1}{36}$
	$\frac{36}{36} = 1.$

The column headed $f(x)$ gives the probabilities associated with the values x. For example $f(3) = \frac{2}{36}$, $f(7) = \frac{6}{36}$. $f(x)$ is called the *frequency function* of x and is defined as the function that yields the probability that the random variable x will assume any particular value in its range.

A frequency function often consists of a simple tabulation like the one above. There are, however, a number of important frequency functions which are expressed as formulae rather than tabulations. We shall study several such functions in the next chapter, among them the frequency function for the Binomial Distribution, which we derive now as an exercise.

Consider the following problem. Two coins are tossed 100 times. What is the probability that exactly x tosses will result in 'Heads', 'Heads' (written HH)? Here x is a random variable which can take values from 0 to 100, each value having an associated probability, the sum of these probabilities equalling 1. We begin by considering a fixed order in which the required event can occur. Let us suppose that the first x throws result in HH and the remaining $100-x$ throws are not HH. We regard HH as a success and 'not HH' as a failure. We can write the above statement as

$$\underbrace{S\,S\,S\,...\,S}_{x}\ \underbrace{F\,F\,...\,F}_{(100-x)}.$$

We now find the probability of the above sequence. We have $P(S) = P(HH) = \frac{1}{4}$ (from Rule 6) and $P(F) = \frac{3}{4}$, therefore a sequence of x S's and $(100-x)$ F's has probability $(\frac{1}{4})^x(\frac{3}{4})^{100-x}$ (again from Rule 6). For every sequence of x successes and $(100-x)$ failures this probability is the same, all we need to know, therefore, is the total number of different sequences. The number of different sequences is equal to the number of permutations of 100 objects when x of one kind and $100-x$ of another kind are alike, i.e.

$$\frac{100!}{x!(100-x)!}$$

The probability that exactly x of the 100 throws will result in HH is therefore

$$\frac{100!}{x!(100-x)!}\,(\tfrac{1}{4})^x(\tfrac{3}{4})^{100-x}.$$

This expression can easily be generalised to the case where n trials are made and p is the probability of success at each trial. The probability of achieving x successes is given by the function

$$f(x) = \frac{n!}{x!(n-x)!}\,p^x(1-p)^{n-x}.$$

We will investigate the properties of this function in Chapter 7.

Prove your prowess

6.1. Three dice are thrown. What is the probability of throwing:

(a) Three sixes?

(b) 4 5 6?

(c) A total of 12?

(d) A total greater than 5?

6.2. Compare the chances of tossing Head with one coin, 2 Heads with two coins and 3 Heads with three coins. Generalise this to n Heads with n coins.

6.3. In how many different ways can the letters in the word MANAGER be arranged? How many of these begin with MAN?

6.4. Four cards are drawn from a standard pack of 52. What is the probability that 2 Red and 2 Black Picture cards are drawn?

6.5. What is the probability of drawing a black Ace or a red King or a black Queen or a red Jack in one draw from a standard pack of cards?

6.6. What is the probability of throwing either an even number or a number greater than 3 with one throw of a die?

6.7. If the probability that a married man will vote in a General Election is 0·7 and the probability that a married woman will vote provided that her husband votes is 0·8, what is the probability that a married couple will both vote?

6.8. A customer in a glassware store accidentally breaks an expensive ornament. The store manager demands the full retail price in payment for the breakage. The customer offers to pay the cost to the store, plus the expected profit during the period in which the store is awaiting replacement. Assuming that only one of this particular ornament is kept in stock, the replacement will take one week, and the average rate of sale is one ornament per month, calculate the approximate amount that the customer will have to pay if his offer is accepted. The cost to the store is £25, the selling price £52.

6.9. A production process produces components with a defect rate of 5 per cent. (a) What is the probability that a box of 200 components will contain exactly 10 defectives? (b) What is the probability that the box contains no defectives?

6.10. A colour television set is constructed from 500 sub-assemblies. If the chance of each sub-assembly failing to meet its specification without further adjustment is 1 in 1000, what is the probability that at final inspection stage the set will not need adjustment?

REFERENCES

All introductory texts on mathematical statistics include at least one chapter on probability. Among these can be particularly recommended:

HOEL, P. G. 1962. *Introduction to Mathematical Statistics*, 3rd Edition. New York, John Wiley.

The reader interested in an experimental approach to the subject should consult:

KERRICH, J. E. 1946. *An Experimental Introduction to the Theory of Probability*. Munksgaard.

The reader who would really like to get to grips with the subject should try:

FELLER, W. 1957. *An Introduction to Probability Theory and its Application*, 2nd Edition. New York, John Wiley.

PARZEN, E. 1960. *Modern Probability Theory and its Application*. New York, John Wiley.

7: SUITS 'OFF THE PEG'

The Binomial, Poisson and Normal Distributions

INTRODUCTION

LET us recapitulate a little. The reader is by now very familiar with the idea of a distribution table for a variable. In Chapter 2 he was made to ponder at some length over a distribution table for salaries, and saw, in Fig. 2.2, that the curve of the distribution looked like an admiral's hat.

It so happens that many of the distributions that are encountered in industrial and commercial life are of this variety. Sometimes the curve is lopsided to the right or left, or, to express it in the more technical terms learned earlier, the curve has a positive or negative skew. This much the reader knows.

But strictly speaking of course, graphs of this type of distribution will only look like an admiral's hat when they are of a continuous variable. A histogram of a discrete distribution which has the same kind of frequency pattern as the continuous distribution above, will look like an 'up and down' staircase, e.g. a distribution of the number of points scored in the several Rugby (or Ivy) League games on any one Saturday.

Another common type of distribution is one which deals with the occurrence of rare events. A frivolous (but apt) example is the probability that none, or one or more players, will be sent off the field by the referee during a professional football match. The greatest likelihood is that none will be sent off, with much smaller and diminishing probabilities that one then two, etc., will be sent to the dressing room.

This chapter looks closely at each of these distributions, although for reasons which will emerge, not in the order in which they have been mentioned above. In making our investigations we are aided by a fortunate coincidence. Each one of the above types of distribution is similar in form to a particular theoretical distribution. A great deal is known about these theoretical distributions, such as their means and variances. Once it can be shown, therefore, that there is a reasonable similarity between a set of observations and one of the theoretical distributions, the useful properties of the latter can be applied to the observed numerical data. A process which tests 'the goodness of fit' as it is called, between the observations and the theoretical distribution is dealt with in Chapter 8. It is rather like having a standard fitting suit; providing the client who wears it does not look too ridiculously gawky, we can make a number of assumptions, about his height, his girth, and so on because we know the standards from which the suit was cut.

There are other theoretical distributions, but we have not included them in this book, mainly because they demand a mathematical knowledge which it would be unwise to expect from our readers.

BINOMIAL DISTRIBUTION

THE CONCEPT—BERNOULLI'S THEOREM

Let us try to translate this introduction into practical terms. Consider the somewhat bizarre proposition that exactly equal numbers of men and women have booked to see the annual Promenade Concerts. Towards the end of the season the three booking clerks, to while away the tedium of their occupation, compose a game. The idea of the game is for each clerk to note the sex of the first person to call at his booking office the following day, after having guessed what the sex combination of their three 'first-callers' was going to be. Their immediate problem is to establish the odds to place on the possible outcomes.

Assuming that men and women appear in random fashion, there are eight possible outcomes. Each of these is equally likely, so the probability that each will occur is 1/8. A table of the possible outcomes would appear as follows (M = Male; F = Female):

Outcome	Booking Clerks A	B	C	Total	Probability of occurrence
1	M	M	M	3 M	$\frac{1}{8}$
2	M	F	M	2 M, 1 F	$\frac{1}{8}$
3	M	M	F	2 M, 1 F	$\frac{1}{8}$ $\Big\}\frac{3}{8}$
4	F	M	M	2 M, 1 F	$\frac{1}{8}$
5	M	F	F	1 M, 2 F	$\frac{1}{8}$
6	F	M	F	1 M, 2 F	$\frac{1}{8}$ $\Big\}\frac{3}{8}$
7	F	F	M	1 M, 2 F	$\frac{1}{8}$
8	F	F	F	3 F	$\frac{1}{8}$
					1

This table of probabilities can be re-written as:

Sex-combination	Probability
3 Males	$\frac{1}{8}$
2 Males, 1 female	$\frac{3}{8}$
1 Male, 2 females	$\frac{3}{8}$
3 Females	$\frac{1}{8}$
	$\overline{1}$

This result could have been obtained theoretically. There can be only one way for three men to arrive together or consecutively at the three booking offices, and the same applies to the arrival of three women, but the number of ways that two men and one woman can arrive requires a little thought. There is only one way for them to arrive at the offices *in a specified order*, e.g. Booking Office A, male; Office B, male; Office C, female. Since however we are concerned only with a *total* of two males and one female, irrespective of the office at which they call, we must, in assigning a probability to the combination, remember that it can occur in *three* ways, i.e. MMF, MFM, and FMM, counting the booking offices A, B, C from left to right.

This result was demonstrated in Chapter 6, in which the number of ways in which two items can be selected from three was theoretically shown to be $\binom{3}{2}$ or $\frac{3!}{2!1!}$, i.e. 3. The same argument applies to the arrival of one male and two females.

At this stage it is helpful if we re-write the table of probabilities somewhat differently:

Sex-combination	Probability
3 Males	$\frac{1}{8}$ or $(\frac{1}{2})^3$
2 Males, 1 female	$\frac{3}{8}$ or $3(\frac{1}{2})^2(\frac{1}{2})$
1 Male, 2 females	$\frac{3}{8}$ or $3(\frac{1}{2})(\frac{1}{2})^2$
3 Females	$\frac{1}{8}$ or $(\frac{1}{2})^3$

The probabilities in the third column are the same as those in the second, but expressed in a different algebraic form. The perspicacious reader will have noticed that the indices of the powers of the fraction $\frac{1}{2}$ in the last column represent the sex combination, e.g. the probability that two males and one female will appear is $3(\frac{1}{2})^2(\frac{1}{2})$. It so happens that we have, in a simple empirical way, arrived at a particular case of *Bernoulli's Theorem*, which can be stated as follows:

'If the probability of an event occurring is p and the probability of it not occurring is $q = 1 - p$, then the probabilities of 0, 1, 2, ..., n successes in a set of n trials, are the first, second third, ..., $(n+1)$th terms of the expansion $(q+p)^n$. It follows that if N sets of such trials are made, the number of sets containing 0, 1, 2, 3, ..., n successes are the successive terms of the expansion $N(q+p)^n$.

Because of the form it takes, the expansion $(q+p)^n$ is more familiarly known as the *binomial expansion*.

Relating this theorem to the booking office problem, the possible outcomes are determined by the terms of the expansion $(\frac{1}{2}+\frac{1}{2})^3$— the probability of a male arriving is the same as that of a female, so $p = q = \frac{1}{2}$.

Now

$$(\tfrac{1}{2}+\tfrac{1}{2})^3 = \binom{3}{0}(\tfrac{1}{2})^3 + \binom{3}{1}(\tfrac{1}{2})^2(\tfrac{1}{2}) + \binom{3}{2}(\tfrac{1}{2})(\tfrac{1}{2})^2 + \binom{3}{3}(\tfrac{1}{2})^3$$

$$= \tfrac{1}{8} + \tfrac{3}{8} + \tfrac{3}{8} + \tfrac{1}{8} = 1$$

and as can be seen, the successive terms are those of the probabilities in the table. It is repeated, because it is important, that the theorem stipulates the existence of a known and measurable probability of failure as well as success. It also deals with specific and separate events, i.e. with a discrete variable.

To bring the application of the theorem a little nearer the commercial world, consider now a machine manufacturing springs which have to possess a certain coiled length while under a fixed tension. Samples of five are tested and the number of defectives, i.e. those falling outside specified limits, are counted. Of a large number produced, $\frac{1}{10}$ are known (from accumulated experience) to be defective. Given this, the probability that there will be 0, 1, 2, 3, 4, 5 defectives in a sample of 5, is given by the successive terms of the expansion $(0\cdot9+0\cdot1)^5$.

In 100 such samples of 5 one would expect to find the frequencies of 0, 1, 2, 3, 4, 5 defectives in the samples, from the successive terms of the expansion $100(0\cdot9+0\cdot1)^5$, i.e. 59, 33, 7, 1, 0, 0 which add to 100.

A manufacturer concerned with a problem of this kind would place the frequency distribution obtained by his inspectors beside that derived theoretically from the binomial expansion—the 'standard-fitting suit' that was mentioned earlier—and compare them with each other. He would then apply the 'goodness-of-fit' test to help him decide if his results fit the binomial 'suit'.

However, it is not uncommon in industrial situations for the exact value of q to be unknown. This would be the case for example when a new machine was being tested prior to the fixing of piece-work rates. In these circumstances we have to use the proportion of defectives which our limited machine experience provides. Suppose that in 1000 samples of 5 products, the number of samples of 5 which yielded 0, 1, 2, 3, 4, 5 defectives was 575, 300, 100, 20, 5, 0, respectively. The total number of defectives would then be

$$(575 \times 0) + (300 \times 1) + (100 \times 2) + (20 \times 3) + (5 \times 4) + (0 \times 5)$$

$$= \quad 0 \quad + \quad 300 \quad + \quad 200 \quad + \quad 60 \quad + \quad 20 \quad + \quad 0 \quad = 580$$

and the proportion of defectives is therefore $580/5000 = 0.116$ or 0.12 approximately.

We can see how close the observed distribution is to a binomial distribution by comparing it with the expansion $1000(0.88 + 0.12)^5$. Now

$1000(0.88 + 0.12)^5$

$$= 1000 \left[0.88^5 + \binom{5}{1} 0.88^4 \times 0.12 + \binom{5}{2} 0.88^3 \times 0.12^2 \right.$$

$$\left. + \binom{5}{3} 0.88^2 \times 0.12^3 + \binom{5}{4} 0.88 \times 0.12^4 + \binom{5}{5} 0.12^5 \right]$$

$$= 1000[0.527 + 0.360 + 0.098 + 0.014 + 0.001]$$

$$= \quad 527 + 360 + 98 + 14 + 1 \quad = 1000.$$

The fit does not seem to be a good one; a goodness-of-fit test will tell us just how close it is.

THE EFFECT OF CHANGING THE VALUE OF p OR q OR n

A histogram of the distribution $100(0.9 + 0.1)^5$ is drawn in Fig. 7.1, which also illustrates the histograms of the expansions $100(0.75 + 0.25)^5$, $100(0.6 + 0.4)^5$ and $100(0.5 + 0.5)^5$. The reader will see

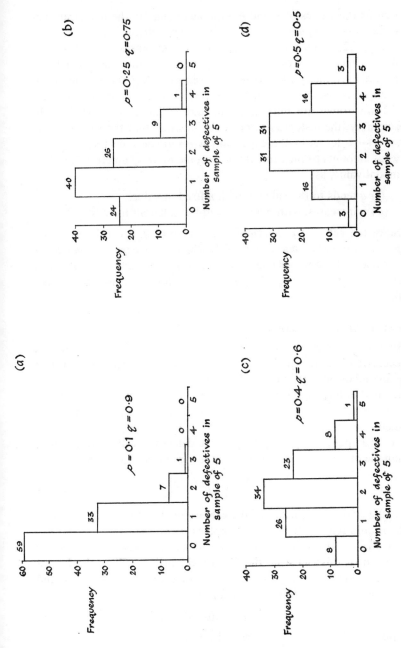

Fig. 7.1. Histograms of binomial expansions, probable frequency of 0, 1, 2, 3, 4, 5 defectives in 100 samples of five.

that as the values of p and q approach equality, the histogram becomes less skew, and when $p = q$ it is symmetrical.

Fig. 7.2 shows the histograms for the expansion $100(0 \cdot 9 + 0 \cdot 1)^n$ for various values of n. It is apparent that the histogram tends towards symmetry as n increases, even though p and q remain unchanged in value. The numerical results from which Figs. 7.1 and 7.2 are drawn are important, and are collated in Table 7.1. In summary, the histograms and table illustrate that the values of the successive terms in the expansion $N(q + p)^n$ become less skew, or, what amounts to the same thing, become more symmetrical about a modal value, as

  (*i*) q tends to the value p, with n fixed in value,
  (*ii*) n increases, with p and q fixed but different in value.

Since the value of p determines the value of q, it can be said that p and n are the two *parameters of the distribution, p* being a *continuous* parameter since it can take any value between 0 and 1, while n is a *discrete* parameter, capable only of assuming the isolated values 0, 1, 2, 3, ..., n.

CALCULATING THE MODAL FREQUENCY

In the examples included in Table 7.1 the number of defectives occurring with the greatest frequency was ascertained only through some laborious arithmetic. It would be helpful to have a shorter method of finding this modal value. It can be calculated as follows:

Consider the binomial expansion $(q + p)^n$ where p, q, and n take their usual meanings, $(p + q = 1)$. Then, in the context of the preceding examples, the probability that x defectives will be found in a sample of n articles is expressed by the frequency function

$$f(x) = \binom{n}{x} q^{n-x} p^x.$$

(If the student is startled by the use of $f(x)$ for 'frequency function' he should refer to the end of Chapter 6.)

Calculate the value of $(n + 1)p$ and let m be the whole number portion of this: let e be the fractional portion. Then the largest value of $f(x)$ occurs when x is made equal to m, e.g. if $n = 5$, and $p = 0 \cdot 25$ as in expansion 2 of Table 7.1, then:

$$(n + 1)p = 6(0 \cdot 25) = 1 \cdot 5,$$

thus $m = 1$ and $e = 0 \cdot 5.$

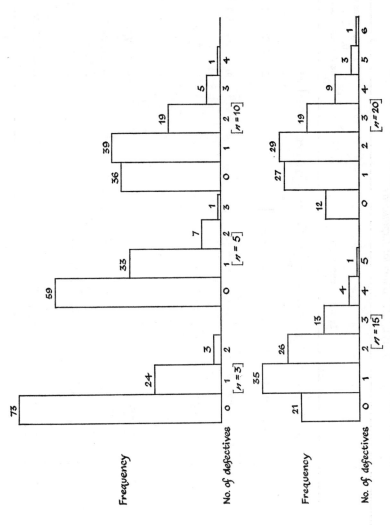

Fig. 7.2. Histograms of binomial expansion $100(0.9+0.1)^n$ for varying values of n.

TABLE 7.1. *Binomial Expansions*

Expansion	Terms of expansion	Frequencies of 0, 1, 2, ... defectives in samples	Total
1. $100(0.9+0.1)^5$	$100\left[0.9^5 + \binom{5}{1}0.9^4\times0.1 + \binom{5}{2}0.9^3\times0.1^2 + \binom{5}{3}0.9^2\times0.1^3 + \binom{5}{4}0.9\times0.1^4+0.1^5\right]$	59, 33, 7, 1	100
2. $100(0.75+0.25)^5$	$100\left[0.75^5 + \binom{5}{1}0.75^4\times0.25 + \ldots \binom{5}{4}0.75\times0.25^4+0.25^5\right]$	24, 40, 26, 9, 1	100
3. $100(0.6+0.4)^5$	$100\left[0.6^5 + \binom{5}{1}0.6^4\times0.4 + \ldots \binom{5}{4}0.6\times0.4^4+0.4^5\right]$	8, 26, 34, 23, 8, 1	100
4. $100(0.5+0.5)^5$	$100\left[0.5^5 + \binom{5}{1}0.5^4\times0.5 + \ldots \binom{5}{4}0.5\times0.5^4+0.5^5\right]$	3, 16, 31, 31, 16, 3	100
5. $100(0.9+0.1)^3$	$100\left[0.9^3 + \binom{3}{1}0.9^2\times0.1 + \binom{3}{2}0.9\times0.1^2+0.1^3\right]$	73, 24, 3	100
6. $100(0.9+0.1)^5$	As for (1) above.	59, 33, 7, 1	100
7. $100(0.9+0.1)^{10}$	$100\left[0.9^{10} + \binom{10}{1}0.9^9\times0.1 + \binom{10}{2}0.9^8\times0.1^2+\ldots \binom{10}{9}0.9\times0.1^9+0.1^{10}\right]$	36, 39, 19, 5, 1	100
8. $100(0.9+0.1)^{15}$	$100\left[0.9^{15} + \binom{15}{1}0.9^{14}\times0.1 + \ldots \binom{15}{14}0.9\times0.1^{14}+0.1^{15}\right]$	21, 35, 26, 13, 4, 1	100
9. $100(0.9+0.1)^{20}$	$100\left[0.9^{20} + \binom{20}{1}0.9^{19}\times0.1 + \ldots \binom{20}{19}0.9\times0.1^{19}+0.1^{20}\right]$	12, 27, 29, 19, 9, 3, 1	100

Then, in a large number of samples of 5 articles, manufactured so that the likelihood of an article being defective is 25%, more samples will have one defective than any other number of defectives. This conforms with the result from expansion 2, where the frequencies for 0, 1, 2, 3, 4, 5 defectives were 24, 40, 26, 9, 1, 0.

BINOMIAL COEFFICIENTS

There is also a simple way of determining the values of the coefficients of a binomial expansion. It is by means of *Pascal's Triangle*.

Pascal's triangle

The coefficients of the expansion $(q+p)^4$ can be read off in the triangle against the value $n = 4$: 1, 4, 6, 4, 1. They can be proved by working the expansion through.

$$(q+p)^4 = \binom{4}{0}q^4 + \binom{4}{1}q^3p + \binom{4}{2}q^2p^2 + \binom{4}{3}qp^3 + \binom{4}{4}p^4$$

$$= 1 . q^4 + 4 . q^3p + 6 . q^2p + 4 . qp^3 + 1 . p^4.$$

The coefficients of any line are derived from those in the line above it. The triangles show how this is done: by adding adjacent coefficients, that which falls between them in the line immediately below is obtained.

THE MEAN AND VARIANCE OF A BINOMIAL DISTRIBUTION

Remembering the work done on location and dispersion in Chapter 2, the reader may be wondering how to calculate the mean and variance of the distribution we have so far been investigating in this chapter. We shall be satisfied for the time being with quoting the results.

With n, p and q taking their usual meanings,

the mean of a binomial distribution is np,
the variance of a binomial distribution is npq,
the standard deviation of a binomial distribution is $\sqrt{(npq)}$.

In attempting to interpret these results through one of the expansions of Table 7.1, one must remember that in that table p represents a 'success', or appearance of a defective article. If one wished to calculate the mean number of defectives, the appropriate expression to quantify would be np. Thus, in expansion 1, the mean number of defectives per sample, using the common method of calculation, is:

$$\tfrac{1}{100}\{(59 \times 0)+(33 \times 1)+(7 \times 2)+(1 \times 3)\} = \tfrac{1}{100}(50) = 0\cdot5.$$

Now, $np = 5 \times 0\cdot1 = 0\cdot5$, which agrees.
Again, quantifying the variance from Table 7.1, expansion 1:

x	Frequency f	$f(x)$	$f(x^2)$
0	59	0	0
1	33	33	33
2	7	14	28
3	1	3	9
	100	50	70

Variance $\sigma^2 = 70/100-(50/100)^2 = 45/100 = 0\cdot45$.
Now, $npq = 5 \times 0\cdot9 \times 0\cdot1 = 0\cdot45$ which agrees.
The formulae for the mean and variance are derived in the Appendix at the end of this chapter.

SUMMARY

An understanding of the binomial expansion and the use of Bernoulli's Theorem is so important to what follows, that our thoughts and findings so far are worth summarising.

We have been discussing situations in which a recognisable event will occur with a known degree of probability, e.g. the probability that we will throw a 6 with a single throw of an unbiased die. In each of these situations the *failure* of the event to occur is also recognisable. In this case the failure to throw a 6 is represented by the throwing of a 1, 2, 3, 4, or 5, and since the sum of all probabilities must equal one, the probability of the failure to throw a 6 is $\tfrac{5}{6}$.

Supposing now we were to ask the question 'When I throw the

die 20 times, what is the probability that a 6 will turn up four times?'
One practical way of answering this would be to throw the die 20
times and count the number of times a 6 appears. Obviously, to
obtain an answer in which we have confidence, we must perform
this throwing operation many times over, and record the number of
6's thrown after each throw. But this would be a tiresome exercise.
Having the results of this chapter so far in mind, we would do better
to recognise that the 'standard-fitting suit' for the problem is the
binomial distribution. We can therefore apply Bernoulli's Theorem.
By this we know that the probability that a 6 will turn up four
times when a regular die is thrown 20 times is the value of the *fifth*
term in the expansion

$$\left(\frac{5}{6} + \frac{1}{6}\right)^{20}, \text{ i.e. } \binom{20}{4}\left(\frac{5}{6}\right)^{16}\left(\frac{1}{6}\right)^{4}.$$

If, however, the problem were stated in the form 'Am I using an
unbiased die?', it would be necessary to run a sequence of 'throwing'
trials, and relate the results to the frequencies derived theoretically
from Bernoulli's Theorem, as set out in Chapter 8. The important
thing to recognise is that the form of the problem permits the use of
the binomial 'standard-fitting suit'. The subject moves into the
commercial world when the question becomes something like 'Does
the number of defective items being recorded conform to the assumption of 10% defectives?' In this case the 'standard-fitting suit'
would be the expansion $(0 \cdot 9 + 0 \cdot 1)^{n}$ where n is the number tested in
a single sample, and the frequencies it yielded would be placed beside
those obtained on the bench. An unsatisfactory comparison would
call for a machine adjustment, or, if it were acceptable, a change in
the value of p. The comparison is effected by the 'goodness-of-fit'
test of Chapter 8.

POISSON DISTRIBUTION

THE CONCEPT

In the binomial distribution we knew the probability of success and
of failure, and knew what we meant by both. We were also aware
of the number of items in a sample or in a series of events. Remembering this 'identi-kit' of the binomial distribution, let us now turn
to a situation which has a binomial flavour but which is different
in a number of important respects.

In a factory there are a few accidents each year. The opportunities for an accident to occur are endless, but the skill and common sense of the employees, combined with the factory's safety precautions reduce the actual number of accidents to a minimum. In the conceptual sense there are many 'accident opportunities' but the probability of an accident occurring is very small indeed. In binomial terms, in the expansion $(q+p)^n$, p (the probability that an accident will happen) is an extremely small fraction, while n (the number of 'accident opportunities') is a very large number. But q (the probability that an accident will *not* occur) has no understandable meaning except that it is measurable as $(1-p)$: a non-accident is not a discernible event.

However, if sufficient experience over a long period is available, the average number of accidents per year can be calculated. This, we have said, is equal to np for a binomial distribution. But if we wish to calculate the probabilities that 0, 1, 2, etc., accidents per year will occur, we are up against a fundamental difficulty because the separate values of n and p are unknown, so $(q+p)^n$ cannot be expanded.

It so happens that there is a discrete distribution which approximates closely to the binomial for small p and large n, the terms of which are independent of p and n taken individually: its terms do however include the mean or expected value of the distribution, i.e. the product np. The expanded terms of the distribution will give the probabilities of 0, 1, 2, ... accidents per year occurring.

The discrete distribution to which we are referring will now be derived from the binomial expansion.

DERIVING THE POISSON DISTRIBUTION FROM THE BINOMIAL EXPANSION

In $(q+p)^n$ let p be very small and n very large, so that $np = \mu$ which is not extremely small. Then

$$(q+p)^n = \binom{n}{0} q^n + \binom{n}{1} q^{n-1}p + \binom{n}{2} q^{n-2}p^2 + \ldots$$

$$= q^n \left(1 + n\frac{p}{q} + \frac{n(n-1)}{2!} \frac{p^2}{q^2} + \frac{n(n-1)(n-2)}{3!} \frac{p^3}{q^3} + \ldots\right)$$

$$= (1-p)^n \left(1 + n\frac{p}{q} + \frac{n(n-1)}{2!} \frac{p^2}{q^2} + \ldots\right)$$

Now $(1-p)^n = \left(1 - \dfrac{\mu}{n}\right)^n$ which tends to $e^{-\mu}$ when n becomes very

large, and $\left(1+n\dfrac{p}{q} + \dfrac{n(n-1)}{2!}\dfrac{p^2}{q^2} +...\right)$ tends to

$$\left(1+np+ \dfrac{n^2}{2!} p^2 +...\right).$$

Hence $(q+p)^n$ tends to

$$e^{-\mu}\left(1+\mu+ \dfrac{\mu^2}{2!} +...\right) = e^{-\mu}.e^{\mu} = 1.$$

This is known as the *Poisson distribution* after the Frenchman who published it in the first half of the nineteenth century. Its frequency function is $e^{-\mu}\dfrac{\mu^x}{x!}$ with μ being the average number of times the event will occur in one selected unit of time.

WHEN THE POISSON MAY BE USED

The reader will find it rewarding to draw a series of histograms of binomial expansions for varying values of small p and large n where np is constant. (The arithmetic is done in Table 7.2.) If he compares them with a histogram of the frequencies or probabilities determined from the Poisson distribution where $\mu = np$, he will find that the two distributions become almost coincident when $n \geqslant 50$ and $p \leqslant 0.1$.

E.g. when $n = 100$ and $p = 0.01$, frequencies for 0, 1, 2 ..., successes are 37, 37, 19, 6, 1 (Table 7.2). Also, $np = \mu = 1$, and

$$e^{-1}\left(1+1+\dfrac{1}{2!}+\dfrac{1}{3!}+...\right)$$ gives frequencies of 37, 37, 18, 6, 2

($e = 2.7183$). The Poisson can therefore be used as a close approximation of the binomial when these limits are observed, although circumstances often demand some flexibility. A further point to note is that since the general term for the Poisson contains the single parameter μ, it follows that ideally the distribution may be used only when the value of μ remains unchanged from trial to trial. Within the example we have so far used, the expected number of accidents in a factory in a year must remain unchanged. There must not be a significant change in the type of machinery used or the strictness or character of the safey precautions laid down. As an illustration of this point, a printing house which this year discarded

TABLE 7.2. *Binomial Expansions with np constant*

Expansion	Terms of expansion	Frequencies of 0, 1, 2, ... defectives in samples	Total
1. $100(0.5+0.5)^2$	$100\left[0.5^2+\binom{2}{1}0.5\times0.5+0.5^2\right]$	25, 50, 25	100
2. $100(0.75+0.25)^4$	$100\left[0.75^4+\binom{4}{1}0.75^3\times0.75+\ldots\binom{4}{3}0.75\times0.25^3+0.25^4\right]$	21, 44, 21, 4	100
3. $100(0.8+0.2)^5$	$100\left[0.8^5+\binom{5}{1}0.8^4\times0.2+\ldots\binom{5}{4}0.8\times0.2^4+0.2^5\right]$	33, 41, 20, 5, 1	100
4. $100(0.9+0.1)^{10}$	$100\left[0.9^{10}+\binom{10}{1}0.9^9\times0.1+\binom{10}{2}0.9^8\times0.1^2+\ldots\binom{10}{9}0.9\times0.1^9+0.1^{10}\right]$	36, 39, 19, 5, 1	100
5. $100(0.99+0.01)^{100}$	$100\left[0.99^{100}+\binom{100}{1}0.99^{99}\times0.01+\ldots\binom{100}{99}0.99\times0.01^{99}+0.01^{100}\right]$	37, 37, 19, 6, 1	100
6. $100(0.999+0.001)^{1000}$	$100\left[0.999^{1000}+\binom{1000}{1}0.999^{999}\times0.001+\ldots\binom{1000}{999}0.999\times0.001^{999}+0.001^{1000}\right]$	37, 37, 18, 7, 2	100

a set of paper guillotines and installed in their place a battery of burster or 'paper-tugger' machines operating on preperforated sheets, could not incorporate this year's accidents with the aggregate for previous years. The value for μ would have changed.

There are two other conditions necessary for the use of the Poisson distribution to be fully justified. There must not be circumstances present tending to give rise to multiple accidents or make any kind of accident more likely to occur once another has happened. Subject to these conditions being satisfied, our findings can be summarised in the statement:

If the probability of success in a single trial (p) approaches zero while the number of trials (n) becomes infinite in such a way that the product $np = \mu$ remains finite and constant, then the binomial distribution tends to the Poisson with mean μ.

SOME APPLICATIONS OF THE POISSON DISTRIBUTION AND THE USE OF POISSON SUMMATION CURVES

Some work done by R. D. Clarke during the Second World War gives an application of the Poisson distribution and at the same time demonstrates its use as a test of the random distribution of a particular variate.

The government required to know if the V2 rockets were falling on London in an indiscriminate pattern, or whether they were so clustered as to suggest that they were being aimed at specific target areas. An area of 144 square kilometres was divided into 576 equal squares and the number of rockets which had fallen in each, counted. The distribution was:

Number of rockets in square	Number of squares	Total number of rockets	Poisson distribution
0	229	0	227
1	211	211	211
2	93	186	98
3	35	105	30
4	7	28	7
5	1	5	1
	576	535	574

$\mu = 535/576 = 0.93$ rockets per square.

Table 7.3 lists the value of $e^{-\mu}$ for various values of μ. With the help of this table and the appropriate Poisson distribution, which in this instance is

$$e^{-0.93}\left(1+0.93+\frac{0.93^2}{2!}+\frac{0.93^3}{3!}+\frac{0.93^4}{4!}+\frac{0.93^5}{5!}+\ldots\right)$$

$$= 0.3946(1+0.93+\ldots),$$

the reader can calculate for himself the frequencies that relate to 0, 1, 2, 3, 4, 5 rockets in a square, and check his arithmetic by comparing his results with the figures in the fourth column in the above table. The 'fit', it can be seen, is a very good one, and, at the time, gave statistical blessing to the philosophy of the Londoner who had already decided to sit it out as long as his chances were no worse than his neighbour's.

However convenient the Poisson distribution may be, it cannot be

TABLE 7.3. (a) Values of $e^{-\mu}$ (μ lying between 0 and +1)

μ	0	1	2	3	4	5	6	7	8	9
0·0	1·000	·9900	·9802	·9704	·9608	·9512	·9418	·9324	·9231	·9139
0·1	·9048	·8958	·8869	·8781	·8694	·8607	·8521	·8437	·8353	·8270
0·2	·8187	·8106	·8025	·7945	·7866	·7788	·7711	·7634	·7558	·7483
0·3	·7408	·7334	·7261	·7189	·7118	·7047	·6977	·6907	·6839	·6771
0·4	·6703	·6636	·6570	·6505	·6440	·6376	·6313	·6250	·6188	·6126
0·5	·6065	·6005	·5945	·5886	·5827	·5770	·5712	·5655	·5599	·5543
0·6	·5488	·5434	·5379	·5326	·5273	·5220	·5169	·5117	·5066	·5016
0·7	·4966	·4916	·4868	·4819	·4771	·4724	·4677	·4630	·4584	·4538
0·8	·4493	·4449	·4404	·4360	·4317	·4274	·4232	·4190	·4148	·4107
0·9	·4066	·4025	·3985	·3946	·3906	·3867	·3829	·3791	·3753	·3716

Example $e^{-0.57} = 0.5655$.

(b) Values of $e^{-\mu}$ (μ lying between +1 and +10)

μ	1	2	3	4	5
$e^{-\mu}$	0·367 88	0·135 34	0·049 79	0·018 32	0·006 738

μ	6	7	8	9	10
$e^{-\mu}$	0·002 479	0·000 912	0·000 335	0·000 123	0·000 045

Example $e^{-2.57} = e^{-2.0} \times e^{-0.57} = 0.135\ 34 \times 0.5655 = 0.076\ 53$.

denied that the arithmetic involved in its use, even with the help of Table 7.3, is often very lengthy. This can be reduced, however, with the employment of *Poisson Summation Curves* (Fig. 7.3) from which the probability that an event will occur x times at least, can be ascertained for any particular value of μ. For example, in Fig. 7.3:

For $\mu = 0.5$

The probability that an event will occur *at least*	The probability that the event will occur
	never $= (1\text{-}0.4) = 0.6$
once $= 0.4$	once $= (0.4\text{-}0.09) = 0.31$
twice $= 0.09$	twice $= (0.09\text{-}0.02) = 0.07$
3 times $= 0.02$	3 times $= (0.02\text{-}0.003) = 0.017$
4 times $= 0.003$	

When the probabilities in the second column are multiplied by the known number of events, such as the number of accidents or number of geographical squares, the Poisson distribution for the study in question is derived.

An obvious corollary to this is that a series of observations converted to the appropriate probability equivalents can be plotted on Poisson Summation Curves: if they all lie on a vertical line, the frequencies conform to the Poisson distribution with the value for μ indicated on the appropriate axis.

Looking back at the figures for the 'rocket' plots:

Number of rockets in square (1)	Number of squares (2)	'At least' frequency (3)	Probability equivalents of (3) (4)
0	229	576	1.0
1	211	347	0.604
2	93	136	0.236
3	35	43	0.075
4	7	8	0.014
5	1	1	0.000

The probability values of column (4) fall on the vertical line for $\mu = 0.93$ so that the original use of the Poisson is vindicated.

The axes of the Poisson Summation Curve are not drawn to an

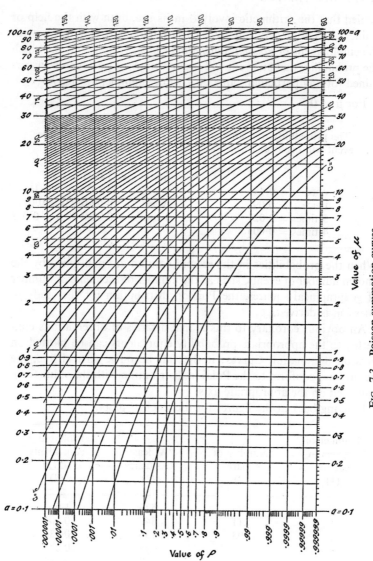

FIG. 7.3. Poisson summation curves.

Value of μ

Value of P

Probability that event will occur at least c times

arithmetic scale; the probability figures running up the side of the graph are drawn to a *probability scale*, and the reader will be quick to notice that they run in the reverse order to what is usual. The scale running along the foot of the graph is logarithmic, and on it are drawn the values for μ, the mean or expected value of the variate.

The construction of a Poisson Summation Curve is tedious but not difficult. Remembering that the Poisson distribution is

$$e^{-\mu}\left(1+\mu+\frac{\mu^2}{2!}+\frac{\mu^3}{3!}+\dots\right)$$

consider function $P(x)$, defined as the probability that, for a given value of μ, the event will occur x times *or more*.

$$\begin{aligned} \text{When } x = 0 \quad & P(0) = 1 \\ x = 1 \quad & P(1) = 1 - e^{-\mu} \\ x = 2 \quad & P(2) = 1 - e^{-\mu}(1+\mu) \\ x = 3 \quad & P(3) = 1 - e^{-\mu}\left(1+\mu+\frac{\mu^2}{2!}\right) \end{aligned}$$

and so on.

It is now possible for a table, of which Table 7.4 is but a section, to be drawn up. When the full results are plotted on axes drawn to the correct scales and the appropriate points joined, a set of Poisson Summation Curves emerge.

A practical application of the Poisson distribution is given in *Analysing Qualitative Data* by A. E. Maxwell (Methuen). It refers to some data obtained by L. Stein and others when sampling *spells* of sickness among workpeople.

The distribution is:

Number of spells (1)	Frequency of spells (2)	Expected frequency (3)	Product (1)×(2) (4)
0	537	494·6	0
1	229	291·7	229
2	95	86·0	190
3	19	16·9	57
4	10 ⎰	2·8	40 ⎰
5+	2 ⎱		10 ⎱
	892	892·0	526

A *spell* of sickness is a period of 4 days or more.

The mean number of spells per employee (μ) is

$$\frac{\text{Total number of spells}}{\text{Total number of employees at risk}} = \frac{\text{Column (4)}}{\text{Column (2)}} = \frac{526}{892} = 0.59.$$

The expected number of spells, i.e. the numbers of employees absent for 0, 1, 2, 3, ... spells, is obtained from the successive terms of the Poisson distribution.

TABLE 7.4. *Plot Values for Poisson Summation Curves*

The table presents the probabilities that an event will occur one or more, three or more, five or more times, for specific values of μ.

μ		0·5	1	2
x				
1	...	0·394	0·632	0·865
3	...	0·015	0·084	0·325
5	...	0·001	0·008	0·055

Explanation of column for $\mu = 2$.

General expansion is

$$e^{-\mu}\left(1 + \mu + \frac{\mu^2}{2!} + \frac{\mu^3}{3!} + \frac{\mu^4}{4!} + ... \right)$$

Expressing probabilities for occurrences:

For column $\mu = 2$

$$\begin{array}{ccccccccc} & 0 & , & 1 & , & 2 & , & 3 & , & 4 & , \\ = 0.135(& 1 & + & 2 & + & \frac{4}{2} & + & \frac{8}{6} & + & \frac{16}{24} & + ...) \end{array}$$

$$= \quad 0.135 + 0.270 + 0.270 + 0.180 + 0.090$$

Aggregating

$$= \quad 0.135 , 0.405 , 0.675 , 0.855 , 0.945$$

Subtracting from 1·0

$$= \quad 0.865 , \quad ... \quad , 0.325 , \quad ... \quad , 0.055$$

For $\mu = 0.59$ the series becomes

$$e^{-0.59}\left(1 + 0.59 + \frac{0.59^2}{2!} + \frac{0.59^3}{3!} + ... \right)$$

$$= 0.554\,51\left(1 + 0.59 + \frac{0.3477}{2} + \frac{0.2051}{6} + \frac{0.1209}{24} + ... \right)$$

$$= 0.554\,51 + 0.326\,99 + 0.096\,41 + 0.018\,95 + 0.002\,79 + 0.000\,35.$$

When the separate probabilities are multiplied by the total number of employees, 892, the frequency distribution given in Column (3) is obtained. A 'goodness-of-fit' test will confirm whether or not the fit is close enough to justify the belief that the distribution is of the Poisson type.

Some interesting work involving the use of the Poisson has been done by R. E. Beard in relation to the number of motor accident insurance claims arising from a known number of cars at risk in a fixed period. For a portfolio of 10 000 insured cars the average number of claims per year is said to be 2000 or 0·2 per insured car. The distribution was employed to calculate the frequency with which 0, 1, 2, ... claims per insured car could be expected.

TABLE 7.5. *Insurance Claims arising from Motor Accidents*

Number of claims (1)	Anticipated frequency (2)	Total Number of claims (3)	Poisson frequency (4)	Total Number of claims (5)	Negative binomial frequency (6)	Total Number of claims (7)
0	8 306	0	8 187	0	8 304	0
1	1 433	1433	1 637	1637	1 437	1437
2	222	444	164	328	221	442
3	34	102	11	33	33	99
4	5	20	1	4	5	20
	10 000	1999	10 000	2002	10 000	1998

These figures were thus obtained from the values of the successive terms of the expansion:

$$10\,000\left[e^{-0\cdot2}\left(1+0\cdot2+\frac{0\cdot2^2}{2!}+\frac{0\cdot2^3}{3!}+...\right)\right]$$

$$= 10\,000[0\cdot8187+0\cdot163\,74+0\cdot016\,374+0\cdot001\,064+0\cdot000\,49]$$

$$= \qquad 8187+\ 1637\ +\ 164\ \ +\ 11\ \ \ +\ 1 = 10\,000.$$

This result is tabulated in columns (1) and (4) of Table 7.5 with the associated number of claims in column (5). The interesting, if digressionary, aspect of this study is the affinity which these figures bear to those in column (2) which were calculated on the basis that

G

the number of claims arising from each insured vehicle was directly proportional to the mileage travelled in the period under review.

However, a theoretical objection could be lodged against the use of the Poisson in the claims situation. It is that the value for μ changes from year to year, because the risk of a claim alters with the changing traffic situation; and indeed if one wished to adopt a wholly purist attitude it could be pointed out that multiple accidents do occur, and that from time to time conditions on our roads are conducive to further accidents if one has already happened. It will not be long before the expression 'pile-up' is socially sanctified by the lexicographers. In spite of all this the Poisson fit to the hypothetical claims distribution in column (2) is fairly close.

RELATIONSHIP BETWEEN THE POISSON AND THE NEGATIVE BINOMIAL DISTRIBUTIONS

It so happens that a result following from the *negative binomial distribution* is suited to a situation which would be counted as Poisson except that the mean μ varies. The probability that an event will occur 0, 1, 2, ... times is given by the value of the successive terms in the series

$$\left(\frac{c}{c+1}\right)^p \left\{1, \frac{p}{(c+1)}, \frac{p(p+1)}{2!(c+1)^2}, \ldots\right\}$$

where
$$\mu = \frac{p}{c}; \ \mu + \frac{\mu}{c} = \sigma^2.$$

If the reader will calculate the mean and variance (μ, σ^2) from column (1) and the 'anticipated frequency' column (2) in Table 7.5, he will find that $\mu = 0\cdot2$ and $\sigma^2 = 0\cdot231$.

p and c are then calculated:

$$0\cdot2 + \frac{0\cdot2}{c} = 0\cdot231 \text{ so that } c = \frac{0\cdot2}{0\cdot031} = 6\cdot451$$

$$0\cdot2 = \frac{p}{6\cdot451} \text{ so that } p = 6\cdot451 \times 0\cdot2 = 1\cdot290.$$

The numbers of 0, 1, 2, ... claims are then derived from the consecutive terms of the series

$$\left(\frac{6\cdot451}{7\cdot451}\right)^{1\cdot290} \left\{1, \frac{1\cdot290}{7\cdot451}, \frac{1\cdot290 \times 2\cdot290}{2 \times 7\cdot451^2}, \ldots\right\}.$$

This could provide the reader with an exercise in the use of logarithms: the frequencies which are thus obtained appear in column (6) of Table 7.5. There is a remarkable closeness of fit between the frequencies obtained from the negative binomial and those based on the assumption of a direct relationship between distance travelled and claims incidence: they can be compared in columns (6) and (2).

THE MEAN AND VARIANCE OF THE POISSON DISTRIBUTION

The mean of the distribution is $np = \mu$ by definition. The variance is $npq = \mu q = \mu$ when q becomes nearly equal to 1. The mean and variance of the Poisson distribution are thus both equal to μ. The Poisson is skew for small values of μ, but becomes almost symmetrical when $\mu = 6$.

So far, in dealing with the binomial, Poisson and negative binomial distributions, we have been dealing solely with discrete distributions. We will now move on to a study of the distribution of a continuous variate.

THE NORMAL DISTRIBUTION

THE CONCEPT

The binomial distribution is like an affectionate stray dog: one had better make friends with it because it remains under the heels no matter how often the foot is raised to it.

The reader will recall that in the first four binomial expansions of Table 7.1 for $n = 5$, the histograms moved towards symmetry as p approached q in value, and finally became wholly symmetrical when $p = q = 0.5$, although the 'staircase' appearance was very marked. However as n increased this became less noticeable, and in fact, when n is infinitely large the outline of the histogram becomes a smooth curve. Fig. 7.4 illustrates two early phases of this process. The smooth curve is known as the *Normal Curve* because it was at one time thought that many of the natural phenomena of life conformed to it.

The Normal Curve is also called the *Gaussian Curve*. It was discovered in 1753 by De Moivre (an Englishman!) as the limiting form of the binomial, and M. G. Kendall in his work *The Advanced Theory of Statistics* gives a brief historical survey of the part it has played (and is playing) in statistical theory. From this it is interesting to note that whatever domestic ruin our titled betters of the

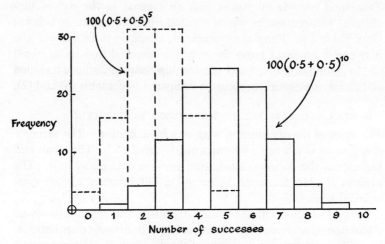

FIG. 7.4. Comparative appearance of histograms for $100(0.5+0.5)^5$
and $100(0.5+0.5)^{10}$.

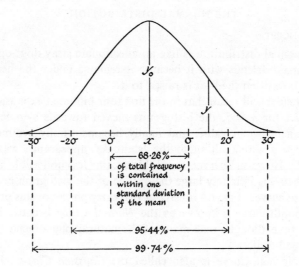

FIG. 7.5. The normal curve.

$$y = \frac{1}{\sigma\sqrt{(2\pi)}} \cdot exp\left(\frac{-\{x - \bar{x}\}^2}{2\sigma^2}\right).$$

Note.—The area contained within the x-axis and the
curve represents the total frequency.

eighteenth century called about their ears in settlement of their gambling honour, the pastime titillated research into the theory of probability, to our present benefit. The current eminence of the Normal distribution rests on its usefulness in the theory of sampling.

But to return to consideration of the curve. It is an interesting fact that the binomial distribution approximates very closely to the Normal distribution when n is large, even if p does not equal 0·5, provided it is not extremely small. Once again, this is a result we should expect from our studies of the histograms of the binomial expansion in Tables 7.1 and 7.2. The mathematical equation of the Normal curve has a snarling appearance and had better be faced four-square, so that the strength of its malevolence can be tested. It is

$$y = \frac{1}{\sigma\sqrt{(2\pi)}} \cdot e^{-\frac{(x-\bar{x})^2}{2\sigma^2}}$$

and its shape is shown in Fig. 7.5. Its derivation from $(p+q)^n$ when n is increased indefinitely, involves some mathematics beyond the scope of this book, and the reader will have to take it on trust. In the above equation:

σ is the standard deviation of the distribution;

e is 'exponential e' and is the value of the series

$$1+1+ \frac{1}{2!} + \frac{1}{3!} +....$$

which forms the base of Napierian logarithms. Its value is 2·7183.

x is any value of the variate;

$\bar{x}$ is the arithmetic mean of the distribution;

y is the height of the ordinate on the curve which relates to any chosen value for x;

$\dfrac{1}{\sigma\sqrt{(2\pi)}}$ is that part of the equation which is introduced to make the total area under the curve equal unity.

The equation thus introduces two parameters—the mean and the standard deviation—which give each distribution its own individuality.

If we were to express in graphical form a series of Normal distributions whose frequencies had been converted to relative frequencies totalling unity, they would appear as in Fig. 7.6. Those with the same standard deviations but with different means would appear as in (*a*), and those with the same mean but different standard deviations, as in (*b*). The height of the curves in (*b*) increase with diminishing standard deviations because the areas under the curves are the same, namely unity.

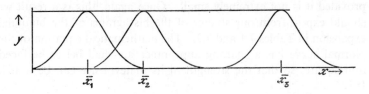

FIG. 7.6(a). Normal curves with similar standard deviations but different means.

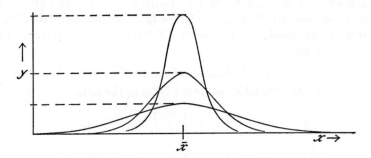

FIG. 7.6(b). Normal curves with same mean but different standard deviations.

THE STANDARDISED FORM

However, if distributions which are thought to be Normal in character are to be compared with each other or with a theoretical Normal distribution, the element of scale must be eliminated from their measurements, so that they become independent of units. This is achieved by expressing all values of the variate in *standardised form*, determined from the formula:

$$z = \frac{x - \bar{x}}{\sigma}$$

where z is the standardised variate and x, $\bar{x}$, and σ take their usual meanings. The formula for the Standardised Normal distribution is

$$y = \frac{1}{\sqrt{(2\pi)}} e^{-z^2/2}.$$

The change in the form of the exponent is derived from a straightforward substitution, and σ has been removed from the denominator in order once again to make the total frequency equal unity.

Readers unfamiliar with mathematical conventions sometimes have difficulty in accepting the concept of an area representing a frequency. It is best tackled from the idea of the histogram, which by definition is an associated series of rectangles, each of which has an area proportional to a class frequency. The sum of all rectangles (i.e. the histogram) therefore represents the total frequency. Moreover if the rectangles are of equal width the height of each rectangle is also proportional to its class frequency.

In discrete distributions the smallest class interval must be unity, whatever its physical scale, but in a continuous distribution it may be as small as we care to make it. It is of course true that the narrower the class intervals are made, the fewer will be the number of items falling into each of them. Theoretically anyway, the class intervals can be made smaller and smaller until the 'staircase' effect at the head of the histogram vanishes and becomes a smooth curve. At that stage, even though the width of the rectangle becomes extremely small, the area of each rectangle continues to represent the class frequency, so that the total area of all rectangles represents the total frequency. This total area is contained within the x-axis and the Normal curve.

But to return to the standardised Normal distribution, which is sometimes written in the mathematical shorthand form of $N(0, 1)$, meaning, specifically, 'this is a standardised Normal distribution which has zero mean and unit standard deviation'. Let us turn it over in our hand and have a good look at it.

TABLES FOR USE WITH THE STANDARDISED NORMAL DISTRIBUTION

The discerning reader will have privately wondered how he was to derive the ordinate values. Repeatedly to calculate an odd negative power of 'e' seems too much to ask, while to determine the total frequency between specified values of the variate appears to be an even more perplexing problem. Fortunately both have been tabulated against standardised values of the variate, as can be seen in Table 7.6. Because the Normal curve is symmetrical it is only necessary to quote values for z lying between zero and 3·99. Measurement of the area under the curve needs some explanation (Table 7.6 (a)). The listed values relate to the area lying between the mean and the selected z reading. The use of this table is best expressed by example.

PROBLEM. What proportion of the total frequency of a standard Normal distribution lies within one, two and three standard deviations of the mean, i.e. between $z = 1, 2, 3$?

> *Situation* 1.   Area between mean and $z = +1$ is 0·3413
>   Area between mean and $z = -1$ is 0·3413
>
> Therefore   Area between $z = +1$ and $z = -1$ is 0·6826

i.e. 68·26% of the total frequency lies within one standard deviation of the mean of the distribution.

> *Situation* 2.   Area between mean and $z = +2$ is 0·4772
>   Area between mean and $z = -2$ is 0·4772
>
> Therefore   Area between $z = +2$ and $z = -2$ is 0·9544

i.e. 95·44% of the total frequency lies within two standard deviations of the mean of the distribution.

> *Situation* 3.   Area between mean and $z = +3$ is 0·4987
>   Area between mean and $z = -3$ is 0·4987
>
> Therefore   Area between $z = +3$ and $z = -3$ is 0·9974

i.e. 99·74% of the total frequency lies within three standard deviations of the mean of the distribution.

The reverse of this process is also helpful. Let us derive the z values, lying symmetrically about the mean, within which 95% and 99% of the total frequency falls.

Situation 4. If 95% of the total frequency falls within certain values of z which are symmetrical about the mean, the table will record the z value containing 47·5% of the total frequency, or, since the total frequency under the curve is unity, 0·4750.

TABLE 7.6(a). *Areas under the Standard Normal Curve*

z	0	1	2	3	4	5	6	7	8	9
0·0	·0000	·0040	·0080	·0120	·0160	·0199	·0239	·0279	·0319	·0359
0·1	·0398	·0438	·0478	·0517	·0557	·0596	·0636	·0675	·0714	·0754
0·2	·0793	·0832	·0871	·0910	·0948	·0987	·1026	·1064	·1103	·1141
0·3	·1179	·1217	·1255	·1293	·1331	·1368	·1406	·1443	·1480	·1517
0·4	·1554	·1591	·1628	·1664	·1700	·1736	·1772	·1808	·1844	·1879
0·5	·1915	·1950	·1985	·2019	·2054	·2088	·2123	·2157	·2190	·2224
0·6	·2258	·2291	·2324	·2357	·2389	·2422	·2454	·2486	·2518	·2549
0·7	·2580	·2612	·2642	·2673	·2704	·2734	·2764	·2794	·2823	·2852
0·8	·2881	·2910	·2939	·2967	·2996	·3023	·3051	·3078	·3106	·3133
0·9	·3159	·3186	·3212	·3238	·3264	·3289	·3315	·3340	·3365	·3389
1·0	·3413	·3438	·3461	·3485	·3508	·3531	·3554	·3577	·3599	·3621
1·1	·3643	·3665	·3686	·3708	·3729	·3749	·3770	·3790	·3810	·3830
1·2	·3849	·3869	·3888	·3907	·3925	·3944	·3962	·3980	·3997	·4015
1·3	·4032	·4049	·4066	·4082	·4099	·4115	·4131	·4147	·4162	·4177
1·4	·4192	·4207	·4222	·4236	·4251	·4265	·4279	·4292	·4306	·4319
1·5	·4332	·4345	·4357	·4370	·4382	·4394	·4406	·4418	·4429	·4441
1·6	·4452	·4463	·4474	·4484	·4495	·4505	·4515	·4525	·4535	·4545
1·7	·4554	·4564	·4573	·4582	·4591	·4599	·4608	·4616	·4625	·4633
1·8	·4641	·4649	·4656	·4664	·4671	·4678	·4686	·4693	·4699	·4706
1·9	·4713	·4719	·4726	·4732	·4738	·4744	·4750	·4756	·4761	·4767
2·0	·4772	·4778	·4783	·4788	·4793	·4798	·4803	·4808	·4812	·4817
2·1	·4821	·4826	·4830	·4834	·4838	·4842	·4846	·4850	·4854	·4857
2·2	·4861	·4864	·4868	·4871	·4875	·4878	·4881	·4884	·4887	·4890
2·3	·4893	·4896	·4898	·4901	·4904	·4906	·4909	·4911	·4913	·4916
2·4	·4918	·4920	·4922	·4925	·4927	·4929	·4931	·4932	·4934	·4936
2·5	·4938	·4940	·4941	·4943	·4945	·4946	·4948	·4949	·4951	·4952
2·6	·4953	·4955	·4956	·4957	·4959	·4960	·4961	·4962	·4963	·4964
2·7	·4965	·4966	·4967	·4968	·4969	·4970	·4971	·4972	·4973	·4974
2·8	·4974	·4975	·4976	·4977	·4977	·4978	·4979	·4979	·4980	·4981
2·9	·4981	·4982	·4982	·4983	·4984	·4984	·4985	·4985	·4986	·4986
3·0	·4987	·4987	·4987	·4988	·4988	·4989	·4989	·4989	·4990	·4990
3·1	·4990	·4991	·4991	·4991	·4992	·4992	·4992	·4992	·4993	·4993
3·2	·4993	·4993	·4994	·4994	·4994	·4994	·4994	·4995	·4995	·4995
3·3	·4995	·4995	·4995	·4996	·4996	·4996	·4996	·4996	·4996	·4997
3·4	·4997	·4997	·4997	·4997	·4997	·4997	·4997	·4997	·4997	·4998
3·5	·4998	·4998	·4998	·4998	·4998	·4998	·4998	·4998	·4998	·4998
3·6	·4998	·4998	·4999	·4999	·4999	·4999	·4999	·4999	·4999	·4999
3·7	·4999	·4999	·4999	·4999	·4999	·4999	·4999	·4999	·4999	·4999
3·8	·4999	·4999	·4999	·4999	·4999	·4999	·4999	·4999	·4999	·4999
3·9	·5000	·5000	·5000	·5000	·5000	·5000	·5000	·5000	·5000	·5000

TABLE 7.6(b). *Ordinates of the Standard Normal Curve*

z	0	1	2	3	4	5	6	7	8	9
0·0	·3989	·3989	·3989	·3988	·3986	·3984	·3982	·3980	·3977	·3973
0·1	·3970	·3965	·3961	·3956	·3951	·3945	·3939	·3932	·3925	·3918
0·2	·3910	·3902	·3894	·3885	·3876	·3867	·3857	·3847	·3836	·3825
0·3	·3814	·3802	·3790	·3778	·3765	·3752	·3739	·3725	·3712	·3697
0·4	·3683	·3668	·3653	·3637	·3621	·3605	·3589	·3572	·3555	·3538
0·5	·3521	·3503	·3485	·3467	·3448	·3429	·3410	·3391	·3372	·3352
0·6	·3332	·3312	·3292	·3271	·3251	·3230	·3209	·3187	·3166	·3144
0·7	·3123	·3101	·3079	·3056	·3034	·3011	·2989	·2966	·2943	·2920
0·8	·2897	·2874	·2850	·2827	·2803	·2780	·2756	·2732	·2709	·2685
0·9	·2661	·2637	·2613	·2589	·2565	·2541	·2516	·2492	·2468	·2444
1·0	·2420	·2396	·2371	·2347	·2323	·2299	·2275	·2251	·2227	·2203
1·1	·2179	·2155	·2131	·2107	·2083	·2059	·2036	·2012	·1989	·1965
1·2	·1942	·1919	·1895	·1872	·1849	·1826	·1804	·1781	·1758	·1736
1·3	·1714	·1691	·1669	·1647	·1626	·1604	·1582	·1561	·1539	·1518
1·4	·1497	·1476	·1456	·1435	·1415	·1394	·1374	·1354	·1334	·1315
1·5	·1295	·1276	·1257	·1238	·1219	·1200	·1182	·1163	·1145	·1127
1·6	·1109	·1092	·1074	·1057	·1040	·1023	·1006	·0989	·0973	·0957
1·7	·0940	·0925	·0909	·0893	·0878	·0863	·0848	·0833	·0818	·0804
1·8	·0790	·0775	·0761	·0748	·0734	·0721	·0707	·0694	·0681	·0669
1·9	·0656	·0644	·0632	·0620	·0608	·0596	·0584	·0573	·0562	·0551
2·0	·0540	·0529	·0519	·0508	·0498	·0488	·0478	·0468	·0459	·0449
2·1	·0440	·0431	·0422	·0413	·0404	·0396	·0387	·0379	·0371	·0363
2·2	·0355	·0347	·0339	·0332	·0325	·0317	·0310	·0303	·0297	·0290
2·3	·0283	·0277	·0270	·0264	·0258	·0252	·0246	·0241	·0235	·0229
2·4	·0224	·0219	·0213	·0208	·0203	·0198	·0194	·0189	·0184	·0180
2·5	·0175	·0171	·0167	·0163	·0158	·0154	·0151	·0147	·0143	·0139
2·6	·0136	·0132	·0129	·0126	·0122	·0119	·0116	·0113	·0110	·0107
2·7	·0104	·0101	·0099	·0096	·0093	·0091	·0088	·0086	·0084	·0081
2·8	·0079	·0077	·0075	·0073	·0071	·0069	·0067	·0065	·0063	·0061
2·9	·0060	·0058	·0056	·0055	·0053	·0051	·0050	·0048	·0047	·0046
3·0	·0044	·0043	·0042	·0040	·0039	·0038	·0037	·0036	·0035	·0034
3·1	·0033	·0032	·0031	·0030	·0029	·0028	·0027	·0026	·0025	·0025
3·2	·0024	·0023	·0022	·0022	·0021	·0020	·0020	·0019	·0018	·0018
3·3	·0017	·0017	·0016	·0016	·0015	·0015	·0014	·0014	·0013	·0013
3·4	·0012	·0012	·0012	·0011	·0011	·0010	·0010	·0010	·0009	·0009
3·5	·0009	·0008	·0008	·0008	·0008	·0007	·0007	·0007	·0007	·0006
3·6	·0006	·0006	·0006	·0005	·0005	·0005	·0005	·0005	·0005	·0004
3·7	·0004	·0004	·0004	·0004	·0004	·0004	·0003	·0003	·0003	·0003
3·8	·0003	·0003	·0003	·0003	·0003	·0002	·0002	·0002	·0002	·0002
3·9	·0002	·0002	·0002	·0002	·0002	·0002	·0002	·0002	·0001	·0001

The value of z in the table appropriate to 0·4750 is 1·96,

i.e. 95% of the total frequency of a Normal distribution is contained within 1·96 standard derivations of the mean.

Situation 5. With the same reasoning as that put forward in Section 4:

The value of z in the table appropriate to 0·4949 is 2·570.
The value of z in the table appropriate to 0·4951 is 2·580.

Therefore

The value of z in the table appropriate to 0·4950 is 2·575,

i.e. 99% of the total frequency of a Normal distribution is contained within 2·575 standard deviations of the mean.

In order to emphasise the full use of the table, the following further examples are worked.

PROBLEM. Find the proportion of the total frequency contained within $z = +1·5$ and $z = -0·4$.

Solution. Proportion required is that contained within $z = 0$ and $z = +1·5$ and $z = 0$ and $z = -0·4$.
This is $0·4332 + 0·1554 = 0·5886$ or 58·86%.

PROBLEM. Find the proportion of the total frequency contained within $z = +1·5$ and $z = +0·4$.

Solution. Using the figures from the problem above, and looking at Table 7.7(c), the proportion required is $0·4332 - 0·1554 = 0·2778$ or 27·78%.

PROBLEM. Find the proportion of the total frequency which has a z value greater than $+1·8$.

Solution. Proportion lying between $z = 0$ and $z = +1·8$ is 0·4641. Proportion lying between $z = +1·8$ and all greater values of z is $0·5 - 0·4641 = 0·0359$ or 3·59%.

PROBLEM. Find the proportion of the total frequency of a standard Normal distribution which lies outside z values $\pm 2·5$.

Solution. Proportion required $= 2(0·5 - 0·4938)$

$$= 2(0·0062) = 0·0124 \text{ or } 1·24\%.$$

PROBLEM. 1000 students sat for an examination. The mean mark was 49 and the distribution of marks had a standard deviation of 6.

Assuming the marks to be Normally distributed, what proportion of students scored more than 55 marks.

Solution. $z = \dfrac{55-49}{6} = 1.$

Now the proportion of students scoring more than 49 marks (or zero on the standard scale) is 0·5. The proportion of students scoring between 49 and 55 (or between zero and 1 on the standard scale) is 0·3413.

Therefore the proportion scoring more than 55 marks is

$$0.5 - 0.3413 = 0.1587 \text{ or } 15.87\%.$$

PROBLEM. In the same examination, Grade 'A' is to be given to students scoring more than 64 marks. What proportion of the candidates will receive Grade 'A'?

Solution. $z = \dfrac{64-49}{6} = \dfrac{15}{6} = 2.5.$

A proportion of 0·5 score more than 49. Proportion of 0·4938 score between 49 and 64 (i.e. between $z = 0$ and $z = 2.5$).

Therefore proportion $0.5 - 0.4938 = 0.0062$ score more than 64 marks, i.e. 0·62% or, in this instance, since there are 1000 candidates, 6 candidates will receive Grade 'A'.

The reader is advised to perform a number of similar exercises so that he can handle the table with ease. It is an extremely important concept and time spent in mastering it is well employed.

COMPARING A DISTRIBUTION WITH A NORMAL CURVE

Using Table 7.6(a). There is an extension to this technique which compares a given distribution with a Normal distribution. The technique is illustrated in Table 7.7 for a contrived distribution.

The procedure is as follows:

1. Calculate $\bar{x}$ and σ for the distribution. (The problem can now be posed as 'Find a Normal distribution with $\bar{x} = $; $\sigma = $; $N = $ '.)

2. Determine the true class boundaries. In the given data the variate is clearly rounded to the nearest whole number, so the true boundaries lie 0·5 units below and above the recorded lower and upper class boundaries respectively.

3. Calculate $z = \dfrac{x - \bar{x}}{\sigma}$ for each class boundary.

TABLE 7.7(a). *New Business written by* 100 *Salesmen*

New business (£00)	Frequency f	Mid class	x	xf	x^2f
50–52	10	51	−6	−60	360
53–55	20	54	−3	−60	180
56–58	33	57	0	0	0
59–61	22	60	3	+66	198
62–64	15	63	6	+90	540
	100			+36	1278

$$\bar{x} = 100 \left(57 + \frac{36}{100} \right) = £5736.$$

$$\sigma^2 = 100^2 \left[\frac{1278}{100} - \left(\frac{36}{100} \right)^2 \right] = 100^2[12\cdot78 - 0\cdot1296] = 12\cdot65[100^2]$$

$$\sigma = 100(3\cdot56) = 356.$$

4. Refer to Table 7.6(a) and obtain the proportions of the total frequency which lie between the zero mean and each z value. The table records only those proportions which lie on one side of the mean, and one should remember that when the sign of z changes, the part of the Normal curve to which the table relates 'crosses' the zero mean line. See 5 below.

5. Subtract from each other the consecutive values obtained in (4). The frequency proportions thus derived will be those relating to the several class intervals.

TABLE 7.7(b). *Fitting a Normal Curve to the above Data*

New business £00	Class boundaries (x)	z for class boundaries	Area under N curve from 0 to z	Area for each class	Expected[1] frequency	Observed frequency
50–52	49·50	−2·20	·4861			10
53–55	52·50	−1·36	·4131	·0730	7	20
56–58	55·50	−0·52	·1985	·2146	22	33
59–61	58·50	0·32	·1255	·3240	32	22
62–64	61·50	1·16	·3770	·2515	25	15
	64·50	2·00	·4772	·1002	10	

[1] These frequencies add to 96 only, because the z values stop at −2·2 and 2·0. The terminal frequencies can be converted to 9 and 12. See page 249.

A word is required about the addition of the values 0·1950 and 0·1217. They correspond to z values of −0·52 and +0·32, that is, to z values which span the mean, as they do in Table 7.7(c). The proportion of total frequency which lies *between* these 2 values must therefore be the addition of the Table 7.6(a) figures.

TABLE 7.7(c).   Portions of Normal curve
tabulated in Table 7.7(b).

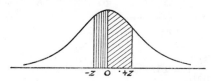

6. Since the standard Normal curve subtends unit area, the class proportions in (4) must be multiplied by the observed total frequency before they are compared with the observed class frequencies. The 'goodness of fit' test described earlier can now be applied to the expected and observed frequencies.

Using Table 7.6(b). The subject could have been tackled in another—but less informative—way by using the ordinates of the Normal curve. To do this the normal equation must be rewritten as

$$y = \frac{Nc}{\sigma \sqrt{(2\pi)}} \cdot e^{-\frac{(x-\bar{x})^2}{2\sigma^2}}$$

where N = Total frequency,

$\quad\quad c$ = width of the class intervals of the distribution,

$\quad\quad \sigma$ = the standard deviation of the distribution.

If the data in Table 7.7 is again used, the calculations necessary are shown in Table 7.8. The process is as follows:

1. Compute the length of the ordinate y_0 appropriate to the mean $\bar{x}$. Here we have; $N = 100$; $c = 300$; $\sigma = 356$; $\bar{x} = 5736$; so

$$y_0 = \frac{100 \times 300}{356} \cdot \frac{1}{2\cdot5066} \cdot e^{-\frac{(5736-5736)^2}{2 \times 356^2}} .$$

Now the value of

$$\frac{100 \times 300}{356} = 84\cdot3$$

TABLE 7.8. *New Business written by* 100 *Salesmen*

New business £	z	Ordinate	84·3 × ordinate
4668	−3·0	0·0044	0·37
4846	−2·5	0·0175	1·48
5024	−2·0	0·0540	4·55
5202	−1·5	0·1295	10·9
5380	−1·0	0·2420	20·4
5558	−0·5	0·3521	29·7
5736	0	0·3989	33·6
5914	+0·5	0·3521	29·7
6092	1·0	0·2420	20·4
6270	1·5	0·1295	10·9
6448	2·0	0·0540	4·55
6626	2·5	0·0175	1·48
6804	3·0	0·0044	0·37

NOTE. Column (1) is constructed as follows:

$$\text{Mean} = 5736,$$
$$\text{Standard deviation} = 356.$$

Therefore new business relevant to $z = -1·0$ is

$$5736 - 356 = 5380$$

and so on.

$$\left(z = \frac{x - \bar{x}}{\sigma} \right).$$

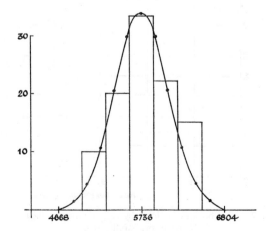

and the value of

$$\frac{1}{2\cdot5066}\cdot e^{-\frac{(5736-5736)^2}{2\times356^2}}$$

is contained within Table 7.6(b) for $z = 0$, and equals 0·3989. ($e^0 = 1$) Therefore $y_0 = 84\cdot3 \times 0\cdot3989$.

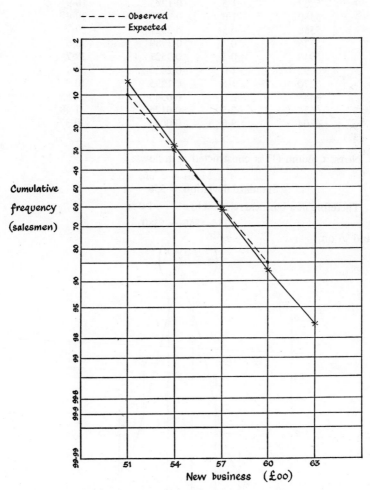

FIG. 7.7. New business written by salesmen. Testing normality by using probability scale.

2. Compute the lengths of ordinates (y_c) for z values, 0·5, 1·0, 1·5, 2·0, 2·5, 3·0 by multiplying the factor 84·3 by the appropriate entry in Table 7.6(b). The figures for the ordinates relating to Table 7.7 data appear in Table 7.8.

3. Join the 'tops' of the ordinates to form the curve.

The histogram for this data is drawn beneath the curve to illustrate the relationship.

Using probability graph paper. It has been shown earlier that a series of numbers increasing in arithmetic progression become a straight line when plotted on an absolute scale: those increasing in geometrical progression become a straight line when plotted on a ratio scale. In a similar way a Normal distribution plotted on *probability paper* becomes a straight line. This must therefore be counted as a third method of testing whether data is Normally distributed. The data for Table 7.7 is plotted on probability paper in Fig. 7.7. If the straight-line plot of the first four cumulative frequencies is continued, it cuts the vertical for £6300 at 97: this is not too far wrong.

Although this is a rough test for Normality, the method is of little real value because it cannot provide a 'goodness of fit' test as can the 'area' method.

APPENDIX: CALCULATING THE MEAN AND VARIANCE OF A BINOMIAL DISTRIBUTION

The mean of a distribution is

$$\bar{x} = \frac{\sum\limits_{x = -\infty}^{\infty} fx}{\Sigma f}.$$

In a relative frequency table where the proportional frequencies sum to unity

$$\bar{x} = \sum_{x = -\infty}^{\infty} x . \text{Prob}(x)$$

which is sometimes written as

$$\bar{x} = \sum_{x = -\infty}^{\infty} x . f(x).$$

Now, in a binomial distribution, which is discrete, and where x can take any integral value between 0 and n

$$\bar{x} = \sum_{x=1}^{n} x \cdot f(x) = \sum_{x=1}^{n} x \cdot \frac{n!}{x!(n-x)!} \cdot p^x \cdot q^{n-x}$$

$$= np \sum_{x=1}^{n} \frac{(n-1)!}{(x-1)!(n-x)!} \cdot p^{x-1} \cdot q^{n-x}$$

$$= np(p+q)^{n-1} = np \quad \text{since} \quad (p+q) = 1.$$

The mean of a binomial distribution is thus np.

The calculation of the variance of a binomial distribution is approached in the same way.

$$\text{Variance} = \sigma^2 = \sum_{x=0}^{n} (x-\bar{x})^2 f(x) = \sum_{x=0}^{n} (x^2 - 2\bar{x} \cdot x + \bar{x}^2) \cdot f(x)$$

$$= \sum_{x=0}^{n} x^2 f(x) - 2\bar{x} \sum_{x=0}^{n} x f(x) + \bar{x}^2 \sum_{x=0}^{n} f(x). \tag{1}$$

Now

$$\sum_{x=0}^{n} x^2 f(x) = \sum_{x=1}^{n} x^2 \frac{n! \, p^x q^{n-x}}{x!(n-x)!}$$

$$= \sum_{x=1}^{n} \{x(x-1) + x\} \frac{n! \, p^x q^{n-x}}{x!(n-x)!}$$

$$= \sum_{x=1}^{n} x(x-1) \frac{n! \, p^x q^{n-x}}{x!(n-x)!} + \sum_{x=1}^{n} x \frac{n! \, p^x q^{n-x}}{x!(n-x)!}$$

$$= n(n-1)p^2 \sum_{x=2}^{n} \frac{(n-2)! \, p^{x-2} q^{n-x}}{(x-2)!(n-x)!} + \sum_{x=1}^{n} x \frac{n! \, p^x q^{n-x}}{x!(n-x)!}$$

$$= n(n-1)p^2 (p+q)^{n-2} + np$$

$$= n(n-1)p^2 + np \quad \text{because} \quad (p+q) = 1.$$

Reverting to equation (1).

$$\sum_{x=0}^{n} x^2 f(x) - 2\bar{x} \sum_{x=0}^{n} x \cdot f(x) + \bar{x}^2 \sum_{x=0}^{n} f(x)$$

$$= n(n-1)p^2 + np - 2np \cdot np + (np)^2 \cdot (1)$$

$$= n^2 p^2 - np^2 + np - 2n^2 p^2 + n^2 p^2$$

$$= np(1-p) = npq.$$

The variance of a binomial distribution is thus npq and the standard deviation is $\sqrt{(npq)}$.

Prove your prowess [1]

7.1.* Fit a Normal frequency curve to the following distribution:

Weight of package contents (ounces)	Number of packages
3·98 and up to but less than 3·99	3
3·99 and up to but less than 4·00	6
4·00 and up to but less than 4·01	37
4·01 and up to but less than 4·02	29
4·02 and up to but less than 4·03	16
4·03 and up to but less than 4·04	9
Total	100

From your fitted curve estimate the proportion of the population of packages with a weight of less than 4 ounces.   (I.O.S., Part II)

7.2.* According to a recent survey, $\frac{7}{8}$ of all students who enter universities receive degrees. What is the probability that out of 4 students chosen at random, (a) 1, (b) 3, (c) at most 2, (d) all 4, will graduate.

7.3.* The probability that a patient will suffer an adverse reaction from a new drug is 0·001.

What is the probability that, of 1000 patients who have received the treatment, (a) 2, (b) at most 3, (c) 4 patients suffer a reaction?

7.4. During the 'running in' period for a new machine, the following measurements were taken for 55 rods which it produced:

Length (cm)	Number
4·9 less than 4·95	2
4·95 less than 5·00	12
5·00 less than 5·05	32
5·05 less than 5·10	6
5·10 less than 5·15	3

Fit a Normal curve to this frequency table.

[1] An asterisk against the exercise number indicates that the solution is given in detail.

Repeat the process for the distribution:

Length (cm)	Number
48 to 50	2
51 to 53	12
54 to 56	32
57 to 59	6
60 to 62	3

(This is an unusually contrived example because a machine would not be expected to vary within such wide limits: but it serves to make a point about the treatment of class boundaries in problems of this kind. That lets the cat out of the bag.)

7.5. The mean weight of an army draft of 500 men is 150 lb, and the standard deviation is 15 lb. Assuming the distribution of weights to be Normal, how many soldiers weigh between (a) 120 lb and 145 lb, (b) 140 lb and 165 lb, (c) 155 lb and 180 lb?

7.6. During a season lasting 25 weeks, 50 footballers are sent off the field by referees. Assuming that 100 games are played each week, what are the probabilities that 0, 1, 2 footballers are sent off in any one game. (Ignore the varying probability induced by the heat of local Derbys.)

7.7. 10% of the photographs printed by a new and cheap printing process are defective. What is the probability that in one edition of a newspaper containing ten photographs, three of them will be defective?

(a) Use the binomial distribution. (b) Use the Poisson distribution.

REFERENCES

BROOKES, B. C., and DICK, W. F. L. 1953. *Introduction to Statistical Method.* London, Heinemann.

DAVID, F. N. 1953. *A Statistical Primer.* London, Griffin.

WILKS, S. S. 1948. *Elementary Statistical Analysis.* Princeton, University Press.

8: MAKING GOOD GUESSES
Estimation and Tests of Hypotheses

IN this chapter we shall be concerned with the problems that arise when we seek to infer knowledge about a population (or populations) from evidence provided by a sample (or samples). The sort of problems which arise are illustrated in the following examples:

1. What is the average life of a particular make and model of car battery?
2. What proportion of 'jumping crackers' produced by a certain fireworks factory is defective?
3. A particular kind of bottling machine produces 5% defectives. A rival machine is claimed to produce only 2% defectives. How long do we have to test this machine before we can decide whether it really is better than the old one?
4. We wish to test whether the values in a certain population are Normally distributed. We select a sample from the population and fit a Normal distribution to our observations. How big must the deviations of actual from theoretical frequencies be for us to reject the idea of a Normally-distributed population?

The factor common to all the above problems is that we have to estimate the population value from a sample. In example 2 the reason for this is quite obvious; in the other examples consideration of time and cost suggest a sample rather than an attempt at a complete enumeration. Time and cost, however, are rather negative recommendations; a more positive one is that in a great many cases a sample provides information reliable enough for the purpose for which it is required, and there is therefore no point in attempting a complete enumeration. We shall pursue this idea in more detail in Chapter 11, where we shall be concerned mainly with the sampling of human population; for now we will merely state that there are many occasions on which decisions have to be based on information gained from samples, and that the statistician can do a great deal towards avoiding wrong decisions.

One vital qualification must be made when talking about samples in this chapter. When we refer to a sample we mean a *random* sample. A *random sample* is one in which every unit in the population has a known and non-zero probability of selection. It is *not*, necessarily, as is frequently supposed, a sample in which every unit in the population has an *equal* chance of selection. Such a sample is certainly a random sample, but the requirement of equal probabilities is not necessary for a sample to be random. In this chapter we shall in fact assume that sampling has been carried out with equal probabilities of selection; this assumption simplifies the calculations and for many practical applications is a valid assumption anyway.

The opposite of random sampling is, naturally enough, *non-random sampling*, i.e. a method of sampling where selection is made according to somebody's judgement, or where those units which are most readily available are selected. The pitfalls of this form of sampling will be discussed in Chapter 11, as will the practical problems of actually selecting a random sample.

Reverting to the subject of the present chapter, we begin our investigations of the problems outlined above by considering what happens when we take samples from a population. The basic idea involved is that of the sampling distribution. (*Note.* Throughout this chapter we shall use μ and σ^2 to represent the *population* mean and variance; $\bar{x}$ and s^2 to represent the *sample* mean and variance.)

SAMPLING DISTRIBUTIONS

Following our usual practice we will introduce the concept by means of an example. Consider the case of a biscuit manufacturer making a particular type of biscuit for which the specified weight is $\frac{1}{2}$ oz. The biscuits are sold in 8-oz packs, the packing being done by a machine which receives batches of 16 biscuits and wraps them. We would not expect to find that every biscuit weighed exactly $\frac{1}{2}$ oz. Random effects arising from the raw materials, the method of blending them, oven temperatures and baking times will result in a distribution of values for the biscuits. This distribution would probably be a Normal distribution: we will assume that it is. If the process is under control, and if the manufacturer is aiming at a biscuit weight of $\frac{1}{2}$ oz we will find that the actual weight of biscuits is distributed Normally with a mean of $\frac{1}{2}$ oz.

Now, the maker would like to produce 8-oz packs which weigh 8 oz *exactly*. If they weigh more he is losing money; if they weigh less he is likely to fall foul of the Inspectors of Weights and Measures.

We can express the manufacturer's problem in a slightly different way by saying that his aim is to produce batches of biscuits with an average weight per biscuit of exactly $\frac{1}{2}$ oz. We have already seen, however, that the weight of individual biscuits varies. We can therefore pose a slightly different question, namely: 'Given that the distribution of biscuit-weight is Normal, with a mean of $\frac{1}{2}$ oz, how would we expect the average weight per biscuit of batches of 16 biscuits to vary?'

We could answer the question empirically by weighing a large number of reputedly 8-oz packs, calculating the average weight per biscuit, and forming a frequency distribution of the resulting means. Expressing this operation in statistical terminology, we can say that we are selecting random samples of 16 biscuits, calculating the mean of each sample, recording the value, selecting another sample, and so on. To be completely rigorous, we would have to add that we are assuming that a given pack consists of 16 biscuits selected at random, and, furthermore, that after weighing each sample we return the 16 biscuits to the population of biscuits (the maker's total stock) before selecting the next sample. With regard to the last point, however, the population of biscuits will be so large compared with the sample size of 16, that we can ignore it.

Now, we know that the population of biscuits is Normally distributed with a mean of $\frac{1}{2}$ oz but we have said nothing yet about its variance. For the time being let us call this σ^2 so that the standard deviation will be σ. If we now perform our sampling operation say 100 times, we will have a frequency distribution consisting of 100 values of the mean weight per biscuit of batches of 16 biscuits. If we calculated the mean and variance of this frequency distribution we would find that the mean was $\frac{1}{2}$ oz or thereabouts, and the variance was $\sigma^2/16$, i.e. the population variance divided by the sample size. The reader is urged to study this example until the principle is quite clear to him, because the fundamental ideas of sampling are contained in it.

In the example we were concerned with the distribution of a set of mean weights of 16 biscuits. We could equally well have concerned ourselves with the distribution of the variance, the standard deviation, the median or any other statistic which could be obtained from a set of samples. The procedure is the same in every case. We select a sample, calculate a chosen statistic from it, record the value of the statistic, return the sample to the population, select another sample of the same size, repeat the calculation, and so on. The frequency distribution constructed from the values of the statistic derived from the samples is known as the *sampling distribution* of that statistic. If we imagine the sampling operation carried out an extremely large number of times, the resulting sampling distribution would tend to the theoretical or 'ideal' distribution for that particular statistic.

We conclude our discussion of sampling distributions by noting that there is a definite relationship between the distribution of individual values in the population and the distribution of a given

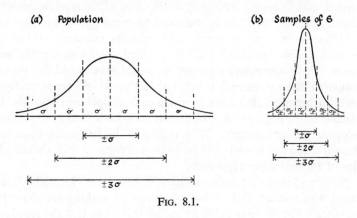

FIG. 8.1.

statistic calculated from samples selected from that population. We mentioned above in our biscuit example that the sampling distribution of the mean had the same mean as the population, but with variance σ^2/n (where n is the sample size). What is much more remarkable is that the sampling distribution of the mean is itself Normal. We summarise this important property as follows:

Given a population which is $N(\mu, \sigma^2)$, then

$$\bar{x} \text{ is } N(\mu, \sigma^2/n)$$

where n is the sample size.

This relationship can be clarified with the aid of a diagram.

Fig. 8.1(a) represents the distribution of the individual values of the variate in the population, while (b) represents the sampling distribution of means from samples of size 6. The distributions are both Normal and both have mean μ, but $\sigma_{\bar{x}}$ is only $\frac{2}{5}$ths the size of

$\sigma, \left(\dfrac{1}{\sqrt{6}} = \dfrac{2}{5} \text{ approx.} \right)$. The greater height of the distribution in (b) (assuming that they are drawn to the same scale) is explained by the fact that with a smaller variance, values will be clustered more closely around μ, and the height of the curve, of course, is proportional to the number of observations at that point.

If we had been interested in the sampling distribution of the *median*, we would have found it distributed as $N\left(\mu, \dfrac{1 \cdot 57 \sigma^2}{n} \right)$ approximately, i.e. with the same mean as the sampling distribution of the mean, but with a larger variance. (This is the reason why the sample mean gives the better estimate of the population mean than the sample median.)

In the next section of this chapter, where we shall be concerned with estimating population means, we present, without proof, the fundamental theorem relating the sampling distribution of the mean to the population distribution.

ESTIMATION

ESTIMATING POPULATION MEANS

The actual estimation of the population mean from a random sample is an extremely simple operation: we simply calculate the sample mean and take this as our estimate of the population mean. This is not quite so self-evident as it sounds. It does not always follow that when we wish to estimate a population value we apply exactly the same formula to our sample, but in the case of the mean this is so.

Having estimated the population mean, we are now concerned with calculating the limits around the estimate between which the true value is likely to lie. We cannot be absolutely certain that our calculated limits contain the true value; in general, the more certain we wish to be, the wider will be the limits. Expressing this idea in a different way, the narrower the limits we fix, the less certain are we that these limits contain the true value.

To derive a method for attaching limits to our estimate we must know something about the sampling distribution of the mean. We saw in the previous section that the sampling distribution of means based on random samples from a Normal distribution is itself

Normal, with the same mean as the population, but with variance σ^2/n. We can extend this result to the case when the population distribution is not Normal. We have the fundamental theorem of random sampling, which we state without proof:

Theorem. As n becomes large, the sampling distribution of $\bar{x}$ tends to Normality with mean μ and variance σ^2/n irrespective of the shape of the population distribution.

This very remarkable theorem, known as the *Central Limit Theorem*, provides us with a method for calculating the probabilities of obtaining various values of $\bar{x}$. Two questions at once spring to mind:

    (a) How large must n be before the sampling distribution approximates to the Normal?

    (b) What inferences can be drawn from the sample when the population is itself Normally distributed?

The answer to the first question depends to a large extent on the shape of the population distribution, but unless this is extremely skew, the approximation is valid for n greater than about 20. The answer to the second is that when the population is distributed Normally the sampling distribution of $\bar{x}$ is also Normal, i.e. there is no element of approximation no matter how small n is.

We will now use our knowledge of the Normal distribution to see how the Central Limit Theorem enables us to attach limits to an estimate.

EXAMPLE 1. We turn right back to the data of Table 1.1, the frequency distribution of the salaries of 1000 salesmen. We regard this as our population and imagine that we wish to estimate the population mean from a random sample of 50 salesmen. At this stage we also make the unlikely assumption that we know the value of the population variance even though we do not know the value of the population mean.

Suppose our sample of 50 gave us a value of $\bar{x} = £1500$.

We also know that the population variance $\sigma^2 = 102\ 400$ (see Table 2.1, where $\sigma^2 = (320)^2$).

From the Central Limit Theorem we know that our sample mean is distributed approximately Normally with estimated mean $\bar{x} = £1500$ and with variance $102\ 400/50 = 2048$. Hence the standard deviation of $\bar{x} = \sqrt{2048} = £45$ approximately.

The standard deviation of a statistic is such an important concept that it is given a special name. It is referred to as the *standard error* of the statistic. Its population value, i.e. its calculated value when the population variance is known, is denoted by $\sigma_{\bar{x}}$; when it is estimated from a sample it is denoted by $s_{\bar{x}}$.

Now from our knowledge of the Normal distribution we know that 68·3%, 95·4% and 99·7% of the observations lie in the interval $\mu \pm \sigma$, $\mu \pm 2\sigma$, $\mu \pm 3\sigma$ respectively (see Fig. 7.5).

Similarly, from Table 7.6 (a) we can find the interval enclosing any given percentage of the distribution. An interval of $\mu \pm 1\cdot96\sigma$, for example, includes exactly 95% of the observations.

Applying this idea to our sampling distribution of $\bar{x}$ we can now say that 95% of all values of $\bar{x}$ will lie in the interval £1500 ± 1·96

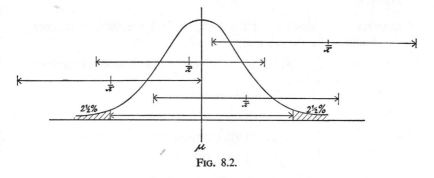

FIG. 8.2.

times £45, i.e. $\bar{x} \pm 1\cdot96\sqrt{(\sigma^2/n)}$, = £1500 ± £88·20 = £1411·80 to £1588·20 and we say that these limits are the 95% *confidence limits* to our estimate, i.e. that the true value lies between £1411·80 and £1588·20. We note that this interval does in fact include the true mean £1420, but we only know this because of our highly artificial example.

The interpretation of this statement is rather subtle, and is a fruitful source of confusion. When we state that the interval includes the true value of the mean, the statement is either correct or incorrect, it cannot be '95% correct'. If, however, we repeated the sampling experiment a large number of times, each time taking a random sample of 50 observations, each time constructing an interval like the one above, and each time stating that the interval included the true value of the population mean, our statements would, in the long run, be correct 95 times out of 100. This is what is meant by '95% confidence limits'. Fig. 8.2 will help to make this clear.

In Fig. 8.2 the curve represents the sampling distribution of the mean. The mean of this distribution is μ and the standard deviation is $\sqrt{(\sigma^2/n)}$. We know that 95% of the sample means fall in the interval $\mu \pm 1\cdot96\sqrt{\sigma^2/n}$. This interval is indicated by the horizontal line $\leftarrow\mu\rightarrow$. The remaining 5% of observations fall outside this interval, $2\frac{1}{2}$% occurring in each tail of the distribution. Whatever value of $\bar{x}$ we obtain from the sample, we fit limits of $\pm 1\cdot96\sqrt{(\sigma^2/n)}$. These limits are also shown in the diagram for several possible values of $\bar{x}$. It is clear from the graph that only when $\bar{x}$ falls into either tail of the distribution do the confidence limits fail to include μ, i.e. in only 5% of cases. Hence the interval $\bar{x} \pm 1\cdot96\sqrt{\sigma^2/n}$ fails to include μ only 5% of the time. To stress the point to the edge of tedium—this is why we call it a 95% confidence interval.

EXAMPLE 2. Calculate (a) the 90% and (b) the 99% confidence limits in the above example.

(a) From Table 7.6(a), 90% of a Normal distribution is contained in the interval $\mu \pm 1\cdot65\sigma$.

Therefore 90% confidence limits are

$$\bar{x} \pm 1\cdot65\sqrt{(\sigma^2/n)},$$

i.e. £1500 ± £(1·65)(45),

i.e. £1500 ± £74·20

or £1425·80 to £1574·20.

(b) For 99% confidence limits the appropriate constant from Table 7.6(a) is 2·58 approximately.

Therefore the 99% confidence limits are

$$\bar{x} \pm 2\cdot58\sqrt{(\sigma^2/n)}$$

i.e. £1500 ± £116·10

or £1383·90 to £1616·10.

We note that in the case of the 90% limits we are wrong, the true value of the mean being £1420. The 99% confidence limits illustrate the fact that the more certain we wish to be, the wider must be the interval in which we claim μ to lie.

In the above examples we made the assumption that we knew the value of the population variance. In the great majority of applications, however, we will not have this information, but provided the sample is large, we can estimate it from the sample. Thus we now not only estimate μ from our sample, but also σ^2, the parameter which is the basis of our confidence limits for μ. We estimate σ^2 by taking the sample variance, denoted by s^2, but our formula is

$$s^2 = \frac{1}{n-1} \Sigma(x-\bar{x})^2$$

and *not*

$$s^2 = \frac{1}{n} \Sigma(x-\bar{x})^2.$$

The reason for this change in the divisor is that dividing by n introduces an element of bias into the estimate. This bias can be removed by simply replacing n by $(n-1)$. Naturally, the larger the sample the less difference this makes.

Note. The proof of the above result is a simple exercise in mathematical expectation.

Thus, when the population is not known, the 95% confidence limits for the mean are given by the expression

$$\bar{x} \pm 1{\cdot}96\sqrt{(s^2/n)}$$

and it can be shown that our previous interpretation of confidence limits still holds good.

EXAMPLE 3. A random sample of 21 salesmen, selected from the population of Table 1.1, yields the following observations: £2150, £1325, £850, £1150, £1950, £1690, £1420, £2150, £1500, £1500, £1250, £1800, £1200, £1375, £1325, £1750, £1500, £1250, £1560, £1350, £1050. Estimate the population mean and fit (a) 95% and (b) 80% confidence limits to the estimate.

Our sample is just about large enough to estimate σ^2 by s^2: we have

$$\bar{x} = £31\ 095/21 = £1481$$

and

$$s^2 = \frac{1}{n-1} \Sigma(x-\bar{x})^2 = 115\ 008.$$

Therefore (a) 95% confidence limits are

$$£1481 \pm £(1{\cdot}96)(74)$$

i.e. £1481 ± £145.

Note that this interval includes the true value of the mean, £1420.

(b) 80% confidence limits are

$$£1481 \pm £(1\cdot28)(74)$$

i.e. $£1481 \pm £95$

which, in this example, also includes the true mean.

We must now consider the case where we do not know the value of σ^2 and our sample size is too small to estimate it reliably from our sample. In this situation we can no longer rely on the properties of the Normal distribution—the errors in estimating σ^2 would destroy the validity of the interval $\bar{x} \pm 1\cdot96\sqrt{(s^2/n)}$. The solution to the problem lies in the use of a theoretical distribution which we have not met before. This distribution has the rather odd name of 'Student's-t distribution', 'Student' being the pen-name of the researcher who first derived it, and 't' referring to the statistic whose distribution this is

The statistic t is defined as

$$t = \frac{\bar{x} - \mu}{s/\sqrt{n}}$$

$$= \frac{\bar{x} - \mu}{s} \cdot \sqrt{n}.$$

The distribution is, therefore, a standardised one very similar indeed to the standardised Normal variable z, where

$$z = \frac{\bar{x} - \mu}{\sigma} \cdot \sqrt{n}.$$

All we have done is to substitute s for σ.

The distribution of t is more complicated than the distribution of z but since its area is tabulated for us we do not need to bother with the equation. Compared with the Normal distribution, Student's-t distribution (which is also symmetrical) has a rather larger variance, but a more important difference is that its equation depends on the size of the sample, or rather on the value $n-1$, a value which is referred to as the number of degrees of freedom. Why it should be called this is a complicated question; we can offer a hint here in that the expression

$$s^2 = \frac{1}{n-1} \Sigma(x - \bar{x})^2$$

involves n values of $(x-\bar{x})$. Since $\Sigma(x-\bar{x}) = 0$, $n-1$ of these can take any value but the nth must be such that the total is 0; hence '$n-1$ degrees of freedom'.

Values of t are tabulated in Table 8.1, but only those values of t above which 10, 5, $2\frac{1}{2}$, 1 and $\frac{1}{2}\%$ of the area lies are tabulated for any given number of degrees of freedom since otherwise a whole table would be required for every number of degrees of freedom. These values of t are denoted by $t_{.100}$, $t_{.050}$, $t_{.025}$, etc., where $t_{.025}$, for example, is the value of t which is exceeded by only $2\frac{1}{2}\%$ of all values of t.

Looking down Table 8.1 the reader will see that as the number of degrees of freedom increases the values of t approximate closer and closer to the values of the standardised Normal distribution until with an infinite number of degrees of freedom they are the same as the Normal, e.g. $t_{.025} = 1\cdot96$. The use of Student's-t distribution in setting confidence limits to $\bar{x}$ is precisely the same as the use of the Normal distribution. If we are interested in a 95% confidence interval, we know that 95% of the standardised distribution lies between the values $-t_{.025}$ and $t_{.025}$; we can therefore construct an interval

$$\bar{x} \pm t_{.025}\sqrt{(s^2/n)}$$

and this will be our 95% confidence interval. The appropriate value of $t_{.025}$ is looked up in Table 8.1 with degrees of freedom (d.f.) equal to $n-1$. If we compare this expression with the corresponding expression for a large sample with σ^2 unknown, we find that the only difference is that the value for $t_{.025}$ replaces $1\cdot96$. Similarly, for 99%, 90% etc., confidence intervals, $t_{.005}$, $t_{.050}$, etc., replace $2\cdot58$, $1\cdot65$, etc.

EXAMPLE 4. A random sample of 10 salesmen, selected from the population of Table 1.1 yields the following observations: £1050, £1325, £1150, £2150, £1500, £1200, £1325, £1750, £1500, £1560. Estimate the population mean and fit (a) 90%, (b) 95% and (c) 99% confidence limits to the estimate.

We have

$$n-1 = v = 9$$

$$\bar{x} = \frac{£14\,510}{10} = £1451$$

$$s^2 = 104\,538.$$

TABLE 8.1

Probability Points of the t-Distribution (Single-sided)

v	P				
	0·1	0·05	0·025	0·01	0·005
1	3·08	6·31	12·7	31·8	63·7
2	1·89	2·92	4·30	6·96	9·92
3	1·64	2·35	3·18	4·54	5·84
4	1·53	2·13	2·78	3·75	4·60
5	1·48	2·01	2·57	3·36	4·03
6	1·44	1·94	2·45	3·14	3·71
7	1·42	1·89	2·36	3·00	3·50
8	1·40	1·86	2·31	2·90	3·36
9	1·38	1·83	2·26	2·82	3·25
10	1·37	1·81	2·23	2·76	3·17
11	1·36	1·80	2·20	2·72	3·11
12	1·36	1·78	2·18	2·68	3·05
13	1·35	1·77	2·16	2·65	3·01
14	1·34	1·76	2·14	2·62	2·98
15	1·34	1·75	2·13	2·60	2·95
16	1·34	1·75	2·12	2·58	2·92
17	1·33	1·74	2·11	2·57	2·90
18	1·33	1·73	2·10	2·55	2·88
19	1·33	1·73	2·09	2·54	2·86
20	1·32	1·72	2·09	2·53	2·85
21	1·32	1·72	2·08	2·52	2·83
22	1·32	1·72	2·07	2·51	2·82
23	1·32	1·71	2·07	2·50	2·81
24	1·32	1·71	2·06	2·49	2·80
25	1·32	1·71	2·06	2·48	2·79
26	1·32	1·71	2·06	2·48	2·78
27	1·31	1·70	2·05	2·47	2·77
28	1·31	1·70	2·05	2·47	2·76
29	1·31	1·70	2·05	2·46	2·76
30	1·31	1·70	2·04	2·46	2·75
40	1·30	1·68	2·02	2·42	2·70
60	1·30	1·67	2·00	2·39	2·66
120	1·29	1·66	1·98	2·36	2·62
∞	1·28	1·64	1·96	2·33	2·58

The standard error of $\bar{x}$ is therefore:

$$\sqrt{(s^2/n)} = \sqrt{(104\,538/10)} = \sqrt{10\,454} = \text{£}102 \text{ approximately.}$$

Hence (a) 90% confidence limits are given by:

$$\text{£}1451 \pm \text{£}(1\cdot83)\,(102)$$

i.e. £1451 ± £187.

(b) 95% confidence limits are given by:

$$\text{£}1451 \pm \text{£}(2\cdot26)\,(102)$$

i.e. £1451 ± £231.

(c) 99% confidence limits are given by:

$$\text{£}1451 \pm \text{£}(3\cdot25)\,(102)$$

i.e. £1451 ± £332.

Thus, with 9 d.f., 1·83, 2·26 and 3·25 replace the corresponding Normal values of 1·65, 1·96 and 2·58. The confidence interval is, therefore, much wider than it would have been had we used the Normal approximation and shows that the error resulting from the misuse of the Normal approximation leads to spuriously precise estimates. The wide intervals in the above example are, of course, only to be expected with samples as small as 10.

ESTIMATING POPULATION PROPORTIONS

The best estimate of π, the population proportion, is p, the sample proportion where

$$p = \frac{x}{n},$$

i.e. $\dfrac{\text{the number of observations with the required characteristic}}{\text{the total number of observations in the sample}}$,

EXAMPLE 5. A batch of 10 000 photographic flash bulbs is manufactured on a particular machine. As a test of the efficiency of the machine an estimate of the proportion defective is required. A sample of 100 bulbs is selected and tested. 6 bulbs in the sample fail to flash, therefore

$$p = 6/100 = 0\cdot06$$

and this is our estimate of π, the proportion defective in the population of 10 000.

The question might have been put in the form 'How many bulbs

H

in the batch of 10 000 will be defective?' To obtain the answer to this question we simply calculate p as before, then take $p \times$ (population size) as our estimate. In the above example we have $p = 0.06$, therefore the number defective is $0.06 \times 10\ 000 = 600$.

We can fit confidence limits to our estimate of p by using the Normal approximation provided that the value of p is not close to 0 or 1 and provided that the population from which the sample is drawn is large. As a rough guide, we can use the Normal approximation provided np and $n(1-p)$ are both greater than 5, and the sample includes less than 10% of the population. When these conditions are fulfilled, the sampling distribution of $\frac{x}{n}$ is approximately Normal with

$$\mu = \pi \text{ and } \sigma^2 = \frac{\pi(1-\pi)}{n}.$$

We will not set out the steps of the argument as we did in the previous section, but content ourselves with quoting the final expression.

We use our statistic p to estimate π in the above expression and, taking the 95% level for our example, write our confidence limits for π as

$$p \pm 1.96 \sqrt{\left\{ \frac{p(1-p)}{n} \right\}}.$$

From our flash bulb example above we have

$$N = 10\ 000, p = 0.06, (1-p) = 0.94 \text{ and } n = 100$$

Since $np = 6$, $n(1-p) = 94$ and $\frac{n}{N} \times 100 = 1\%$, we can use the Normal approximation and say that 95% confidence limits for π are given by

$$0.06 \pm 1.96 \sqrt{\frac{0.06\ (0.94)}{100}}$$

i.e. 0.06 ± 0.05

i.e. 0.01 to 0.11.

Confidence limits at different confidence levels are calculated by replacing 1.96 by the appropriate value from Table 7.6(a).

An alternative method of fitting confidence limits to p, that eliminates the need for the arithmetic carried out above is to use specially constructed charts and read off the required limits, Fig. 8.3 illustrates two such charts, one for the 95% confidence level and one for the 99% level.

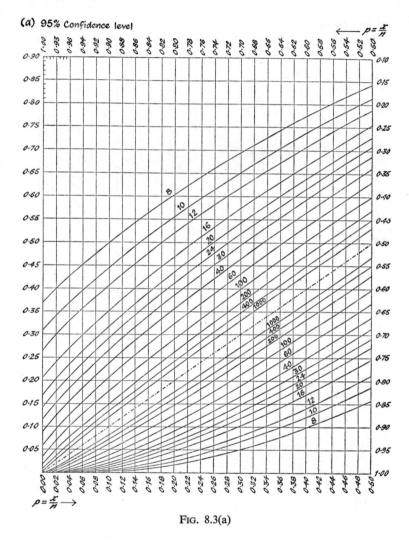

(a) 95% Confidence level

FIG. 8.3(a)

With our flash bulb example, we have $p = 0.06$ and $n = 100$. All we do is to read off 0·06 along the horizontal scale of the chart and then read vertically from this point until we see two curves labelled 100. The intersections of these two curves with the vertical line for the value 0·06 gives us our confidence limits. They are read from the left-hand scale.

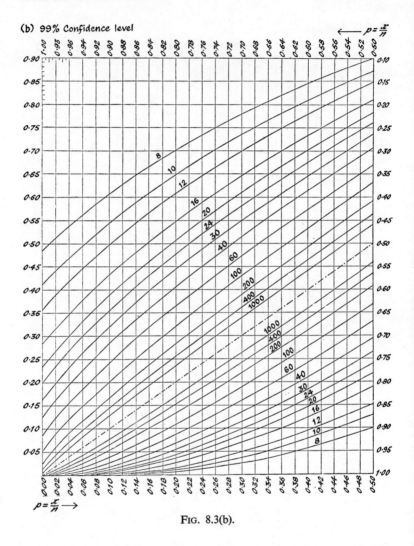

FIG. 8.3(b).

The accuracy of this method obviously depends on how big the chart is; how carefully it is drawn and how carefully we read it. If we wish to be very accurate in our calculations we have to perform the arithmetic; in many cases, however, the rough and ready values provided by the chart will be good enough for our purpose. We need, of course, a separate chart for every confidence level we wish to use.

ESTIMATING POPULATION TOTALS

The object of a statistical investigation is often to estimate population totals rather than population means. We saw an instance of this with the flash-bulb example in the previous section. In an earlier section we were concerned with estimating the average salary of 1000 salesmen from a sample of 50. This investigation might be prompted by a desire to know the total salary bill of the salesmen, and confidence limits might be required for this estimate.

The simplest way of calculating confidence limits for a total, assuming that confidence limits for the mean have already been calculated, is a matter of trivial arithmetic. The estimate of the total is

$$N\bar{x}$$

where N is the population size. In our salesmen example (Example 3, p. 207) we found

$$\bar{x} = \pounds1481 \quad n = 21 \quad N = 1000$$

therefore   Total salaries $= \pounds1\ 481\ 000$.

Now, the 95% confidence limits for the mean are $\bar{x} \pm \pounds145$, hence if the population mean lies in the interval £1336 to £1626 the population total must lie in the interval £1 336 000 to £1 626 000.

Hence, the standard error of the total is simply $Ns_{\bar{x}}$.

If we are not interested in $\bar{x}$, we can proceed to a direct estimate of T the population total. The sample total, in the above example, is £31 095. T is therefore

$$\pounds31\ 095 \times \frac{N}{n},$$

i.e. £31 095 000/21 $= \pounds1\ 480\ 714$, to the nearest £, which is rather more accurate, arithmetically, than £1481 × 1000, and the 95% confidence limits are

$$T \pm 1 \cdot 96 N s_{\bar{x}}$$

i.e. £1 480 714 $\pm 1 \cdot 96\ \pounds(74 \times 1000)$

i.e. £1 480 714 $\pm \pounds145\ 040$

which, again, is marginally more accurate than using $\bar{x}$.

We can formalise the above theory for sample totals as follows:

For large values of n, the sampling distribution of sample totals is approximately Normal with mean T and variance $N^2\sigma^2$ (where T is the population total).

Note. The factor $\dfrac{N}{n}$ is known as the *'raising factor'*. It is the factor by which we must multiply the sample total and the sample size to estimate the population total and to get the population size. Its reciprocal, which is a more familiar concept, is $\dfrac{n}{N}$, the *sampling fraction*.

ESTIMATING OTHER PARAMETERS

As we said at the beginning of the chapter, there are a number of statistics which can be calculated from a sample, and confidence limits can be attached to all of them. We have dealt with the three most common of them: the *mean, proportion* and *total*. The only other one we will mention is the *variance*, with which we shall include the *standard deviation*.

The fitting of confidence limits to σ^2 and σ is based on the sampling distribution of s^2. We shall demonstrate two alternative methods, the first an exact method the second a large-sample approximation based, as usual, on the Normal distribution.

Method 1. We consider the sampling distribution of the ratio s^2/σ^2, which conforms to a theoretical distribution known as the χ^2/d.f. distribution. (This is pronounced 'ki-square over degrees of freedom' with 'ki' as in 'kite'.) The distribution is continuous and positively skewed. Percentiles of the distribution are tabulated in Table 8.2. The χ^2/d.f. distribution is closely related to the χ^2 (without the 'd.f.') distribution and Student's-t distribution. Like both of these, it depends on the number of degrees of freedom, which in the present application is always $n-1$.

EXAMPLE 6. Returning once more to our salesmen example, in Example 3 (p. 207), we estimated their average salary from a sample of 21 observations. We also calculated s^2, the sample variance, obtaining the value $s^2 = 115\,008$. We fit 95% confidence limits to this value as follows:

We have 20 d.f. so, from Table 8.2, we have $p_{0.025} = 1.71$. and $p_{0.975} = 0.480$

(*Note.* $P_{0.975}$ is the value of $\chi^2/d.f.$, which is exceeded by 97·5% of the observations, $p_{0.025}$ is the value exceeded by 2·5% of the observations, therefore 95% of observations fall within the interval $p_{0.975}$ to $p_{0.025}$.) Hence

$$0.480 < \frac{s^2}{\sigma^2} < 1.71$$

therefore

$$\frac{1}{0.480} > \frac{\sigma^2}{s^2} > \frac{1}{1.71}.$$

(*Note.* We have simply taken the inverse of all three terms in the inequality. This has the effect of reversing the inequality signs, as any simple numerical example will demonstrate.)

TABLE 8.2. *Percentiles of the $\chi^2/d.f.$ Distribution*

ν	0·995	0·99	0·975	0·95	0·90	0·75	0·50	0·25	0·10	0·05	0·025	0·01	0·005	0·001	ν
1	..	..	..	..	·016	·102	·455	1·32	2·71	3·84	5·02	6·64	7·88	10·83	1
2	·005	·010	·025	·052	·106	·288	·693	1·39	2·30	3·00	3·69	4·61	5·30	6·91	2
3	·024	·038	·072	·117	·195	·403	·789	1·37	2·08	2·60	3·12	3·78	4·28	5·42	3
4	·052	·074	·121	·178	·266	·480	·839	1·35	1·94	2·37	2·70	3·32	3·72	4·62	4
5	·082	·111	·166	·229	·322	·534	·870	1·33	1·85	2·21	2·57	3·02	3·35	4·10	5
6	·113	·145	·206	·272	·367	·575	·891	1·31	1·77	2·10	2·41	2·80	3·09	3·74	6
7	·141	·177	·241	·310	·405	·607	·907	1·29	1·72	2·01	2·29	2·64	2·90	3·47	7
8	·168	·206	·272	·342	·436	·634	·918	1·28	1·67	1·94	2·19	2·51	2·74	3·27	8
9	·193	·232	·300	·369	·463	·656	·927	1·27	1·63	1·88	2·11	2·41	2·62	3·10	9
10	·216	·256	·325	·394	·487	·674	·934	1·25	1·60	1·83	2·05	2·32	2·52	2·96	10
11	·237	·278	·347	·416	·507	·689	·940	1·25	1·57	1·79	1·99	2·25	2·43	2·84	11
12	·256	·298	·367	·436	·525	·703	·945	1·23	1·55	1·75	1·94	2·18	2·36	2·74	12
13	·274	·316	·385	·453	·542	·715	·949	1·23	1·52	1·72	1·90	2·13	2·29	2·66	13
14	·291	·333	·402	·469	·556	·729	·953	1·22	1·50	1·69	1·87	2·08	2·24	2·58	14
15	·307	·349	·418	·484	·570	·733	·956	1·21	1·49	1·67	1·83	2·04	2·19	2·51	15
16	·321	·363	·432	·498	·582	·744	·959	1·21	1·47	1·64	1·80	2·00	2·14	2·45	16
17	·335	·377	·445	·510	·593	·753	·961	1·21	1·46	1·62	1·78	1·97	2·10	2·40	17
18	·348	·390	·457	·522	·604	·761	·963	1·20	1·44	1·60	1·75	1·93	2·06	2·35	18
19	·360	·402	·469	·532	·613	·768	·965	1·19	1·43	1·59	1·73	1·90	2·03	2·31	19
20	·372	·413	·480	·543	·622	·775	·967	1·19	1·42	1·57	1·71	1·88	2·00	2·27	20
21	·382	·424	·490	·552	·629	·776	·968	1·19	1·41	1·56	1·69	1·85	1·97	2·23	21
22	·393	·434	·499	·561	·638	·782	·970	1·18	1·40	1·54	1·67	1·83	1·95	2·19	22
23	·403	·443	·509	·570	·643	·787	·971	1·18	1·39	1·53	1·66	1·81	1·92	2·16	23
24	·412	·452	·517	·577	·652	·792	·972	1·17	1·38	1·52	1·64	1·79	1·90	2·13	24
25	·420	·460	·524	·584	·660	·796	·972	1·17	1·38	1·51	1·62	1·77	1·88	2·10	25
26	·429	·469	·532	·592	·665	·800	·974	1·17	1·37	1·50	1·61	1·76	1·86	2·08	26
27	·437	·478	·541	·600	·670	·804	·975	1·17	1·36	1·49	1·60	1·74	1·84	2·06	27
28	·445	·484	·547	·605	·676	·811	·976	1·16	1·35	1·48	1·59	1·72	1·82	2·03	28
29	·452	·493	·552	·610	·683	·814	·976	1·16	1·35	1·47	1·58	1·71	1·80	2·01	29
30	·460	·498	·560	·616	·687	·817	·978	1·16	1·34	1·46	1·57	1·70	1·79	1·99	30

The values in this table are obtained by dividing the values in Table 8.3 by the appropriate number of d.f.

ν is the number of degrees of freedom.

TABLE 8.3. *Probability Points of the χ^2 Distribution*

ν	0·995	0·99	0·975	0·95	0·90	0·75	0·50	0·25	0·10	0·05	0·025	0·01	0·005	0·001	ν
1	...	...	...	...	·016	·102	·455	1·32	2·71	3·84	5·02	6·63	7·88	10·8	1
2	·010	·020	·051	·103	·211	·575	1·39	2·77	4·61	5·99	7·38	9·21	10·6	13·8	2
3	·072	·115	·216	·352	·584	1·21	2·37	4·11	6·25	7·81	9·35	11·3	12·8	16·3	3
4	·207	·297	·484	·711	1·06	1·92	3·36	5·39	7·78	9·49	11·1	13·3	14·9	18·5	4
5	·412	·554	·831	1·15	1·61	2·67	4·35	6·63	9·24	11·1	12·8	15·1	16·7	20·5	5
6	·676	·872	1·24	1·64	2·20	3·45	5·35	7·84	10·6	12·6	14·4	16·8	18·5	22·5	6
7	·989	1·24	1·69	2·17	2·83	4·25	6·35	9·04	12·0	14·1	16·0	18·5	20·3	24·3	7
8	1·34	1·65	2·18	2·73	3·49	5·07	7·34	10·2	13·4	15·5	17·5	20·1	22·0	26·1	8
9	1·73	2·09	2·70	3·33	4·17	5·90	8·34	11·4	14·7	16·9	19·0	21·7	23·6	27·9	9
10	2·16	2·56	3·25	3·94	4·87	6·74	9·34	12·5	16·0	18·3	20·5	23·2	25·2	29·6	10
11	2·60	3·05	3·82	4·57	5·58	7·58	10·3	13·7	17·3	19·7	21·9	24·7	26·8	31·3	11
12	3·07	3·57	4·40	5·23	6·30	8·44	11·3	14·8	18·5	21·0	23·3	26·2	28·3	32·9	12
13	3·57	4·11	5·01	5·89	7·04	9·30	12·3	16·0	19·8	22·4	24·7	27·7	29·8	34·5	13
14	4·07	4·66	5·63	6·57	7·79	10·2	13·3	17·1	21·1	23·7	26·1	29·1	31·3	36·1	14
15	4·60	5·23	6·26	7·26	8·55	11·0	14·3	18·2	22·3	25·0	27·5	30·6	32·8	37·7	15
16	5·14	5·81	6·91	7·96	9·31	11·9	15·3	19·4	23·5	26·3	28·8	32·0	34·3	39·3	16
17	5·70	6·41	7·56	8·67	10·1	12·8	16·3	20·5	24·8	27·6	30·2	33·4	35·7	40·8	17
18	6·26	7·01	8·23	9·39	10·9	13·7	17·3	21·6	26·0	28·9	31·5	34·8	37·2	42·3	18
19	6·84	7·63	8·91	10·1	11·7	14·6	18·3	22·7	27·2	30·1	32·9	36·2	38·6	43·8	19
20	7·43	8·26	9·59	10·9	12·4	15·5	19·3	23·8	28·4	31·4	34·2	37·6	40·0	45·3	20
21	8·03	8·90	10·3	11·6	13·2	16·3	20·3	24·9	29·6	32·7	35·5	38·9	41·4	46·8	21
22	8·64	9·54	11·0	12·3	14·0	17·2	21·3	26·0	30·8	33·9	36·8	40·3	42·8	48·3	22
23	9·26	10·2	11·7	13·1	14·8	18·1	22·3	27·1	32·0	35·2	38·1	41·6	44·2	49·7	23
24	9·89	10·9	12·4	13·8	15·7	19·0	23·3	28·2	33·2	36·4	39·4	43·0	45·6	51·2	24
25	10·5	11·5	13·1	14·6	16·5	19·9	24·3	29·3	34·4	37·7	40·6	44·3	46·9	52·6	25
26	11·2	12·2	13·8	15·4	17·3	20·8	25·3	30·4	35·6	38·9	41·9	45·6	48·3	54·1	26
27	11·8	12·9	14·6	16·2	18·1	21·7	26·3	31·5	36·7	40·1	43·2	47·0	49·6	55·5	27
28	12·5	13·6	15·3	16·9	18·9	22·7	27·3	32·6	37·9	41·3	44·5	48·3	51·0	56·9	28
29	13·1	14·3	16·0	17·7	19·8	23·6	28·3	33·7	39·1	42·6	45·7	49·6	52·3	58·3	29
30	13·8	15·0	16·8	18·5	20·6	24·5	29·3	34·8	40·3	43·8	47·0	50·9	53·7	59·7	30

If we now multiply through by s^2 we have

$$\frac{s^2}{0\cdot480} > \sigma^2 > \frac{s^2}{1\cdot71}$$

i.e.

$$\frac{115\,008}{0\cdot480} > \sigma^2 > \frac{115\,008}{1\cdot71}$$

i.e. $239\,600 > \sigma^2 > 67\,256.$

In practice, of course, we go straight to the expression

$$\frac{s^2}{p_{0\cdot975}} > \sigma^2 > \frac{s^2}{p_{0\cdot025}}$$

Confidence limits for σ may be obtained directly from the confidence limits for σ^2 by taking the square root of the three terms in the inequality.

Thus $\sqrt{239\,600} > \sigma > \sqrt{67\,256}.$

i.e.   £489$>\sigma>$£259 approx.

Method 2. This method provides us with approximate confidence limits for σ. It is rather less reliable for σ^2.

When n is large the distribution of s, the sample standard deviation, is approximately

$$N\left(\sigma, \frac{\sigma^2}{2n}\right).$$

(We state this result without proof).

Hence, 95% confidence limits are given by the following expression:

$$s - 1\cdot96\,\frac{\sigma}{\sqrt{2n}} < \sigma < s + 1\cdot96\,\frac{\sigma}{\sqrt{2n}}.$$

A certain amount of algebraic manipulation yields the alternative form:

$$\frac{s}{1 + \dfrac{1\cdot96}{\sqrt{2n}}} < \sigma < \frac{s}{1 - \dfrac{1\cdot96}{\sqrt{2n}}}$$

Other confidence levels are obtained by replacing 1·96 by the appropriate factors.

EXAMPLE 7. We recalculate 95% confidence limits for the example in Method 1 above. We have

$$s = \sqrt{115\,008} = £339 \text{ approx.}$$

Therefore

$$\frac{339}{1+\dfrac{1{\cdot}96}{\sqrt{42}}} < \sigma < \frac{339}{1-\dfrac{1{\cdot}96}{\sqrt{42}}}$$

from which $£260 < \sigma < £486$

The 95% confidence limits for σ^2, if required, are then

$$260^2 < \sigma^2 < 486^2$$

i.e. $67\,600 < \sigma^2 < 236\,196$

Comparison of the two methods shows only small differences.

SAMPLING FROM SMALL POPULATIONS

The methods of fitting confidence limits discussed so far assume that the populations from which we selected the samples were very large in relation to the sample. This assumption was made so that we could further assume that the probability of selection was so small that, as each unit was selected, no difference was made to the probabilities of selection of the remaining members of the population. The importance of this assumption can be illustrated by an example.

If we wished to select a sample of 20 from a population of 100, the probability of selection of the first member selected would be 1/100. Unless this member was returned to the population before another member was selected, the second member would be selected with probability 1/99, and by the time the 20th member was selected, this probability would be 1/81. Thus, members of the sample would *not* be selected with equal probabilities, in which case the methods considered so far would be invalid.

A second point is that where an appreciable proportion of the population is included in the sample, the methods used so far overestimate the sampling errors. We can correct this by introducing a *finite population correcting factor* (f.p.c. for short). These matters we shall be considering in Chapter 11, when we are dealing with sampling and sample surveys in greater detail.

TESTS OF HYPOTHESES

In this section we consider how to use the evidence provided by samples for testing hypotheses. We can think of a hypothesis in general as an assumption concerning a frequency distribution (or distributions). For example, a firm may decide to launch a certain new product provided it finds acceptance with at least 20% of housewives. A sample survey is carried out among 1000 housewives; each housewife being classified as an 'acceptance' or a 'rejection'. We will not pursue the question of how valid this classification system is; what concerns us is that only 18% of the housewives in the sample find the product acceptable. Should the firm put the new product on the market?

The hypothesis we are testing in this example is that the proportion of housewives who find the product acceptable is 20% *or more*. This hypothesis is called by statisticians the *null hypothesis* and is denoted by H_0. (This curious expression originates from the testing of differences between observed values of statistics and hypothetical values. The difference under test is assumed to be *zero*, hence the term 'null' hypothesis.) Every null hypothesis is tested against an *alternative hypothesis*, which is usually denoted by H_1. In this example the alternative hypothesis is that the proportion is less than 20%. The precise form of the alternative hypothesis has a marked effect on the construction of the test; we shall observe this effect as we study various examples.

The question we now have to answer, with regard to our null hypothesis is 'What evidence have we got for either accepting or rejecting the hypothesis?' On the face of it we ought to reject the hypothesis straight away, because the observed proportion in favour is only 18%. But this proportion was calculated from a sample and is therefore only an estimate of the population proportion. Can we really conclude that the population proportion is less than 20%?

We now construct a test of the null hypothesis by stating the problem in the following terms.

Assuming that the population proportion in favour of the product is 20%, what is the probability that in a random sample of 1000 housewives an observed proportion of 18% will be obtained?

If this probability is very small we reject the hypothesis that the population proportion is 20%; if the probability is large then we accept the null hypothesis (we shall consider later on what we mean by 'small' and 'large'). Applying this procedure to our example we have:

$$H_0 : \pi = 0 \cdot 20$$
$$H_1 : \pi < 0 \cdot 20$$

Our sample gives us:

$$n = 1000$$

$$p = 0.18.$$

Using our old friend the Normal approximation we calculate the standardised Normal variate, z:

$$z = \frac{p - \pi}{\sqrt{\{\pi(1-\pi)/n\}}}$$

$$= \frac{0.18 - 0.20}{\sqrt{\{(0.20)(0.80)/1000\}}}.$$

From Table 7.6(a) we see that the probability of obtaining a value of -1.54 or less from a standardised Normal distribution is 0.06 approx. Should we reject H_0?

In order to answer this question we must go back to our example and settle something which we left undecided when we constructed our test. We have to say in advance how small our 'small' probability must be before we reject H_0. There are no hard and fast rules regarding this question. Much depends on the alternative hypothesis and the consequences of making a wrong decision. In our case we presume that the criterion of 20% has been set bearing in mind the expected profitability arising from a 20% market share and the required rate of return on capital. It would probably be not too disastrous if the actual proportion of housewives favouring the product turned out to be 19% or even a bit less. Let us assume the directors of the firm will accept H_0 provided that the probability of being wrong is not more than 0.10. From Table 7.6(a) we see that the value of z such that 10% of the distribution has values smaller than z is -1.28. Our observed value of z is -1.54 and since this is smaller than -1.28 we reject the hypothesis that $H_0 = 0.20$ and cancel the launch.

The reader will probably be wondering why we bothered with the 10% value -1.28, since we had already calculated the probability of our actual value of $z = -1.54$ as 0.06 approx. Since 0.06 is less than 0.10 we can reject H_0 without going any further. In fact we used both methods merely to demonstrate the two approaches.

We can illustrate the ideas in the above example, and introduce some of the nomenclature of hypothesis testing by means of a diagram.

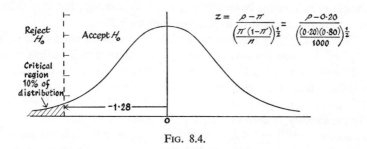

FIG. 8.4.

The distribution above is of the standardised variate

$$z = \frac{p-\pi}{\sqrt{\{\pi(1-\pi)/n\}}}$$

Under the null hypothesis H_0, $\pi = 0.20$, therefore

$$z = \frac{p-0.20}{\sqrt{\{(0.20)(0.80)/1000\}}},$$

Also under H_0, values of z less than -1.28 will occur only 10% of the time. This 10% tail is indicated by the shaded area in the diagram, and is called the *critical region* of the test statistic. The diagram can be drawn as soon as we set H_0 and decide on a probability level for our critical region; all that remains to be done is the sampling operation and the calculation of p from the sample results. Once p is obtained we simply calculate the value of z under the null hypothesis. If this value falls in the critical region of the test we reject H_0, if it falls outside the critical region we accept H_0. In our example the boundary of the critical region was -1.28 and our observed value of z was -1.54. Hence we rejected H_0, i.e. we rejected the hypothesis that our sample came from a population in which $\pi = 0.20$.

It is worth going a little deeper into the logic of our decision. In order to test whether $\pi \geqq 0.20$ against the hypothesis that $\pi < 0.20$ we assumed that $\pi = 0.20$. We then said, in effect, 'We have obtained an estimate $p = 0.18$ from our sample of 1000. Assuming that $\pi = 0.20$ what is the probability of obtaining a value of $p = 0.18$ from a sample of 1000'. If this probability is less than some predetermined value, in our case 0.10, then *either* we have obtained a rather rare value from our sampling distribution *or* $\pi \neq 0.20$. We choose to believe the latter and reject the hypothesis that $\pi = 0.20$, accepting the alternative hypothesis that $\pi < 0.20$.

Now, in applying this test we are likely to make a wrong decision 10% of the time because 10% of the values of z will be less than -1.28 when $\pi = 0.20$. From the expression for z we can easily calculate that $z < -1.28$ when $p < 0.183$, thus the probability is 0.10 that the sample proportion will be less than 0.183. Therefore the probability is 0.10 that under the criterion we have set up for ourselves we will reject H_0 when in fact H_0 is true. This is known as the *Type I error* and the probability of making this error is called the *size* of the error. If the size of the Type I error is too big, we can easily reduce it. All we have to do is to make the critical region smaller, e.g. in the example above, if we required the Type I error to be no larger than 0.05 we would simply shift the boundary of the critical region to the left until only 5% of the distribution lay outside. We would do this by reading off from Table 7.6(a), the value of z which cut off a 5% tail. This value is $z = 1.65$. If we adopt this value as our critical region boundary, and go back to our observed value of $z = -1.54$ we find that we now accept the hypothesis that $\pi = 0.20$, or rather we say that there is insufficient evidence on which to reject the hypothesis.

But now it occurs to us that we are likely to be committing another type of error. We have reduced the probability of our Type I error by reducing the size of the critical region. The Type I error occurs when we *reject H_0* when it is *true*. A different error, known as the *Type II error*, occurs when we *accept H_0* when it is *false*. Clearly, a reduction in the size of the Type I error, other things being equal, can only be obtained at the expense of increasing the size of the Type II error. The only way in which we can reduce the size of both types or error simultaneously is by increasing the sample size.

We can describe the effect of Type I and Type II errors with the aid of the following table:

| | Decision | |
Situation	Accept H_0	Reject H_0
H_0 true	Correct	Type I error
H_0 false	Type II error	Correct

The two errors are denoted by α and β respectively.

The discussion can be illustrated diagrammatically:

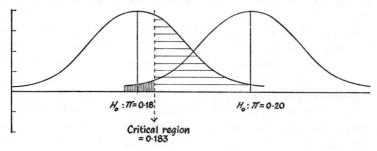

$H_0 : \pi = 0.18$ $H_0 : \pi = 0.20$

Critical region
= 0.183

FIG. 8.5.

In Fig. 8.5, the right-hand curve represents the sampling distribution of π under the null hypothesis $H_0 : \pi = 0.20$. The left-hand curve represents the sampling distribution of π under the assumption that $\pi = 0.18$. (As we have said elsewhere, the size of the Type II error depends on the particular value of the parameter being estimated. In order to illustrate both types of error we have assumed that $\pi = 0.18$, which is the value of our sample proportion p.) Under the null hypothesis, we establish the 10% critical region as that part of the distribution to the left of the value 0.183. This is the small area vertically shaded. We have, therefore, constructed a test with a Type I error of 10%.

Now, if the true value of π is 0.18, the 'true' sampling distribution of π overlaps the distribution under the null hypothesis, i.e. it overlaps the critical region boundary 0.183. This overlap is represented in the diagram by the horizontally shaded area. We immediately see that even if the true value of π is 0.18, a large proportion of sample values p will exceed 0.183 and the null hypothesis $H_0 : \pi = 0.20$ will be accepted incorrectly. Thus, having fixed the size of the Type I error, the size of the Type II error is determined by:

(*a*) the true value of π and
(*b*) the spread of the two sampling distributions.

If we wish to vary the size of either error, we simply shift our critical region boundary—left to decrease the Type I error and right to decrease the Type II error. But we can only decrease the size of one type of error at the cost of increasing the size of the other type. The only way in which we can decrease the size of both errors (or hold

one constant while decreasing the size of the other) is by increasing
the sample size. If we do this, we immediately reduce the variance
of p, as can be seen from Fig. 8.6

In Fig. 8.6 we have increased sample size to such an extent that
the sampling distributions are much more tightly clustered—so much
so that with the same critical region boundary as before we have
virtually removed the Type I error and at the same time greatly
reduced the size of the Type II error. But given such a sampling
distribution, under the null hypothesis $H_0 : \pi = 0.20$ we would
re-calculate the critical region boundary. As can be seen from the
diagram, in this new situation a critical region boundary of about
0.19 would make both types of error extremely small.

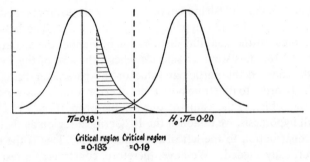

$\pi = 0.18$

$H_o : \pi = 0.20$

Critical region
$= 0.183$

Critical region
$= 0.19$

.Fig. 8.6.

Comparing Fig. 8.6 with Fig. 8.5 we can see that by increasing
sample size, thereby reducing the standard error of p, and by careful
choice of the critical region, we can set both types of error to whatever size we please until, ultimately, when we have included the entire
population in our 'sample' the sampling distributions disappear and
we simply calculate the population value from our 100% 'sample'.

The size of the Type I error can be specified in advance, as soon
as we have formulated H_0. The size of the Type II error, however,
depends on what in fact is the true value of the parameter. We
illustrate this point with our example above.

Let us assume that H_0 and α are as in Fig. 8.5, and let us make the
further assumption that the true value of π is 0.15. With our
observed value $p = 0.18$ we began by asking ourselves 'What is the
probability of obtaining a value of $p = 0.18$ from a sample of 1000
when $\pi = 0.20$?' and ended up with a value of $p = 0.183$ as our

critical value below which we would reject H_0. We now have to ask ourselves 'What is the probability of obtaining a value of $p > 0.183$ from a sample of 1000 when $\pi = 0.15$?' This probability is the size of the Type II error. We calculate the value of the standardised variable

$$z = \frac{0.183 - 0.15}{\sqrt{\{(0.15)(0.85)/1000\}}}$$

$$= 2.92$$

and consult Table 7.6(a) to find the area of the standardised Normal curve lying above this value. This area is 0.0018, hence the probability that $z > 2.92$ is 0.0018, or 0.002 approx. This, therefore, is the size of the Type II error for $\pi = 0.15$.

If now we decide to reduce the size of the Type I error to 0.05 we obtain, under the null hypothesis, a value of $z = -1.65$, which gives us a value of $p = 0.178$. The Type II error, for a true value of $\pi = 0.15$ is now obtained via the standardised variable

$$z = \frac{0.178 - 0.15}{\sqrt{\{(0.15)(0.85)/1000\}}}$$

$$= 2.48$$

which, from Table 7.6(a), gives us a probability of 0.0066. Hence the size of the Type II error is now 0.007 approx. compared with 0.002 when the Type I error was 0.10.

The difficulty of calculating the size of the Type II error arises from the fact that the alternative hypothesis is less specific than the null hypothesis. In the example above we formulated the null hypothesis as $H_0 : \pi = 0.20$ and the alternative hypothesis as $H_1 : \pi < 0.20$, thus we assumed a specific value for H_0 and this enabled us immediately to calculate z and α. This illustrates a general rule for hypothesis testing: we choose a fixed value for the null hypothesis rather than a range of values because only by so doing can we define a critical region for the test.

This rule leads us to turn our questions back to front in many situations. Thus when we really wanted to know whether the proportion of housewives favouring the product in the above example was less than 0.20 we turned the question round, assumed that the proportion *was* 0.20 and tested the probability of this assumption.

Similarly, as we shall see shortly, if we wish to test whether machine-tool A produces fewer defective components than machine-tool B we set up the hypothesis that there is no difference between the two machines, i.e. that machine A has the same (known) rate of defectives as machine B. The formulation of hypotheses in terms of 'no difference', as we have already noted, is the reason why the hypothesis is called a 'null' hypothesis.

The size of α and β that can be tolerated in any given application depends to a large extent on the particular hypothesis being tested. If we were testing a new drug for its efficacy against a certain disease, for example, we would set up the null hypothesis that there was no difference between the new drug and the existing drug. If we were concerned for the public welfare we would want to be fairly certain that the new drug was more efficient, we would therefore specify a very small value for α, possibly 0·001. This, for a given size sample would probably result in a very large value of β. This we might be prepared to tolerate if we felt that the Type II error was less important than the Type I error. Of course, in this situation, as in all others, we want to make the *right* decision, in other words we want both α and β to be very small. The only way of achieving this, however, is to increase the size of the sample. We have, therefore, to weigh our desire for a correct decision against the time and money involved in assessing its correctness.

We now proceed to a second example; this one concerned with a mean rather than a proportion.

EXAMPLE 8. A car-hire company bases its hire rates and replacement policy on the assumption that the average annual mileage per vehicle is 10 000. A sample of 50 vehicles gives an average of 9350 miles with a variance of 1 742 400. Test the null hypothesis $H_0 : \mu = 10000$ against the alternative hypothesis $H_1: \mu \neq 10000$, with a Type I error of 0·05.

Fig. 8.7 illustrates the construction of the test.

Since $n = 50$ we can use the Normal approximation and estimate σ^2 from our sample.

We are testing $H_0 : \mu = 10,000$ against $H_1 : \mu \neq 10\,000$, i.e. we are equally concerned with values of average mileage which are too big or too small to be consistent with H_0. For this reason our critical region consists of both tails of the standardised distribution z, and since we require it to be not greater than 0·05, each tail includes 0·025 of the total distribution. From Table 7.6(a) we see that the required value of z is 1·96, a value which should be growing familiar to the reader.

Our observed value of z (remembering that the standard error of the sampling distribution is $\sqrt{(\sigma^2/n)}$) is:

$$z = \frac{9350 - 10\,000}{\sqrt{\{1\,742\,000/50\}}}$$

$$= \frac{-650}{\sqrt{34\,840}}$$

$$= -3\cdot48.$$

This value is smaller than the critical value $-1\cdot96$ so we reject the null hypothesis. We say, therefore, that the hire-car proprietor is unjustified in treating his cars as though they did average 10 000 miles per annum.

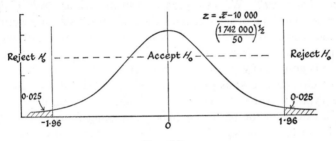

FIG. 8.7.

ONE-TAIL AND TWO-TAIL TESTS

In the first example we were interested in testing $H_0 : \pi = 0\cdot20$ against the alternative hypothesis $H_1 : \pi < 0\cdot20$. Our critical region was the left-hand 5% tail of the sampling distribution of p. In the second example we were interested in testing $H_0 : \mu = 10\,000$ against the alternative hypothesis $H_1 : \mu \neq 10\,000$. Our critical region consisted of both $2\frac{1}{2}$% tails of the sampling distribution.

These two different test situations are described, for obvious reasons, as *one-tail* and *two-tail* tests. In both examples the size of the critical region was 0·05 (5%) but in the first example this 5% was concentrated at one end of the distribution, in the second case the 5% was divided into two equal parts, each accounting for $2\frac{1}{2}$% of the distribution.

The question of whether to use a one-tail or a two-tail test can be answered very simply by looking at H_1. In our first example we had $H_1 : \pi < 0.20$. We were only interested in rejecting H_0 if the evidence suggested that the true value of π might be less than 0·20. We were not testing the hypothesis that π might be greater than 0·20; we would be only too delighted if it were. In our second example, however, our alternative hypothesis was $H_1 : \mu \neq 10\,000$, i.e. we were just as interested in rejecting H_0 if our sample mean $\bar{x}$ were too big as in rejecting it if it were too small.

Thus the question of a one-tail or two-tail test is settled as soon as we have framed our hypotheses H_0 and H_1.

STATISTICAL SIGNIFICANCE

Tests of hypotheses are often referred to as *significance tests* because in carrying out the tests we are comparing an observed value with the value under H_0 and asking ourselves: 'Is the difference between them *significant*?' We also speak of tests at various *levels of significance*. The level of significance is the same as the size of the critical region (and also of the Type I error). The level of significance is usually fixed in advance: the most commonly used level is 5%, but there is no reason why a higher or lower level should not be used in given circumstances; as we have already pointed out, the size of α, the Type I error, and therefore the significance level, depend to a large extent on the particular hypothesis being tested. The alternative to fixing the significance level in advance is to calculate the probability of obtaining the particular value which results from our sample. Thus, in our first example, at the 10% significance level, the critical value of z was -1.28, which was chosen because values less than this figure occurred only 10% of the time. Our observed value of z was -1.54 which from Table 7.6(a) we found to occur with a probability of 0·06. If we had not set the significance level in advance, we would have had to consider the value 0·06 and decide whether this was small enough to reject the null hypothesis, or large enough to accept it. So one advantage of setting the significance level in advance is that we only need to compare our observed value with a pre-determined value, and when we are using the standardised Normal distribution these values soon become familiar. A more important advantage, however, is that by setting a significance level in advance we are forced to consider how big a risk we are prepared to run of making a Type II error. Such considerations might lead to a change of plan, e.g. it might show clearly that a much larger sample than originally envisaged will be necessary.

This consideration of risk, and the resulting decision as to the level of significance to be adopted is a management decision, not a technical decision to be taken by the statistician. The manager must decide what risks he is prepared to run in reaching a decision. The statistician can help the manager by controlling and measuring those risks which arise because the evidence is based on a sample rather than on a complete enumeration. In doing so, he often performs the much more valuable service of drawing to the manager's attention problems, and their attendant risks, of which the manager was not previously aware, e.g. the problems involved in collecting valid data by means of a sample.

The fact that we set our significance level in advance does not preclude us from calculating the probability of obtaining our observed value as well. For example, we might accept a hypothesis at the 5% significance level. If we investigated further and found that the probability of our sample was only just greater than the significance level we might proceed with more caution than if we found the probability of the sample value to be much greater than the significance level. Of course, it will be said that if we select a significance level, we should abide by it. This is perfectly true but there is no harm in seeing how close we are to the critical level. It may be that although we accept our null hypothesis, we are so close to the critical value that we would like to make more observations and test again with a larger sample. There is after all an element of arbitrariness in the selection of the significance levels.

The most commonly used significance levels are the 5, 1 and 0·1% levels, in descending order. These levels are used so much that a convention has arisen whereby the results of tests are often presented with one, two, or three asterisks against them, indicating significance at the 5, 1 and 0·1% levels respectively. We will meet this notation in future examples.

Finally, in this section, we must emphasise as strongly as we can that when we talk of significance we mean *statistical significance*. A result is 'statistically significant' if it is unlikely to have arisen by chance. There is a strong tendency to confuse significance with importance. The reader should remember at all times the old aphorism: 'A difference is not a difference unless it makes a difference'. Thus, in Example 2 above, we rejected the hypothesis that our sample of cars came from a population with a mean annual mileage of 10 000 miles. Our observed value was 9350 miles, a difference of 650 miles. Now although this difference is statistically significant we cannot say, without knowing more about the costing system used by the company, whether it is a practical difference. All we can say is that we are fairly certain that 9350 is a better estimate of the current average annual mileage than 10 000.

SIGNIFICANCE TESTS AND CONFIDENCE LIMITS

It has probably occurred to the reader that we could have approached the problem of significance testing by fitting confidence limits to our sample statistics. Thus, in the car-hire example, instead of testing the significance of the difference between the observed value 9350 and the null hypothesis value 10 000 we could have fitted confidence limits to the value 9350. Let us do this now. We have:

$$n = 50, \quad \bar{x} = 9350, \quad s^2 = 1\ 742\ 000.$$

The standard error of $\bar{x}$ is therefore

$$s_{\bar{x}} = \sqrt{\frac{1\ 742\ 000}{50}}$$

$$= 187 \text{ approx.}$$

Therefore 95% confidence limits are given by

$$9350 \pm (1 \cdot 96)\ (187)$$

$$= 9350 \pm 367$$

$$= 8983 \text{ to } 9717.$$

Thus the probability is only 0·025 that the true value of μ is greater than 9717 and we can reject the hypothesis that $\mu = 10\ 000$.

TESTS INVOLVING TWO PROPORTIONS

We are often faced with the problem of deciding whether two sample proportions are significantly different. For example, two types of screw-cutting machine may be compared by running them for one shift, inspecting the entire output and calculating the proportion of defectives for each machine.

Suppose this operation provides us with the following data:

Machine A: Total output = 27 600; number defective = 1932;

Machine B: Total output = 25 400; number defective = 635.

We wish to test the hypothesis that there is no difference between the true rate of defectives for the two machines. We therefore take as our null hypothesis:

$$H_0 : \pi_A = \pi_B = \pi, \text{ say}$$

and for our alternative hypothesis:

$$H_1 : \pi_A \neq \pi_B.$$

Since we do not know the value of π (nor are we assuming a value under our null hypothesis) we estimate π from the two samples combined. We obtain (denoting our estimate by p):

$$p = \frac{1932 + 635}{27\,600 + 25\,400}$$

$$= 0 \cdot 048.$$

We also have

$$p_A = \frac{1932}{27\,600} = 0 \cdot 07.$$

and

$$p_B = \frac{635}{25\,400} = 0 \cdot 025.$$

The sampling distribution in which we are interested is the sampling distribution of the difference between sample proportions. This may seem a rather odd idea but it can easily be illustrated. If we have a population in which a proportion π has a certain characteristic, and if we draw *pairs* of samples (making sure that the samples are drawn independently of each other), we obtain two estimates of π; p_1 and p_2. If we record the value of $p_1 - p_2$ and continue sampling we obtain a distribution of values of $p_1 - p_2$. The mean of this distribution will be 0 and it can be shown that the variance is

$$\sigma^2_{p_1 - p_2} = \pi(1 - \pi)\left(\frac{1}{n_1} + \frac{1}{n_2}\right)$$

where n_1 and n_2 are the size of the two samples (note that the samples do not have to be the same size).

For large samples, the standardised sampling distribution of $p_1 - p_2$ is approximately Normal.

Now, we are testing $H_0 : \pi_A = \pi_B$ against $H_1 : \pi_A \neq \pi_B$; hence we are interested in a two-tail test. If we choose the 5% significance level for our test we calculate

$$z = \frac{p_A - p_B}{\sqrt{\left\{ p(1-p)\left(\dfrac{1}{n_1} + \dfrac{1}{n_2}\right)\right\}}}$$

$$= \frac{0 \cdot 07 - 0 \cdot 025}{\sqrt{\left\{ (0 \cdot 048)(0 \cdot 952)\left(\dfrac{1}{27\ 600} + \dfrac{1}{25\ 400}\right)\right\}}}$$

$$= \frac{0 \cdot 045}{0 \cdot 0019}$$

$$= 23 \cdot 68.$$

Since this is much bigger than the critical value 1·96, the difference between the machines is significant. Whether it is also important is another matter. Machine B, although producing only about one-third the proportion of defectives Machine A produces, nevertheless is a slower machine and still produces fewer usable screws per shift. On the other hand, it wastes less material. The decision on which machine to use therefore depends on a number of factors other than the percentage defective rate (we have not mentioned costs, expected life, power consumption, floor area required, etc.). We say again, statistical significance is not of itself enough to base a decision on.

TESTS INVOLVING TWO MEANS

The procedure is very similar to that of the preceding section: we consider the sampling distribution of the difference between two means obtained from two independent random samples. Expressing this formally:

If $\bar{x}_1$ and $\bar{x}_2$ are the means of two independent random samples of size n_1, and n_2, the sampling distribution of $\bar{x}_1 - \bar{x}_2$ is approximately Normal with $\mu = \mu_1 - \mu_2$ and variance

$$\sigma^2_{\bar{x}_1 - \bar{x}_2} = \frac{\sigma_1^2}{n_1} + \frac{\sigma_2^2}{n_2}.$$

(We do not attempt to prove these statements.)

We now encounter the problem met in testing single means. Unless σ_1^2 and σ_2^2 are known in advance, which hardly ever happens, we have to estimate them from our samples. The estimates are only valid for large samples, however; with small samples the Normal approximation is invalid and we have to use Student's -t distribution.

Large samples (*Normal approximation*)

EXAMPLE 9. Two beer-bottling machines are set to produce bottles containing half a pint (10 fl. oz) of beer. A random sample of 100 filled bottles is selected from the output of each machine. The samples yield the following data:

Machine 1: $n_1 = 100$; $\bar{x}_1 = 9\cdot9$ fl. oz; $s_1^2 = 6\cdot9$ fl. oz;

Machine 2: $n_2 = 100$; $\bar{x}_2 = 10\cdot7$ fl. oz; $s_2^2 = 8\cdot7$ fl. oz.

Can we conclude that Machine 2 puts more beer into the bottles than Machine 1?

We set up the null hypothesis $H_0 : \mu_1 = \mu_2 = \mu$, say, against the alternative hypothesis $H_1 : \mu_1 \neq \mu_2$ and compute:

$$z = \frac{\bar{x}_1 - \bar{x}_2}{\sqrt{\left(\dfrac{s_1^2}{n_1} + \dfrac{s_2^2}{n_2}\right)}} = \frac{-0\cdot80}{\sqrt{\left(\dfrac{6\cdot9}{100} + \dfrac{8\cdot7}{100}\right)}}$$

$$= -2\cdot03.$$

We have not yet mentioned a significance level. This, as we have said before, is bad practice but in this example we shall use Table 7.6(a) to see at what level the observed difference becomes significant. From Table 7.6(a) we see that the value $z = -2\cdot03$ or less includes $0\cdot0212$ of the distribution. Similarly $0\cdot0212$ of the distribution includes values greater than or equal to $z = 2\cdot03$. Hence the difference is significant at the $4\cdot2\%$ level. Using the asterisk notation mentioned earlier we would write

$$z = -2\cdot03*$$

implying that the difference was significant at the 5% level but not at the 1% level.

Small samples (*Student's-t distribution*). For small samples ($n \leq 20$ approx.) we cannot rely on our sample estimates of σ_1^2 and σ_2^2. We therefore use Student's -t distribution instead of the Normal approximation. The particular version of the t-test to be used depends on the variances of the two samples. There are two alternatives:

(*i*) σ_1^2 and σ_2^2 not known but believed to be equal

(*ii*) σ_1^2 and σ_2^2 not known but believed to be unequal.

We illustrate both methods and then introduce a test for the equality of the variances.

(*i*) If we can assume that $\sigma_1^2 = \sigma_2^2$, we base our test on the statistic:

$$t = \frac{\bar{x}_1 - \bar{x}_2}{s_p \sqrt{\left(\dfrac{1}{n_1} + \dfrac{1}{n_2}\right)}}$$

with $n_1 + n_2 - 2$ degrees of freedom where

$$s_p = \sqrt{\left\{\frac{(n_1 - 1)s_1^2 + (n_2 - 1)s_2^2}{n_1 + n_2 - 2}\right\}}.$$

Thus, s_p^2 is a weighted average of s_1^2 and s_2^2 with weights equal to the respective number of degrees of freedom. The '*p*' in s_p stands for 'pooled' and the expression is usually referred to as the pooled estimate of σ^2.

EXAMPLE 10. Two brands of canned fruit are sold in 1 lb 12 oz cans. In an investigation into the solid contents of the cans a random sample of 10 Brand A and 12 Brand B cans was selected, the liquid drained off and the remaining fruit weighed. The weights, in ounces, were as follows:

> *Brand A*: 16·5 20·7 19·4 21·4 21·6 18·0
> 18·3 17·2 21·3 20·1
>
> *Brand B*: 19·1 19·9 17·2 18·0 23·2 24·1
> 23·6 19·6 20·2 21·4 17·8 22·6

Do the two brands differ with respect to average fruit content?

We set up the null hypothesis $H_0 : \mu_1 = \mu_2$ against the alternative $H_1 : \mu_1 \neq \mu_2$ and proceed to calculate $\bar{x}$ and s^2 from the two samples. We have:

$$\text{Brand } A: \bar{x}_A = \frac{194\cdot5}{10} = 19\cdot45 \text{ oz, } s_A^2 = \frac{31\cdot03}{9} = 3\cdot45;$$

$$\text{Brand } B: \bar{x}_B = \frac{246\cdot7}{12} = 20\cdot56 \text{ oz, } s_B^2 = \frac{62\cdot69}{11} = 5\cdot70.$$

Assuming that the variances are not significantly different (we will shortly demonstrate that this is the case) we calculate s_p.

$$s_p = \sqrt{\left\{\frac{9(3\cdot45) + 11(5\cdot70)}{9 + 11 - 2}\right\}}$$

$$= 2\cdot28 \text{ oz.}$$

Hence

$$t = \frac{19 \cdot 45 - 20 \cdot 56}{2 \cdot 28 \sqrt{\left(\frac{1}{10} + \frac{1}{12}\right)}}$$

$$= \frac{-1 \cdot 11}{0 \cdot 97}$$

$$= -1 \cdot 14.$$

We decide to test at the 5% level of significance. From Table 8.1 we see that $t = 2 \cdot 09$ for 20 d.f. Therefore the difference is not significant at the 5% level. In other words the test provides no evidence of a real difference between the mean weights of solid fruit.

(ii) If we cannot assume that σ_1^2 and σ_2^2 are equal, an approximate test can be based on the statistic

$$t = \frac{\bar{x}_1 - \bar{x}_2}{\sqrt{\left(\frac{s_1^2}{n_1} + \frac{s_2^2}{n_2}\right)}}$$

with degrees of freedom given by the expression

$$\frac{1}{f} = \frac{1}{f_1}\left[\frac{s_1^2/n_1}{s_1^2/n_1 + s_2^2/n_2}\right]^2 + \frac{1}{f_2}\left[\frac{s_2^2/n_2}{s_1^2/n_1 + s_2^2/n_2}\right]^2$$

where f_1 and f_2 are the degrees of freedom associated with Sample 1 and Sample 2 respectively.

EXAMPLE 11. The manager of the sales order department of a particular firm was concerned at the amount of time spent by his clerks in telephoning. He also wished to know whether the average length of call differed between male and female clerks. A random sample of telephone calls yielded the following information:

Duration of call (minutes)

Men: 3·5 4·1 2·0 1·5 3·3 4·5 4·8 6·0 5·1 4.1
        4·3 3·6 3·9 4·4 2·7 6·7 4·4 5·3 3·0

Women: 2·0 2·7 3·0 1·7 1·0 0·8 2·9 3·6 2·2 2·2
        1·5 1·2 1·8 2·3 2·4 2·5 3·3

From the samples, the following statistics were computed:

Men: $\bar{x}_M = \dfrac{77 \cdot 2}{19} = 4 \cdot 06$, $s_M^2 = \dfrac{29 \cdot 28}{18} = 1 \cdot 63$, $n_M = 19$;

$Women$: $\bar{x}_W = \dfrac{37 \cdot 1}{17} = 2 \cdot 18$, $s_W^2 = \dfrac{10 \cdot 02}{16} = 0 \cdot 63$, $n_W = 17$.

Thus for men the duration of calls appears to be both longer, on average, and more variable. If we accept for now the fact that the two sample variances are significantly different we can test the significance of the difference between the two means as follows:

$$t = \frac{4 \cdot 06 - 2 \cdot 18}{\sqrt{\left(\dfrac{1 \cdot 63}{19} + \dfrac{0 \cdot 63}{17} \right)}}$$

$$= \frac{1 \cdot 88}{0 \cdot 35}$$

$$= 5 \cdot 37$$

with degrees of freedom given by:

$$\frac{1}{f} = \frac{1}{18} \left[\frac{1 \cdot 63/19}{\dfrac{1 \cdot 63}{19} + \dfrac{0 \cdot 63}{17}} \right]^2 + \frac{1}{16} \left[\frac{0 \cdot 63/17}{\dfrac{1 \cdot 63}{19} + \dfrac{0 \cdot 63}{17}} \right]^2$$

$$= \frac{1}{18} \left[\frac{0 \cdot 0858}{0 \cdot 1229} \right]^2 + \frac{1}{16} \left[\frac{0 \cdot 0371}{0 \cdot 1229} \right]^2.$$

Therefore

$$\frac{1}{f} = \frac{0 \cdot 487}{18} + \frac{0 \cdot 091}{16}$$

$$= 0 \cdot 027 + 0 \cdot 006.$$

Therefore

$$f = \frac{1}{0 \cdot 033}$$

$$= 30 \cdot 3$$

or 30 to the nearest whole number.

(Note that this value is slightly lower than would be the case if σ_M^2 and σ_W^2 were equal, in which case the number of degrees of freedom would be $n_1 + n_2 - 2$, i.e. 34.)

For $f = 30$, and applying a two-tailed test, the appropriate value of t from Table 8.1 is $t_{0.025} = 2.04$. The observed value of 5.37 is thus significant at the 5% level. Since $t_{0.005} = 2.75$ the observed difference is also significant at the 1% level. We can therefore conclude, with a high degree of confidence that, in this particular office, men spend longer on the telephone than women.

TESTING THE EQUALITY OF TWO VARIANCES

We now consider the question, postponed from the previous section, of how we test the significance of a difference between two sample variances. The test is very simple to apply although complicated in derivation. We shall not worry about its derivation, however.

If, from two random samples, we have s_1^2 with n_1 degrees of freedom and s_2^2 with n_2 degrees of freedom, our test statistic is the ratio of the larger value of s^2 to the smaller. This ratio is known as the *variance ratio* or the *F-ratio* and its sampling distribution, when the two population variances are equal, is known as the *F*-distribution. '*F*' stands for 'Fisher'; the distribution was named in his honour by its originator.

The *F* distribution depends upon the number of degrees of freedom in both numerator and denominator of the *F*-ratio. Upper percentage points of the distribution are tabulated in Table 8.4. If the observed ratio exceeds the value tabulated for the given degrees of freedom then the variances are significantly different at that significance level.

EXAMPLE 12. (*i*) In the first example of the preceding section the observed values of s^2 were:

$$s_A^2 = 3.45 \text{ with } 9 \text{ d.f.}$$
$$s_B^2 = 5.70 \text{ with } 11 \text{ d.f.}$$

To test the significance of the difference between these values, at the 5% level, we compute:

$$F = \frac{5.70}{3.45}$$
$$= 1.65.$$

From Table 8.4 we see that $F_{0.05}$ for 11 and 9 d.f. is 3.10; therefore the variances are not significantly different. This is what we assumed in the example anyway, and was the justification for applying the straightforward *t*-test to the difference between the two means.

TABLE 8.4. 5% (Top Line) and 1% (Bottom line) Points for the Distribution of F

f_1 degrees of freedom (for greater mean square)

f_2	1	2	3	4	5	6	7	8	9	10	11	12	14	16	20	24	30	40	50	75	100	200	500	∞	f_2
1	161 / 4052	200 / 4999	216 / 5403	225 / 5625	230 / 5764	234 / 5859	237 / 5928	239 / 5981	241 / 6022	242 / 6056	243 / 6082	244 / 6106	245 / 6142	246 / 6169	248 / 6208	249 / 6234	250 / 6258	251 / 6286	252 / 6302	253 / 6323	253 / 6334	254 / 6352	254 / 6361	254 / 6366	1
2	18·51 / 98·49	19·00 / 99·00	19·16 / 99·17	19·25 / 99·25	19·30 / 99·30	19·33 / 99·33	19·36 / 99·34	19·37 / 99·36	19·38 / 99·38	19·39 / 99·40	19·40 / 99·41	19·41 / 99·42	19·42 / 99·43	19·43 / 99·44	19·44 / 99·45	19·45 / 99·46	19·46 / 99·47	19·47 / 99·48	19·47 / 99·48	19·48 / 99·49	19·49 / 99·49	19·49 / 99·49	19·50 / 99·50	19·50 / 99·50	2
3	10·13 / 34·12	9·55 / 30·82	9·28 / 29·46	9·12 / 28·71	9·01 / 28·24	8·94 / 27·91	8·88 / 27·67	8·84 / 27·49	8·81 / 27·34	8·78 / 27·23	8·76 / 27·13	8·74 / 27·05	8·71 / 26·92	8·69 / 26·83	8·66 / 26·69	8·64 / 26·60	8·62 / 26·50	8·60 / 26·41	8·58 / 26·35	8·57 / 26·27	8·56 / 26·23	8·54 / 26·18	8·54 / 26·14	8·53 / 26·12	3
4	7·71 / 21·20	6·94 / 18·00	6·59 / 16·69	6·39 / 15·98	6·26 / 15·52	6·16 / 15·21	6·09 / 14·98	6·04 / 14·80	6·00 / 14·66	5·96 / 14·54	5·93 / 14·45	5·91 / 14·37	5·87 / 14·24	5·84 / 14·15	5·80 / 14·02	5·77 / 13·93	5·74 / 13·83	5·71 / 13·74	5·70 / 13·69	5·68 / 13·61	5·66 / 13·57	5·65 / 13·52	5·64 / 13·48	5·63 / 13·46	4
5	6·61 / 16·26	5·79 / 13·27	5·41 / 12·06	5·19 / 11·39	5·05 / 10·97	4·95 / 10·67	4·88 / 10·45	4·82 / 10·27	4·78 / 10·15	4·74 / 10·05	4·70 / 9·96	4·68 / 9·89	4·64 / 9·77	4·60 / 9·68	4·56 / 9·55	4·53 / 9·47	4·50 / 9·38	4·46 / 9·29	4·44 / 9·24	4·42 / 9·17	4·40 / 9·13	4·38 / 9·07	4·37 / 9·04	4·36 / 9·02	5
6	5·99 / 13·74	5·14 / 10·92	4·76 / 9·78	4·53 / 9·15	4·39 / 8·75	4·28 / 8·47	4·21 / 8·26	4·15 / 8·10	4·10 / 7·98	4·06 / 7·87	4·03 / 7·79	4·00 / 7·72	3·96 / 7·60	3·92 / 7·52	3·87 / 7·39	3·84 / 7·31	3·81 / 7·23	3·77 / 7·14	3·75 / 7·09	3·72 / 7·02	3·71 / 6·99	3·69 / 6·94	3·68 / 6·90	3·67 / 6·88	6
7	5·59 / 12·25	4·74 / 9·55	4·35 / 8·45	4·12 / 7·85	3·97 / 7·46	3·87 / 7·19	3·79 / 7·00	3·73 / 6·84	3·68 / 6·71	3·63 / 6·62	3·60 / 6·54	3·57 / 6·47	3·52 / 6·35	3·49 / 6·27	3·44 / 6·15	3·41 / 6·07	3·38 / 5·98	3·34 / 5·90	3·32 / 5·85	3·29 / 5·78	3·28 / 5·75	3·25 / 5·70	3·24 / 5·67	3·23 / 5·65	7
8	5·32 / 11·26	4·46 / 8·65	4·07 / 7·59	3·84 / 7·01	3·69 / 6·63	3·58 / 6·37	3·50 / 6·19	3·44 / 6·03	3·39 / 5·91	3·34 / 5·82	3·31 / 5·74	3·28 / 5·67	3·23 / 5·56	3·20 / 5·48	3·15 / 5·36	3·12 / 5·28	3·08 / 5·20	3·05 / 5·11	3·03 / 5·06	3·00 / 5·00	2·98 / 4·96	2·96 / 4·91	2·94 / 4·88	2·93 / 4·86	8
9	5·12 / 10·56	4·26 / 8·02	3·86 / 6·99	3·63 / 6·42	3·48 / 6·06	3·37 / 5·80	3·29 / 5·62	3·23 / 5·47	3·18 / 5·35	3·13 / 5·26	3·10 / 5·18	3·07 / 5·11	3·02 / 5·00	2·98 / 4·92	2·93 / 4·80	2·90 / 4·73	2·86 / 4·64	2·82 / 4·56	2·80 / 4·51	2·77 / 4·45	2·76 / 4·41	2·73 / 4·36	2·72 / 4·33	2·71 / 4·31	9
10	4·96 / 10·04	4·10 / 7·56	3·71 / 6·55	3·48 / 5·99	3·33 / 5·64	3·22 / 5·39	3·14 / 5·21	3·07 / 5·06	3·02 / 4·95	2·97 / 4·85	2·94 / 4·78	2·91 / 4·71	2·86 / 4·60	2·82 / 4·52	2·77 / 4·41	2·74 / 4·33	2·70 / 4·25	2·67 / 4·17	2·64 / 4·12	2·61 / 4·05	2·59 / 4·01	2·56 / 3·96	2·55 / 3·93	2·54 / 3·91	10
11	4·84 / 9·65	3·98 / 7·20	3·59 / 6·22	3·36 / 5·67	3·20 / 5·32	3·09 / 5·07	3·01 / 4·88	2·95 / 4·74	2·90 / 4·63	2·86 / 4·54	2·82 / 4·46	2·79 / 4·40	2·74 / 4·29	2·70 / 4·21	2·65 / 4·10	2·61 / 4·02	2·57 / 3·94	2·53 / 3·86	2·50 / 3·80	2·47 / 3·74	2·45 / 3·70	2·42 / 3·66	2·41 / 3·62	2·40 / 3·60	11
12	4·75 / 9·33	3·88 / 6·93	3·49 / 5·95	3·26 / 5·41	3·11 / 5·06	3·00 / 4·82	2·92 / 4·65	2·85 / 4·50	2·80 / 4·39	2·76 / 4·30	2·72 / 4·22	2·69 / 4·16	2·64 / 4·05	2·60 / 3·98	2·54 / 3·86	2·50 / 3·78	2·46 / 3·70	2·42 / 3·61	2·40 / 3·56	2·36 / 3·49	2·35 / 3·46	2·32 / 3·41	2·31 / 3·38	2·30 / 3·36	12
13	4·67 / 9·07	3·80 / 6·70	3·41 / 5·74	3·18 / 5·20	3·02 / 4·86	2·92 / 4·62	2·84 / 4·44	2·77 / 4·30	2·72 / 4·19	2·67 / 4·10	2·63 / 4·02	2·60 / 3·96	2·55 / 3·85	2·51 / 3·78	2·46 / 3·67	2·42 / 3·59	2·38 / 3·51	2·34 / 3·42	2·32 / 3·37	2·28 / 3·30	2·26 / 3·27	2·24 / 3·21	2·22 / 3·18	2·21 / 3·16	13

TABLE 8.4 (continued)

$f1$ degrees of freedom (for greater mean square)

$f2$	1	2	3	4	5	6	7	8	9	10	11	12	14	16	20	24	30	40	50	75	100	200	500	∞	$f2$
14	4·60 / 8·86	3·74 / 6·51	3·34 / 5·56	3·11 / 5·03	2·96 / 4·69	2·85 / 4·46	2·77 / 4·28	2·70 / 4·14	2·65 / 4·03	2·60 / 3·94	2·56 / 3·86	2·53 / 3·80	2·48 / 3·70	2·44 / 3·62	2·39 / 3·51	2·35 / 3·43	2·31 / 3·34	2·27 / 3·26	2·24 / 3·21	2·21 / 3·14	2·19 / 3·11	2·16 / 3·06	2·14 / 3·02	2·13 / 3·00	14
15	4·54 / 8·68	3·68 / 6·36	3·29 / 5·42	3·06 / 4·89	2·90 / 4·56	2·79 / 4·32	2·70 / 4·14	2·64 / 4·00	2·59 / 3·89	2·55 / 3·80	2·51 / 3·73	2·48 / 3·67	2·43 / 3·56	2·30 / 3·48	2·33 / 3·36	2·29 / 3·29	2·25 / 3·20	2·21 / 3·12	2·18 / 3·07	2·15 / 3·00	2·12 / 2·97	2·10 / 2·92	2·08 / 2·89	2·07 / 2·87	15
16	4·49 / 8·53	3·63 / 6·23	3·24 / 5·29	3·01 / 4·77	2·85 / 4·44	2·74 / 4·20	2·66 / 4·03	2·59 / 3·89	2·54 / 3·78	2·49 / 3·69	2·45 / 3·61	2·42 / 3·55	2·37 / 3·45	2·33 / 3·37	2·28 / 3·25	2·24 / 3·18	2·20 / 3·10	2·16 / 3·01	2·13 / 2·96	2·09 / 2·89	2·07 / 2·86	2·04 / 2·80	2·02 / 2·77	2·01 / 2·75	16
17	4·45 / 8·40	3·59 / 6·11	3·20 / 5·18	2·96 / 4·67	2·81 / 4·34	2·70 / 4·10	2·62 / 3·93	2·55 / 3·79	2·50 / 3·68	2·45 / 3·59	2·41 / 3·52	2·38 / 3·45	2·33 / 3·35	2·29 / 3·27	2·23 / 3·16	2·19 / 3·08	2·15 / 3·00	2·11 / 2·92	2·08 / 2·86	2·04 / 2·79	2·02 / 2·76	1·99 / 2·70	1·97 / 2·67	1·96 / 2·65	17
18	4·41 / 8·28	3·55 / 6·01	3·16 / 5·09	2·93 / 4·58	2·77 / 4·25	2·66 / 4·01	2·58 / 3·85	2·51 / 3·71	2·46 / 3·60	2·41 / 3·51	2·37 / 3·44	2·34 / 3·37	2·29 / 3·27	2·25 / 3·19	2·19 / 3·07	2·15 / 3·00	2·11 / 2·91	2·07 / 2·83	2·04 / 2·78	2·00 / 2·71	1·98 / 2·68	1·95 / 2·62	1·93 / 2·59	1·92 / 2·57	18
19	4·38 / 8·18	3·52 / 5·93	3·13 / 5·01	2·90 / 4·50	2·74 / 4·17	2·63 / 3·94	2·55 / 3·77	2·48 / 3·63	2·43 / 3·52	2·38 / 3·43	2·34 / 3·36	2·31 / 3·30	2·26 / 3·19	2·21 / 3·12	2·15 / 3·00	2·11 / 2·92	2·07 / 2·84	2·02 / 2·76	2·00 / 2·70	1·96 / 2·63	1·94 / 2·60	1·91 / 2·54	1·90 / 2·51	1·88 / 2·49	19
20	4·35 / 8·10	3·49 / 5·85	3·10 / 4·94	2·87 / 4·43	2·71 / 4·10	2·60 / 3·87	2·52 / 3·71	2·45 / 3·56	2·40 / 3·45	2·35 / 3·37	2·31 / 3·30	2·28 / 3·23	2·23 / 3·13	2·18 / 3·05	2·12 / 2·94	2·08 / 2·86	2·04 / 2·77	1·99 / 2·69	1·96 / 2·63	1·92 / 2·56	1·90 / 2·53	1·87 / 2·47	1·85 / 2·44	1·84 / 2·42	20
21	4·32 / 8·02	3·47 / 5·78	3·07 / 4·87	2·84 / 4·37	2·68 / 4·04	2·57 / 3·81	2·49 / 3·65	2·42 / 3·51	2·37 / 3·40	2·32 / 3·31	2·28 / 3·24	2·25 / 3·17	2·20 / 3·07	2·15 / 2·99	2·09 / 2·88	2·05 / 2·80	2·00 / 2·72	1·96 / 2·63	1·93 / 2·58	1·89 / 2·51	1·87 / 2·47	1·84 / 2·42	1·82 / 2·38	1·81 / 2·36	21
22	4·30 / 7·94	3·44 / 5·72	3·05 / 4·82	2·82 / 4·31	2·66 / 3·99	2·55 / 3·76	2·47 / 3·59	2·40 / 3·45	2·35 / 3·35	2·30 / 3·26	2·26 / 3·18	2·23 / 3·12	2·18 / 3·02	2·13 / 2·94	2·07 / 2·83	2·03 / 2·75	1·98 / 2·67	1·93 / 2·58	1·91 / 2·53	1·87 / 2·46	1·84 / 2·42	1·81 / 2·37	1·80 / 2·33	1·78 / 2·31	22
23	4·28 / 7·88	3·42 / 5·66	3·03 / 4·76	2·80 / 4·26	2·64 / 3·94	2·53 / 3·71	2·45 / 3·54	2·38 / 3·41	2·32 / 3·30	2·28 / 3·21	2·24 / 3·14	2·20 / 3·07	2·14 / 2·97	2·10 / 2·89	2·04 / 2·78	2·00 / 2·70	1·96 / 2·62	1·91 / 2·53	1·88 / 2·48	1·84 / 2·41	1·82 / 2·37	1·79 / 2·32	1·77 / 2·28	1·76 / 2·26	23
24	4·26 / 7·82	3·40 / 5·61	3·01 / 4·72	2·78 / 4·22	2·62 / 3·90	2·51 / 3·67	2·43 / 3·50	2·36 / 3·36	2·30 / 3·25	2·26 / 3·17	2·22 / 3·09	2·18 / 3·03	2·13 / 2·93	2·09 / 2·85	2·02 / 2·74	1·98 / 2·66	1·94 / 2·58	1·89 / 2·49	1·86 / 2·44	1·82 / 2·36	1·80 / 2·33	1·76 / 2·27	1·74 / 2·23	1·73 / 2·21	24
25	4·24 / 7·77	3·38 / 5·57	2·99 / 4·68	2·76 / 4·18	2·60 / 3·86	2·49 / 3·63	2·41 / 3·46	2·34 / 3·32	2·28 / 3·21	2·24 / 3·13	2·20 / 3·05	2·16 / 2·99	2·11 / 2·89	2·06 / 2·81	2·00 / 2·70	1·96 / 2·62	1·92 / 2·54	1·87 / 2·45	1·84 / 2·40	1·80 / 2·32	1·77 / 2·29	1·74 / 2·23	1·72 / 2·19	1·71 / 2·17	25
26	4·22 / 7·72	3·37 / 5·53	2·98 / 4·64	2·74 / 4·14	2·59 / 3·82	2·47 / 3·59	2·39 / 3·42	2·32 / 3·29	2·27 / 3·17	2·22 / 3·09	2·18 / 3·02	2·15 / 2·96	2·10 / 2·86	2·05 / 2·77	1·99 / 2·66	1·95 / 2·58	1·90 / 2·50	1·85 / 2·41	1·82 / 2·36	1·78 / 2·28	1·76 / 2·25	1·72 / 2·19	1·70 / 2·15	1·69 / 2·13	26

TABLE 8.4 (continued)

f_1 degrees of freedom (for greater mean square)

f_2	1	2	3	4	5	6	7	8	9	10	11	12	14	16	20	24	30	40	50	75	100	200	500	∞	f_2
27	4·21 7·68	3·35 5·49	2·96 4·60	2·73 4·11	2·57 3·79	2·46 3·56	2·37 3·39	2·30 3·26	2·25 3·14	2·20 3·06	2·16 2·98	2·13 2·93	2·08 2·83	2·03 2·74	1·97 2·63	1·93 2·55	1·88 2·47	1·84 2·38	1·80 2·33	1·76 2·25	1·74 2·21	1·71 2·16	1·68 2·12	1·67 2·10	27
28	4·20 7·64	3·34 5·45	2·95 4·57	2·71 4·07	2·56 3·76	2·44 3·53	2·36 3·36	2·29 3·23	2·24 3·11	2·19 3·03	2·15 2·95	2·12 2·90	2·06 2·80	2·02 2·71	1·96 2·60	1·91 2·52	1·87 2·44	1·81 2·35	1·78 2·30	1·75 2·22	1·72 2·18	1·69 2·13	1·67 2·09	1·65 2·06	28
29	4·18 7·60	3·33 5·42	2·93 4·54	2·70 4·04	2·54 3·73	2·43 3·50	2·35 3·33	2·28 3·20	2·22 3·08	2·18 3·00	2·14 2·92	2·10 2·87	2·05 2·77	2·00 2·68	1·94 2·57	1·90 2·49	1·85 2·41	1·80 2·32	1·77 2·27	1·73 2·19	1·71 2·15	1·68 2·10	1·65 2·06	1·64 2·03	29
30	4·17 7·56	3·32 5·39	2·92 4·51	2·69 4·02	2·53 3·70	2·42 3·47	2·34 3·30	2·27 3·17	2·21 3·06	2·16 2·98	2·12 2·90	2·09 2·84	2·04 2·74	1·99 2·66	1·93 2·55	1·89 2·47	1·84 2·38	1·79 2·29	1·76 2·24	1·72 2·16	1·69 2·13	1·66 2·07	1·64 2·03	1·62 2·01	30
32	4·15 7·50	3·30 5·34	2·90 4·46	2·67 3·97	2·51 3·66	2·40 3·42	2·32 3·25	2·25 3·12	2·19 3·01	2·14 2·94	2·10 2·86	2·07 2·80	2·02 2·70	1·97 2·62	1·91 2·51	1·86 2·42	1·82 2·34	1·76 2·25	1·74 2·20	1·69 2·12	1·67 2·08	1·64 2·02	1·61 1·98	1·59 1·96	32
34	4·13 7·44	3·28 5·29	2·88 4·42	2·65 3·93	2·49 3·61	2·38 3·38	2·30 3·21	2·23 3·08	2·17 2·97	2·12 2·89	2·08 2·82	2·05 2·76	2·00 2·66	1·95 2·58	1·89 2·47	1·84 2·38	1·80 2·30	1·74 2·21	1·71 2·15	1·67 2·08	1·64 2·04	1·61 1·98	1·59 1·94	1·57 1·91	34
36	4·11 7·39	3·26 5·25	2·86 4·38	2·63 3·89	2·48 3·58	2·36 3·35	2·28 3·18	2·21 3·04	2·15 2·94	2·10 2·86	2·06 2·78	2·03 2·72	1·98 2·62	1·93 2·54	1·87 2·43	1·82 2·35	1·78 2·26	1·72 2·17	1·69 2·12	1·65 2·04	1·62 2·00	1·59 1·94	1·56 1·90	1·55 1·87	36
38	4·10 7·35	3·25 5·21	2·85 4·34	2·62 3·86	2·46 3·54	2·35 3·32	2·26 3·15	2·19 3·02	2·14 2·91	2·09 2·82	2·05 2·75	2·02 2·69	1·96 2·59	1·92 2·51	1·85 2·40	1·80 2·32	1·76 2·22	1·71 2·14	1·67 2·08	1·63 2·00	1·60 1·97	1·57 1·90	1·54 1·86	1·53 1·84	38
40	4·08 7·31	3·23 5·18	2·84 4·31	2·61 3·83	2·45 3·51	2·34 3·29	2·25 3·12	2·18 2·99	2·12 2·88	2·07 2·80	2·04 2·73	2·00 2·66	1·95 2·56	1·90 2·49	1·84 2·37	1·79 2·29	1·74 2·20	1·69 2·11	1·66 2·05	1·61 1·97	1·59 1·94	1·55 1·88	1·53 1·84	1·51 1·81	40
42	4·07 7·27	3·22 5·15	2·83 4·29	2·59 3·80	2·44 3·49	2·32 3·26	2·24 3·10	2·17 2·96	2·11 2·86	2·06 2·77	2·02 2·70	1·99 2·64	1·94 2·54	1·89 2·46	1·82 2·35	1·78 2·26	1·73 2·17	1·68 2·08	1·64 2·02	1·60 1·94	1·57 1·91	1·54 1·85	1·51 1·80	1·49 1·78	42
44	4·06 7·24	3·21 5·12	2·82 4·26	2·58 3·78	2·43 3·46	2·31 3·24	2·23 3·07	2·16 2·94	2·10 2·84	2·05 2·75	2·01 2·68	1·98 2·62	1·92 2·52	1·88 2·44	1·81 2·32	1·76 2·24	1·72 2·15	1·66 2·06	1·63 2·00	1·58 1·92	1·56 1·88	1·52 1·82	1·50 1·78	1·48 1·75	44
46	4·05 7·21	3·20 5·10	2·81 4·24	2·57 3·76	2·42 3·44	2·30 3·22	2·22 3·05	2·14 2·92	2·09 2·82	2·04 2·73	2·00 2·66	1·97 2·60	1·91 2·50	1·87 2·42	1·80 2·30	1·75 2·22	1·71 2·13	1·65 2·04	1·62 1·98	1·57 1·90	1·54 1·86	1·51 1·80	1·48 1·76	1·46 1·72	46
48	4·04 7·19	3·19 5·08	2·80 4·22	2·56 3·74	2·41 3·42	2·30 3·20	2·21 3·04	2·14 2·90	2·08 2·80	2·03 2·71	1·99 2·64	1·96 2·58	1·90 2·48	1·86 2·40	1·79 2·28	1·74 2·20	1·70 2·11	1·64 2·02	1·61 1·96	1·56 1·88	1·53 1·84	1·50 1·78	1·47 1·73	1·45 1·70	48

TABLE 8.4 (continued)

I

f^1 degrees of freedom (for greater mean square)

f^2	1	2	3	4	5	6	7	8	9	10	11	12	14	16	20	24	30	40	50	75	100	200	500	∞	f^2
50	4·03 / 7·17	3·18 / 5·06	2·79 / 4·20	2·56 / 3·72	2·40 / 3·41	2·29 / 3·18	2·20 / 3·02	2·13 / 2·88	2·07 / 2·78	2·02 / 2·70	1·98 / 2·62	1·95 / 2·56	1·90 / 2·46	1·85 / 2·39	1·78 / 2·26	1·74 / 2·18	1·60 / 2·10	1·63 / 2·00	1·60 / 1·94	1·55 / 1·86	1·52 / 1·82	1·48 / 1·76	1·46 / 1·71	1·44 / 1·68	50
55	4·02 / 7·12	3·17 / 5·01	2·78 / 4·16	2·54 / 3·68	2·38 / 3·37	2·27 / 3·15	2·18 / 2·98	2·11 / 2·85	2·05 / 2·75	2·00 / 2·66	1·97 / 2·59	1·93 / 2·53	1·88 / 2·43	1·83 / 2·35	1·76 / 2·23	1·72 / 2·15	1·67 / 2·06	1·61 / 1·96	1·58 / 1·90	1·52 / 1·82	1·50 / 1·78	1·46 / 1·71	1·43 / 1·66	1·41 / 1·64	55
60	4·00 / 7·08	3·15 / 4·98	2·76 / 4·13	2·52 / 3·65	2·37 / 3·34	2·25 / 3·12	2·17 / 2·95	2·10 / 2·82	2·04 / 2·72	1·99 / 2·63	1·95 / 2·56	1·92 / 2·50	1·86 / 2·40	1·81 / 2·32	1·75 / 2·20	1·70 / 2·12	1·65 / 2·03	1·59 / 1·93	1·56 / 1·87	1·50 / 1·79	1·48 / 1·74	1·44 / 1·68	1·41 / 1·63	1·39 / 1·60	60
65	3·90 / 7·04	3·14 / 4·95	2·75 / 4·10	2·51 / 3·62	2·36 / 3·31	2·24 / 3·09	2·15 / 2·93	2·08 / 2·79	2·02 / 2·70	1·98 / 2·61	1·94 / 2·54	1·90 / 2·47	1·85 / 2·37	1·80 / 2·30	1·73 / 2·18	1·68 / 2·09	1·63 / 2·00	1·57 / 1·90	1·54 / 1·84	1·49 / 1·76	1·46 / 1·71	1·42 / 1·64	1·39 / 1·60	1·37 / 1·56	65
70	3·98 / 7·01	3·13 / 4·92	2·74 / 4·08	2·50 / 3·60	2·35 / 3·29	2·23 / 3·07	2·14 / 2·91	2·07 / 2·77	2·01 / 2·67	1·97 / 2·59	1·93 / 2·51	1·89 / 2·45	1·84 / 2·35	1·79 / 2·28	1·72 / 2·15	1·67 / 2·07	1·62 / 1·98	1·56 / 1·88	1·53 / 1·82	1·47 / 1·74	1·45 / 1·69	1·40 / 1·62	1·37 / 1·56	1·35 / 1·53	70
80	3·96 / 6·96	3·11 / 4·88	2·72 / 4·04	2·48 / 3·56	2·33 / 3·25	2·21 / 3·04	2·12 / 2·87	2·05 / 2·74	1·99 / 2·64	1·95 / 2·55	1·91 / 2·48	1·88 / 2·41	1·82 / 2·32	1·77 / 2·24	1·70 / 2·11	1·65 / 2·03	1·60 / 1·94	1·54 / 1·84	1·51 / 1·78	1·45 / 1·70	1·42 / 1·65	1·38 / 1·57	1·35 / 1·52	1·32 / 1·49	80
100	3·94 / 6·90	3·00 / 4·82	2·70 / 3·98	2·46 / 3·51	2·30 / 3·20	2·19 / 2·99	2·10 / 2·82	2·03 / 2·69	1·97 / 2·59	1·92 / 2·51	1·88 / 2·43	1·85 / 2·36	1·79 / 2·26	1·75 / 2·19	1·68 / 2·06	1·63 / 1·98	1·57 / 1·89	1·51 / 1·79	1·48 / 1·73	1·42 / 1·64	1·39 / 1·59	1·34 / 1·51	1·30 / 1·46	1·28 / 1·43	100
125	3·92 / 6·84	3·07 / 4·78	2·68 / 3·94	2·44 / 3·47	2·29 / 3·17	2·17 / 2·95	2·08 / 2·79	2·01 / 2·65	1·95 / 2·56	1·90 / 2·47	1·86 / 2·40	1·83 / 2·33	1·77 / 2·23	1·72 / 2·15	1·65 / 2·03	1·60 / 1·94	1·55 / 1·85	1·49 / 1·75	1·45 / 1·68	1·39 / 1·59	1·36 / 1·54	1·31 / 1·46	1·27 / 1·40	1·25 / 1·37	125
150	3·91 / 6·81	3·06 / 4·75	2·67 / 3·91	2·43 / 3·44	2·27 / 3·14	2·16 / 2·92	2·07 / 2·76	2·00 / 2·62	1·94 / 2·53	1·89 / 2·44	1·85 / 2·37	1·82 / 2·30	1·76 / 2·20	1·71 / 2·12	1·64 / 2·00	1·59 / 1·91	1·54 / 1·83	1·47 / 1·72	1·44 / 1·66	1·37 / 1·56	1·34 / 1·51	1·29 / 1·43	1·25 / 1·37	1·22 / 1·33	150
200	3·89 / 6·76	3·04 / 4·71	2·65 / 3·88	2·41 / 3·41	2·26 / 3·11	2·14 / 2·90	2·05 / 2·73	1·98 / 2·60	1·92 / 2·50	1·87 / 2·41	1·83 / 2·34	1·80 / 2·28	1·74 / 2·17	1·69 / 2·09	1·62 / 1·97	1·57 / 1·88	1·52 / 1·79	1·45 / 1·69	1·42 / 1·62	1·35 / 1·53	1·32 / 1·48	1·26 / 1·39	1·22 / 1·33	1·19 / 1·28	200
400	3·86 / 6·70	3·02 / 4·66	2·62 / 3·83	2·38 / 3·36	2·23 / 3·06	2·12 / 2·85	2·03 / 2·69	1·96 / 2·55	1·90 / 2·46	1·85 / 2·37	1·81 / 2·29	1·78 / 2·23	1·72 / 2·12	1·67 / 2·04	1·60 / 1·92	1·54 / 1·84	1·49 / 1·74	1·42 / 1·64	1·38 / 1·57	1·32 / 1·47	1·28 / 1·42	1·22 / 1·32	1·16 / 1·24	1·13 / 1·19	400
1000	3·85 / 6·66	3·00 / 4·62	2·61 / 3·80	2·38 / 3·34	2·22 / 3·04	2·10 / 2·82	2·02 / 2·66	1·95 / 2·53	1·89 / 2·43	1·84 / 2·34	1·80 / 2·26	1·76 / 2·20	1·70 / 2·09	1·65 / 2·01	1·58 / 1·89	1·53 / 1·81	1·47 / 1·71	1·41 / 1·61	1·36 / 1·54	1·30 / 1·44	1·26 / 1·38	1·19 / 1·28	1·13 / 1·19	1·08 / 1·11	1000
∞	3·84 / 6·64	2·99 / 4·60	2·00 / 3·78	2·37 / 3·32	2·21 / 3·02	2·09 / 2·80	2·01 / 2·64	1·94 / 2·51	1·88 / 2·41	1·83 / 2·32	1·79 / 2·24	1·75 / 2·18	1·69 / 2·07	1·64 / 1·99	1·57 / 1·87	1·52 / 1·79	1·46 / 1·69	1·40 / 1·59	1·35 / 1·52	1·28 / 1·41	1·24 / 1·36	1·17 / 1·25	1·11 / 1·15	1·00 / 1·00	∞

(*ii*) In the second example of the preceding section the observed values of s^2 were:

$$s_M^2 = 1\cdot63 \text{ with 18 d.f.}$$

$$s_W^2 = 0\cdot63 \text{ with 16 d.f.}$$

Therefore, testing at the 5% level we have:

$$F = \frac{1\cdot63}{0\cdot63}$$

$$= 2\cdot59.$$

From Table 8.4, $F_{0\cdot05}$ for 18 and 16 d.f. lies between the values 2·33 and 2·28. (Values for 18 d.f. in the numerator are not tabulated so we look at the values for 16 and 20 d.f.) Hence the observed value of F is significant at the 5% level and we reject the hypothesis that the two samples came from populations with the same variance. This means that to test the significance of the difference between the sample means we use the approximate method involving the complicated expression for the number of degrees of freedom.

TESTS INVOLVING SEVERAL PROPORTIONS

We have seen how to test the significance of a sample proportion against a hypothesised value, and how to test the significance of an observed difference between two sample proportions. We now consider how we can test the significance of observed differences between more than two sample proportions. For example, suppose 200 motorists are asked about their preferences between two specified brands of petrol. If at the same time they were also asked into which of three broad age groups they fell, we could analyse brand preferences by age groups. Let us suppose that Table 8.5 was constructed from the sample data.

The proportions preferring Brand A are:

$$17\text{–}34: \qquad 27/41 = 0\cdot66$$

$$35\text{–}44: \qquad 37/75 = 0\cdot49$$

$$45 \text{ and over: } 25/84 = 0\cdot30$$

with an overall proportion of $89/200 = 0\cdot44$.

The hypothesis we wish to test here is $H_0 : \pi_1 = \pi_2 = \pi_3 = \pi$, against the alternative hypothesis that the three π's are different from each other. Assuming that π is unknown we estimate it by p, the overall sample proportion. We have $p = 0.44$. Thus we can calculate for each age group how many motorists we could have expected in each category if the null hypothesis were true. For example, if $\pi = 0.44$ for all three age groups the expected number of preferences for Brand A among the 17–34 age group is 18, i.e. $(0.44)(41)$. This figure is entered in the table beneath the observed value of 27. (The expected value is entered in brackets.) Expected values for the other cell frequencies are calculated in a similar fashion and entered in the table.

TABLE 8.5

Brand of petrol	Age group			Total
	17–34	35–44	45 and over	
Brand A	27 (18)	37 (34)	25 (37)	89
Brand B	14 (23)	38 (41)	59 (47)	111
Total	41	75	84	200

We now compare observed and expected frequencies. If there is little difference between them, we accept H_0. If the differences are large we reject H_0. Our test statistic is

$$\chi^2 = \Sigma \frac{(O-E)^2}{E}$$

where
$$O = \text{observed frequency}$$
$$E = \text{expected frequency}$$

and the summation is over the six cells of the table.

χ^2 is the same symbol that we met in discussing confidence limits for σ^2 in an earlier section of this chapter. Under the null hypothesis the sampling distribution of our test statistic χ^2 approximates to the theoretical distribution known as the χ^2 distribution. This is a skewed, continuous distribution which, like Student's $-t$ distribution, depends on the number of degrees of freedom. Values of the distribution are tabulated in Table 8.3.

The value of χ^2 obtained from our sample is 15·46. The number of degrees of freedom is given by the expression $(r-1)(c-1)$ where r = number of rows and c = number of columns in the table. Hence in this example we have $1 \times 2 = 2$ d.f. If we test the null hypothesis at the 5% significance level we find from Table 8.3 that χ^2 for 2 d.f. = 5·99. Since our observed value of χ^2 is much higher than this value we reject the hypothesis that the three population proportions are equal.

Tables such as Table 8.5 are usually referred to as *contingency tables*. In constructing contingency tables the important things to remember are:

(*a*) The sum of the expected frequencies should equal the sum of the observed frequencies, for all row and column totals. (In calculating the expected frequencies it is customary to round to the nearest integer, when the numbers are very large, or to 1 decimal place.)

(*b*) The expected frequency for any cell should not be less than 5. (This is just a rule of thumb, the actual minimum for any one cell depends on other factors as well, but 5 is a safe guide.)

(*c*) The degrees of freedom are calculated from the expression $(r-1)(c-1)$. This can be explained fairly easily. In any one column, there are r observations which must add to the column total. Hence any column has $(r-1)$ degrees of freedom. Similarly in every row there are c observations which must add up to the row total. Hence every row has $(c-1)$ degrees of freedom. The total number of degrees of freedom is, therefore, $(r-1)(c-1)$.

The table in our example is a 2×3 table. In general a $2 \times k$ table can be used for testing differences between k sample proportions. A 2×2 table, therefore, provides us with an alternative method of testing the significance of the difference between two sample proportions.

EXAMPLE 13. We return to the example of two bottling machines (p. 235).

The data can be set out as follows:

	Machine A	Machine B	Total
Number defective	1 932	635	2 567
	(1 337)	(1 230)	
Number not defective	25 668	24 765	50 433
	(26 263)	(24 170)	
Total	27 600	25 400	53 000

The overall proportion defective is 0·048. Using this value to calculate the expected frequencies we complete the table as above. We find $\chi^2 = 555·43$ with 1 d.f. From Table 8·3 we find that $\chi^2_{0·05}$ for 1 d.f. is 3·84. Hence we reject the hypothesis that the two sample proportions are equal.

We can extend the use of the contingency table by considering an $r \times k$ table instead of the $2 \times k$. This removes the restriction of considering only two alternatives in our samples, e.g. 'Defective', 'Not defective'; 'In Favour', 'Against'. The arithmetic of the $r \times k$ table proceeds exactly as before, and the degrees of freedom are again obtained from the expression $(r-1)(c-1)$.

EXAMPLE 14. A sample survey among business men asked, among other things, for opinions on prospects in the next six months. The answers obtained from members of the Electrical Engineering, Construction, Food and Drink and Chemicals industries were compared. Does the evidence suggest that some industries are more optimistic than others?

The data is set out in the table below:

Prospects in next 6 months	Industry				
	Electrical engineering	Construction	Food and drink	Chemicals	Total
Better	15	8	15	8	46
	(13)	(9)	(18)	(6)	
Same	25	13	25	9	72
	(21)	(14)	(28)	(9)	
Worse	5	9	20	3	37
	(11)	(7)	(14)	(5)	
Total	45	30	60	20	155

The expected frequencies are obtained from the overall proportions $(46/155, 72/155, 37/155)$ applied to the column totals, e.g. $\dfrac{46}{155} \times 45$ gives 13 in the 'Electrical Engineering/Better' cell.

From the expression

$$\chi^2 = \Sigma \frac{(O-E)^2}{E}$$

we obtain a value of $\chi^2 = 9\cdot96$.

We have $(3-1)(4-1) = 6$ d.f.

From Table 8.3 we find that $\chi^2_{.05}$ for 6 d.f. $= 12\cdot6$, hence the data provides no evidence that opinions differ between different industry groups.

The χ^2 test is a very useful tool for testing the association between two sets of attributes but it is a fairly blunt instrument. It tests the contingency table as a whole; we are not able to point to any one cell or group of cells in the table and say 'This is where the significance arises'.

TESTING GOODNESS OF FIT

The χ^2 test is also very important as a means of testing whether a particular set of data conform to a certain theoretical distribution. The procedure is to fit the theoretical distribution to the data and compute $\chi^2 = \Sigma \dfrac{(O-E)^2}{E}$ as before. The theoretical frequencies are the expected values and the original data are, of course, the observed values in the expression for χ^2. The number of degrees of freedom is obtained from the expression $(n-1-k)$ where k is the number of parameters which have to be estimated in fitting the theoretical distribution.

EXAMPLE 15. In Chapter 7 a Normal distribution was fitted to the distribution of new business written by 100 insurance salesmen, with the following results.

New business £00	Observed frequency	Expected frequency
50–52	10	9
53–55	20	22
56–58	33	32
59–61	22	25
62–64	15	12

[Note: In the discrete distribution in Chapter 7 the first and last frequencies are 7 and 10 respectively. In the above table we have included the two tails of the Normal distribution to make the sum of the expected frequencies equal to the sum of the observed frequencies.]

Since all class frequencies exceed 5 we can go ahead with the χ^2 test. If any of the expected class frequencies had been less than 5 we would have combined this class with the smaller of its two neighbouring classes (unless the class in question was an end-class when we would combine it with its sole neighbour).

Applying the formula:

$$\chi^2 = \Sigma \frac{(O-E)^2}{E}$$

we obtain a value of $\chi^2 = 1{\cdot}43$.

The number of degrees of freedom is given by the expression $(n-1-k)$. Now, in order to fit the Normal distribution to the data we had to estimate two parameters; μ and σ^2. Therefore, since we have 5 classes in our table we have $5-1-2 = 2$ d.f.

For 2 d.f., $\chi^2 = 5{\cdot}99$. Since our observed value is smaller than this we can accept the hypothesis that our data came from a Normally distributed population.

Prove your prowess [1]

8.1.* A machine producing ferrite rod in nominal lengths of 6 in is reset twice a day. The first 25 rods produced after one such resetting have a mean of 6·25 in with standard deviation 0·05 in. Is it reasonable to conclude that the machine has been set too high?

8.2. A random sample of 36 children yielded the following 36 values of weight of confectionery consumed in a week (figures are in ounces).

8·5	8·0	9·5	10·0	15·5	9·0
9·0	7·0	9·5	13·0	8·5	9·5
4·5	6·0	7·5	16·0	12·5	8·5
nil	10·5	8·5	14·5	7·5	7·5
5·0	11·0	8·0	12·0	9·0	8·0
5·5	2·0	8·0	8·5	7·0	7·5

Estimate the average weekly consumption of confectionery by children. Fit (a) 95% and (b) 90% confidence limits to your estimate.

[1] An asterisk against the exercise number indicates that the solution is given in detail.

8.3. Suppose that the sample in the previous question had consisted of only 18 children. Using the first 3 rows of the data, repeat the calculation of 8.2.

8.4. In the manufacture of paper rolls for ticket-issuing machines, wooden bobbins, in sets of 10, are threaded on to a rod which is then loaded on to a winding machine. The bobbins are manufactured to a specified length of 2 in with a standard deviation of 0·1 in. The rod on to which they are threaded has a tolerance of ± 0.05 in. If bobbins are delivered in large batches and loaded by hand, on what proportion of occasions will the threaded rod be rejected?

(HINT: Consider the mean length and standard deviation of samples of 10 bobbins.)

8.5.* In a sample of 500 garages it was found that 170 sold tyres at prices below those recommended by the manufacturer.

(a) Estimate the percentage of all garages selling below list price.
(b) Calculate the 95% confidence limits for this estimate and explain briefly what these mean.
(c) What size sample would have to be taken in order to estimate the percentage to within 2%?   (I.C.W.A., Part I)

8.6. A random sample of 1000 households was selected in a city. Each household was asked:

(i) whether it owned or rented the accommodation;
(ii) whether it had exclusive use of an indoor toilet?

The following results were obtained:

	Owned	Rented	Total
Exclusive toilet	358	483	841
Not exclusive toilet	12	147	159
Total	370	630	1000

(a) Estimate the proportion, among households with rented accommodation, of households without exclusive use of an indoor toilet and fit 95% confidence limits to this estimate.
(b) Estimate the proportion of all households which are rented and fit 95% confidence limits to this estimate.

8.7. If the sample in the above example was drawn from a population of 250 000 households what would be your estimate of the total number of households in rented accommodation and what would be the 95% confidence limits to this estimate?

8.8. For the data in 8.2, assume that the sample was drawn from a school of 600 pupils. Estimate the total weekly consumption of confectionery at the school and the 90% confidence limits to this estimate.

8.9. From the data of 8.2, fit 95% confidence limits to the estimate of the population standard deviation using both methods described in the text.

8.10. A promotion campaign is run to produce consumer awareness of a new trademark. The company decides to move their advertising account to a different agency unless independent research shows that at least 25% of consumers are familiar with the mark by the end of the campaign.
A random sample of 800 consumers gives a recognition rate of 21%. Should the account be shifted?
(Use the 5% significance level.)

8.11. A personnel manager claims that 50% of all single women taken on as secretaries or clerks marry and leave work within two years. According to staff records, of 100 such staff 53 did in fact marry and leave within two years. Test the personnel manager's hypothesis against the alternative hypothesis that the proportion is not 50%.
(Use the 5% significance level.)

8.12. A manufacturer of children's clothing assumes that the average waist size of boys aged 12 to 14 is 26 in. A random sample of 100 schoolboys resulted in an average of 27·5 in with a standard deviation of 1·5 in. Test the null hypothesis $H_0 : \mu = 26$ against the alternative hypothesis $H_1 : \mu \neq 26$, with a Type I error of 0·05.

8.13. After corrosion tests, 42 of 536 metal components treated with Primer A and 91 of 759 components treated with Primer B showed signs of rusting. Test the hypothesis that Primer A is superor to Primer B as a rust inhibitor. (I.C.W.A., Part I)

8.14. To decide whether two brands of white paint can be differentiated, 60 individuals were asked to say which of two contiguous white-painted panels was the 'whiter'. 22 individuals chose the left-hand panel, whilst the other 38 chose the right-hand panel. Do these results indicate any significant difference between the whiteness of the two panels?   (I.O.S., Part III)

8.15. In an enquiry into earnings, samples of 50 process workers from each of two steelworks were drawn. The following results were obtained:

	Works A	Works B
Average earnings (weekly)	£27	£22·80
Standard deviation	£1·60	£1·30

Is this difference significant at the 5% level? If not, at what level is it significant?

8.16. In a work-study experiment a certain job was carried out by 11 men using Method A and 14 men using Method B. The results were:

Method A—Mean time = 51 min, with standard deviation 5 min.
Method B—Mean time = 47 min, with standard deviation 4 min.

Is the observed difference significant at the 5% level?

8.17. In 8.10, a random sample of 800 consumers gave an overall recognition rate of 21% for the new trademark. More detailed analysis of the data produced the following table:

| | Social class | | | | |
	AB	C_1	C_2	DE	Total
Good recognition	24	46	20	6	96
Poor recognition	30	19	13	10	72
No recognition	46	195	287	104	632
Total	100	260	320	120	800

Test, at the 5% level the hypothesis that there is no difference between social classes in recognition of the trademark.

8.18. In 8.7, a Normal distribution was fitted to the observed data. Test the goodness of fit of this distribution.

REFERENCES

The approach in this chapter has been from the standpoint of classical hypothesis testing. We have not ventured into the area of modern decision theory and Bayesian statistics. For an elementary treatment of this subject the reader is referred to:

HADLEY, G. 1968. *Introduction to Business Statistics.* New York, Holden-Day.
SCHLAIFER, R. 1959. *Probability and Statistics for Business Decisions.* New York, McGraw-Hill.

Other introductory treatments of estimation and hypotheses testing will be found in:

FREUND, J. E., and WILLIAMS, F. J. 1959. *Modern Business Statistics.* London, Pitman.
WALLIS, W. A., and ROBERTS, H. V. 1956. *Statistics—A New Approach.* Glencoe, Free Press.

and in numerous other texts.

A more mathematical treatment will be found in :

HOEL, P. G. 1962. *Introduction to Mathematical Statistics,* 3rd Edition. New York, John Wiley.

9: CAUSE, EFFECT AND COINCIDENCE
Regression and Correlation

NEARLY all of the statistical methods discussed so far have been concerned with frequency distributions, sampling distributions, etc., of single variables. Many problems in statistics, however, involve relationships and the joint distribution of two or more variables. For example, we might be interested in the relationship between advertising expenditure and sales revenue in a particular industry. A sample of firms from the industry would yield two values for each firm, advertising expenditure and sales revenue, and we would study the joint distribution of these two variables.

Broadly speaking there are two main approaches to such a study. We can either take one of the variables and try to 'explain' it and predict future values in terms of the other variable, or we can study the variables simultaneously to examine the strength of the association between them. These two approaches lead to *regression* and *correlation* methods respectively. They will be illustrated by pursuing the example of advertising expenditure and sales revenue.

REGRESSION

Table 9.1 displays values of advertising expenditure and sales revenue from a sample of 40 retailers. We are interested in the extent to which sales revenue can be explained by advertising revenue and vice versa. (The classic example of this kind of analysis is the relationship between crop yield and rainfall, temperature and amount of fertiliser used.)

The first step in our analysis is to plot the data of Table 9.1 in the form of a *scatter chart*. This has been done in Fig. 9.1 where the two axes of the graph represent the two variables and each firm in the sample is represented by a point corresponding to its values of sales revenue and advertising expenditure. The relationship between the two variables can be gauged very roughly from the pattern of points on the scatter chart. High values of sales seem to be associated with high values of advertising expenditure and this relationship appears to be linear. This, however, is an imprecise, subjective

254

TABLE 9.1. *Advertising Expenditure and Sales Revenue for 40 Retailers*
(£'000)

Adv. exp.	Sales rev.	Adv. exp.	Sales rev.	Adv. exp.	Sales rev.	Adv. exp.	Sales rev.
14	160	41	270	13	100	11	160
25	180	21	130	9	130	26	150
6	130	17	200	20	160	36	230
9	170	39	240	38	170	13	190
37	260	7	80	24	230	10	100
23	150	20	210	29	190	32	170
12	130	32	220	8	110	18	160
8	150	26	210	16	170	28	230
16	140	35	210	40	220	34	250
16	110	21	190	31	250	14	240

impression and something more sophisticated and objective is needed in order to quantify the relationship. What is required, in fact, is an equation relating sales and revenue and advertising expenditure and such an equation can be obtained by the method of least squares which we met in Chapter 4. We will assume at the start that the relationship is linear (it is, in fact) and discuss later how we decide the order of the equation to be fitted to the data.

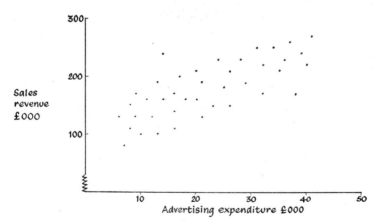

FIG. 9.1. Scatter chart of advertising expenditure and sales revenue for 40 retailers.

Turning back to the data of Table 9.1 the first thing we have to do is decide which of the two variables is the 'explaining' variable and which the 'explained'. Whether we choose to explain sales revenue in terms of advertising expenditure, or advertising expenditure in terms of sales revenue depends on our theory of business behaviour. Statistical methods do not help us in this choice; they enable us to test a hypothesis but not to formulate one.

The question of which variable to use as the 'explaining' variable is not just an academic one. It has great practical importance since the equation relating sales revenue to advertising expenditure will differ from that relating advertising expenditure to sales revenue unless the points on the scatter chart fall on a perfectly straight line. This might seem unlikely but it will be demonstrated in our worked example.

It should be noted here that this problem of choice between the two variables does not always arise. In the classic agricultural example cited above, a simple regression of crop yield on rainfall would raise no doubt as to which variable was the explaining one and which the explained.

We will choose to explain sales revenue in terms of advertising expenditure and we will refer to sales revenue as the *dependent* variable and advertising expenditure as the *independent* or *regressor* variable.

We know from our scatter chart that there is a tendency for sales revenue to increase with advertising expenditure. What we are now asking is: 'by how much, on average, should we expect sales revenue to rise for a given increase in advertising expenditure?' A reliable answer to this question not only improves our understanding of the relationship between the two variables; it also enables us to predict the sales revenue which would result from a given level of advertising expenditure, and thus provides information useful to a manager contemplating an increase in his advertising expenditure.

Since we know that we are dealing with a linear relationship, the problem is to fit a straight line to the data of Table 9.1. We could of course simply take a ruler and draw a reasonable looking line through the points in Fig. 9.1. Such a subjective line, however, would be unlikely to estimate accurately the relationship between the two variables and more important it would not be amenable to arithmetic manipulation for predicting future values and for assessing the reliability of such predictions. What we need is a set of criteria which will enable us to select the line which best fits our data. The theory underlying the choice of criteria can be expressed rather crudely as follows:

We assume that in the total population of retailers there is an inherent linear relationship between the two variables. On the basis of data from a sample of this population we attempt to discover this inherent relationship. Since all sorts of factors other than advertising expenditure affect sales revenue we find that our observations do not fall exactly on a straight line; they spread out in the form of a scatter chart.

From this scatter of observations we have to estimate the parameters of the underlying straight line, in other words we have to estimate a and b in the equation

$$Y' = a+bX$$

in such a way that our estimates meet the criteria of a 'best' line. (We write the prime against Y to show that we mean the regression value, not the actual value.) What are these criteria? The two main ones are that the estimates should be unbiased and as precise as possible; two ideas which will be familiar by now. Armed with these criteria we can go in search of our best line. We find that the least squares line possesses the required properties. It provides unbiased estimates of a and b and no other line can be found which gives the estimates with greater precision. We state these facts without proof: their proof is one of the standard results of advanced statistical theory.

FITTING THE LEAST SQUARES LINE

We can now fit a least squares line to the data of Table 9.1 taking sales revenue as Y and advertising expenditure as X. The method is exactly the same as in Chapter 4; what we are now doing is to show that the least squares line, apart from being a useful piece of mathematics, also possesses optimal properties for estimating population parameters from a sample of observations.

In our present example:

$$\Sigma X = 875, \ \Sigma X^2 = 23541, \ \Sigma Y = 7150, \ \Sigma XY = 171\,380, \ n = 40.$$

Substituting these values in the Normal Equations and applying the method of Chapter 4, we find that the least squares line is:

$$Y' = 104\cdot36+3\cdot40X.$$

This equation is known as 'the *regression equation of Y on X*'. The parameter *b* is known as the *regression coefficient*. In our example the equation tells us that on average an increase in advertising expenditure of £1000 yields an increase in sales revenue of £3400.

Fig. 9.2 shows the scatter chart of the data of Table 9.1, with the regression line drawn in. To illustrate the use of the line for prediction we can ask what the expected level of sales revenue would be for a firm spending £50 000 on advertising. Reading from our

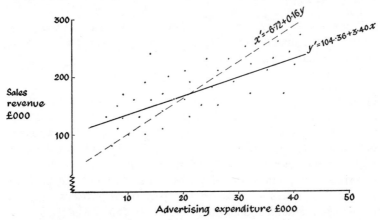

FIG. 9.2. Regression of sales revenue on advertising expenditure and advertising expenditure on sales revenue.

graph the answer is £275 000 approximately. For a more precise estimate we would substitute the value 50 in our regression equation and obtain:

$$= 104 \cdot 36 + 3 \cdot 40 \, (50)$$

$$= 104 \cdot 36 + 170$$

$$= 274 \cdot 36$$

or £274 400 to the nearest £100.

(The reader might be wondering why the equation is called a 'regression' equation. The word 'regression' was first used by Galton, who in the late nineteenth century conducted a series of investigations into the extent to which the height of sons of tall parents reverts or 'regresses' towards the average height of males. Galton fitted least squares lines, calling them regression lines, and the name 'regression' has stuck to this set of techniques ever since.)

THE TWO REGRESSION LINES

We pointed out, in our introduction to regression analysis, that an equation relating Y to X gives a different result from an equation relating X to Y, except when the points fall exactly on a straight line. We now demonstrate the truth of this statement by calculating the regression line of advertising expenditure on sales revenue. Our equation is now $X' = a' + b'Y$ which can also be written

$$X' - \bar{X} = b'(Y - \bar{Y}),$$

The primes against a and b are simply to remind us that they relate X to Y instead of Y to X.

We obtain an expression for b' from that for b by simply changing X's to Y's, and vice versa. This leaves the numerator unchanged while the denominator becomes:

$$n\Sigma Y^2 - (\Sigma Y)^2.$$

We thus have

$$b' = \frac{n\Sigma XY - \Sigma X\Sigma Y}{n\Sigma Y^2 - (\Sigma Y)^2}.$$

Substituting in this expression, and remembering that $\bar{X} = 21\cdot88$, we have

$$X' - 21\cdot88 = 0\cdot16\,(Y - 178\cdot75)$$

or $$X' = -6\cdot72 + 0\cdot16\,Y.$$

This line is also drawn in Fig. 9.2.

The reason why we obtain different equations is simply that we are asking different questions. In the first situation we were asking 'what is the average value of sales revenue for a given level of advertising expenditure?' in the second case the question was 'what is the average value of advertising expenditure for a given level of sales revenue?'

TESTS OF SIGNIFICANCE AND CONFIDENCE LIMITS FOR THE ESTIMATES

In fitting a regression line to a set of data we estimate several parameters, which need to be tested for significance before being accepted. Similarly, in using the regression equation to predict future values of the dependent variable we need to be able to fit confidence limits to our estimates. In this section we present a significance test for b, confidence limits for the regression line and confidence limits for a predicted value of the dependent variable. The basis of all these calculations is the *standard error of estimate of Y from X*. This is the standard deviation of the values of Y about the regression values Y'. It is denoted by $s_{Y.X}$ where:

$$s_{Y.X} = \sqrt{\left\{\frac{\Sigma(Y-Y')^2}{n-2}\right\}}.$$

Note that the divisor is $n-2$ instead of the more familiar $n-1$. This is because in calculating Y' two parameters, a and b are estimated each accounting for 1 degree of freedom.

We do not use the computational form for $s_{Y.X}$ shown above. An alternative form, which does not require all values of Y' to be calculated is

$$s_{Y.X} = \sqrt{\left\{\frac{\Sigma(Y-\bar{Y})^2 - b\Sigma(X-\bar{X})(Y-\bar{Y})}{n-2}\right\}}$$

which is obtained by simply substituting for Y' in the first expression, remembering that $Y' = \bar{Y} - b(X-\bar{X})$. A further reduction, to simplify the arithmetic still further gives us

$$s_{Y.X} = \sqrt{\left\{\frac{n\Sigma Y^2 - (\Sigma Y)^2 - b(n\Sigma XY - \Sigma X\Sigma Y)}{n(n-2)}\right\}}$$

which is the expression we shall use.

All but one of the above values have already been calculated for the data of Table 9.1 (see p. 257). The missing value is $\Sigma Y^2 = 1\ 372\ 500$. Inserting these values in the expressions above and performing the arithmetic we find

$$s_{Y.X} = £33.80.$$

Thus, with our usual assumptions of Normality we can say that 95% of the sales revenue figures in Table 9.1 lie within £66,000 of the regression line, i.e. (1·96)(33·80) in £'000. This is a wide range of uncertainty and indicates that sales revenue cannot be predicted accurately solely on the basis of advertising revenue. Our main reason for calculating $s_{Y.X}$, however, was not for its own sake but in order to construct a confidence interval for b.

Confidence Limits for the Regression Coefficient

The *standard error of b* is given by:

$$s_b = \frac{s_{Y.X}}{\sqrt{\{\Sigma(X-\bar{X})^2\}}}.$$

For large samples (n greater than 20, say) confidence limits can be fitted to b using the Normal approximation. For example, 95% confidence limits are given by

$$b \pm 1 \cdot 96 s_b.$$

In our present example $s_{Y \cdot X} = 33 \cdot 80$ and $\Sigma(X - \bar{X})^2 = 4400$

therefore

$$s_b = \frac{33 \cdot 8}{66 \cdot 3}$$

$$= 0 \cdot 51$$

and the 95% confidence limits are

$$3 \cdot 40 \pm 1 \cdot 00$$

When n is less than 20, the Normal approximation is unreliable and Student's-t distribution must be used. All that is necessary is that for a given confidence level, say 95%, the Normal coefficient $1 \cdot 96$ is replaced by the t value for $n - 2$ d.f. Although this is not necessary in our present example, where $n = 40$, we calculate the t-based confidence limits to demonstrate the method. From Table 8.1 $t_{0 \cdot 025}$ for 38 d.f. is $2 \cdot 02$ approximately (the difference between 38 and 40 d.f. is negligible). Therefore 95% confidence limits for b are given by

$$b \pm 2 \cdot 02 \, s_b$$

i.e. $\qquad\qquad 3 \cdot 40 \pm 1 \cdot 03$

The confidence limits fitted to b also provide a test of significance, since if the limits do not include the value zero, the observed value of b is significantly different from zero.

A more direct test of significance which does not involve confidence limits is provided by the standardised statistic

$$z = \frac{b}{s_b}$$

which, for large samples, approximates to the Normal distribution.

In the above example:

$$z = \frac{3 \cdot 40}{0 \cdot 51}$$

$$= 6 \cdot 67$$

which from Table 7.6(a) is significant at the 0·01% level.

For n less than 20 the statistic

$$t = \frac{b}{s_b} \text{ with } n-2 \text{ d.f.}$$

is used instead of z.

Confidence limits for the regression line. Since the parameters a and b are subject to sampling error, Y', the estimate of Y, is also subject to sampling error. The *standard error of Y'* is given by

$$s_{Y'} = s_{Y.x} \sqrt{\left\{ \frac{1}{n} + \frac{(X-\overline{X})^2}{\Sigma(X-\overline{X})^2} \right\}}.$$

Note that in the second term of the expression under the square root sign, the numerator is $(X-\overline{X})^2$. The value of X to be used is the value corresponding to the particular value of Y'. This implies that $s_{Y'}$ varies with the value of X, and so it does. The confidence limits for the regression line, sometimes called the *confidence band*, are not parallel lines above and below the regression line; they diverge as the value of X moves further away from $\overline{X}$. Thus there is no single value for $s_{Y'}$. Table 9.2 shows calculated values of $s_{Y'}$ for selected

TABLE 9.2. *95% Confidence Limits for Y' and Y*

X	$s_{Y}{'}$	$\pm 1\cdot96 s_{Y}{'}$	$s_{Y-Y'}$	$\pm 1\cdot96 s_{Y-Y'}$
5	10·14	±19·86	35·33	±69·24
10	8·08	±15·84	34·79	±68·19
15	6·39	±12·53	34·44	±67·50
20	5·43	±10·65	34·27	±67·18
25	5·58	±10·94	34·47	±67·55
30	6·77	±13·27	34·51	±67·64
35	8·57	±16·79	34·91	±68·42
40	10·67	±20·92	35·48	±69·55
45	12·94	±25·37	36·23	±71·02
50	15·30	±29·99	37·14	±72·79

values of X together with the 95% confidence limits. These values of $s_{Y'}$ are plotted against the scatter chart and regression line in Fig. 9.3. It will be seen that the further we move from the mean of X (21·88) the wider the confidence band. The reason for this is simply that errors in estimating b have a greater effect the further the line is projected. It is rather like using a garden hose: small movements of the wrist cause under- or over-shooting of the target, and the farther away the target the bigger the error. As before, if the sample size is less than 20, the appropriate t-value is substituted for the Normal value in calculating the confidence limits, again with $n-2$ d.f.

Confidence limits for a predicted value of Y. If we extrapolate the regression line we are predicting a value of Y'. We have just seen how to calculate confidence limits for this estimate. We now consider confidence limits for the predicted value of an individual

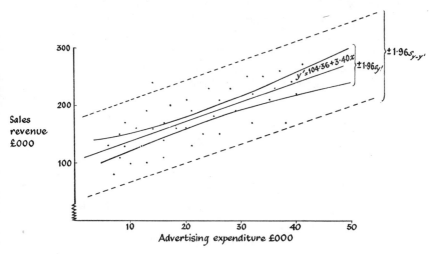

FIG. 9.3. Ninety-five % confidence limits for Y' and Y.

observation Y. These limits will be wider than those for Y' because the regression line is essentially a predictor of trend and Y', in the theoretical model, is the mean of all the observations Y which might correspond to a given value of X. What we need to estimate, in effect, is the standard error of the difference between the individual observations Y and the corresponding computed values Y'. Denoting this by $s_{Y-Y'}$ we have:

$$s_{Y-Y'} = s_{Y.X} \sqrt{\left\{\frac{n+1}{n} + \frac{(X-\overline{X})^2}{\Sigma(X-\overline{X})^2}\right\}}.$$

Table 9.2 also shows $s_{Y-Y'}$ for selected values of X, and 95% confidence limits based on the Normal approximation are shown in dotted form in Fig. 9.3. Once again, if n is less than 20, we use the corresponding t-value with $n-2$ d.f.

We have seen that regression analysis enables us to quantify the relationship between two variables and provides an attractive method of predicting further values of the dependent variable.

We shall have more to say about this use of regression in the chapter on forecasting, where we shall point out the dangers as well as the merits, but for the moment we shall consider another use.

Let us suppose that we are interested in analysing the variation in retail sales revenue and that the sales revenue data of Table 9.1 are the only data we have. We would probably form a frequency distribution of the data and calculate the mean and standard deviation of the figures. We would obtain:

Mean: 178·75
Variance: 2421
Standard Deviation: 49·2

These parameters, while conveniently summarising the data, tell us nothing about the *causes* of the variation.

Let us now consider the variance of our observations applying the usual formula:

$$\frac{1}{n}\,\Sigma(Y-\bar{Y})^2.$$

What we are doing, of course, is to calculate the variation of the points about their mean. We could represent this on our familiar scatter chart of the data of Table 9.1. Fig. 9.4 shows the same scatter chart with a horizontal line drawn through $\bar{Y}$. (The reader is asked to forget for a moment the values of advertising expenditure and consider only those of sales revenue.)

The variance of sales revenue is simply a measure of the scatter of the points about the line $Y = \bar{Y}$, and for a given sample size depends only on $\Sigma(Y-\bar{Y})^2$, the sum of squares about the mean.

Suppose now that we obtain the advertising expenditure data and carry out a regression analysis. The points on our scatter chart cluster much more closely about the regression line than about their mean. We can measure this algebraically as follows:

The sum of squares about the mean, $\Sigma(Y-\bar{Y})^2$, can be broken down ('decomposed' in the jargon) into two components to give

$$\Sigma(Y-\bar{Y})^2 = \Sigma(Y'-\bar{Y})^2 + \Sigma(Y-Y')^2.$$

The equation can be expressed in words as:

Total sum of squares = Sum of squares due to regression (regression sum of squares)

+ Sum of squares about regression (residual sum of squares).

(We have already met the residual sum of squares in the denominator of our significance test for b.)

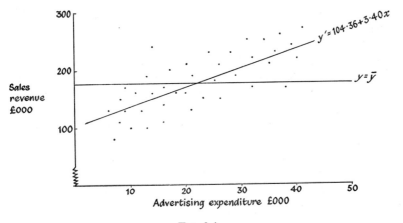

Fig. 9.4.

The interpretation of this decomposition of the total sum of squares is of great interest. The sum of squares due to regression measures the amount of variation there would have been if advertising expenditure had been the only contributing factor to total variation. It is the amount of the total variation explained by the influence of advertising expenditure. The sum of squares about regression, or 'residual sum of squares' measures the amount of variation left unexplained by the variation in advertising expenditure. It is the amount of variation due to other causes (including random influences).

LINEAR AND NON-LINEAR REGRESSION

A certain amount of confusion surrounds the definitions of *linear* and *non-linear* regression. Linear regression is often defined as regression in which the regressor variables do not involve squares or higher powers. This draws a rather false distinction between first and higher order expressions because the theory of regression treats any regressor variable as a set of n values irrespective of the order of the values. A rigorous definition of linearity involves only the regression coefficient b. The regression is linear if b is linear. Thus

$$Y' = a + bX^5 \quad \text{is linear but}$$

$$Y' = X^b \quad \text{is not.}$$

MULTIPLE REGRESSION

The methods described in the preceding sections for explaining one variable by means of a related variable give rather poor results when applied to the sales/advertising data. Only 54% of the total sales variation is explained by variations in advertising expenditure. (See p. 270 for the derivation of this figure.) This is quite a common occurrence with simple regression because a single related variable is often incapable of explaining much of the total variation. Frequently, however, it is possible to find a set of several related variables which together explain a large proportion of the total variation in the dependent variable. These variables can be combined in a *multiple regression equation* of the form:

$$Y' = a + b_1 X_1 + b_2 X_2 + b_3 X_3 + \dots$$

For example, the sales of our 40 retailers might depend as much, if not more, on size of shop and number of people served by the shopping area in which the shop is situated, as on advertising expenditure. If this were the case, any regression equation which omitted these variables would lose in efficiency.

The multiple regression equation is a straightforward generalisation of the simple equation. Estimates of a and the regression coefficients b_1, b_2, etc., are obtained as before from a set of Normal Equations but, as would be expected, these are now more complicated. If we assume a linear regression of k regressor variables the Normal equations are as follows:

$$\Sigma Y = an + b_1 \Sigma X_1 + b_2 \Sigma X_2 + \dots + b_k \Sigma X_k$$

$$\Sigma X_1 Y = a\Sigma X_1 + b_1\Sigma X_1^2 + b_2\Sigma X_2 X_1 + \ldots + b_k\Sigma X_k X_1$$

$$\Sigma X_2 Y = a\Sigma X_2 + b_1\Sigma X_1 X_2 + b_2\Sigma X_2^2 + \ldots + b_k\Sigma X_k X_2$$

$$\cdot \quad \cdot \quad \cdot \quad \cdot \quad \cdot \quad \cdot \quad \cdot \quad \cdot \quad \cdot$$

$$\Sigma X_k Y = a\Sigma X_k + b_1\Sigma X_1 X_k + b_2\Sigma X_2 X_k + \ldots + b_k\Sigma X_k^2.$$

With k regressor variables there are $k+1$ equations. The solution of these equations if k is greater than 2 becomes rather tedious and error-prone if tackled as a straightforward exercise in algebra. Fortunately, a number of special computing schemes (in the current jargon they are called 'algorithms') designed specifically for solving such systems of equations are available. A more fashionable method, of course, is to use a ready-made computer program, feed in the data and wait for an answer.

With two regressor variables the parameters can be obtained by formula although the formulae look rather ugly. To simplify the appearance of the formulae we have expressed all sums of squares and cross-products in terms of deviations about their means. In practice these corrected sums of squares and products are always calculated before proceeding to the formulae (or the algorithm where $k>2$). We have:

$$a = \bar{Y} - b_1\bar{X}_1 - b_2\bar{X}_2$$

$$b_1 = \frac{\Sigma x_1 y \Sigma x_2^2 - \Sigma x_2 y \Sigma x_1 x_2}{\Sigma x_1^2 \Sigma x_2^2 - (\Sigma x_1 x_2)^2}$$

$$b_2 = \frac{\Sigma x_2 y \Sigma x_1^2 - \Sigma x_1 y \Sigma x_1 x_2}{\Sigma x_1^2 \Sigma x_2^2 - (\Sigma x_1 x_2)^2}.$$

Note that the denominators for b_1 and b_2 are the same.

The confusion between linear and non-linear regression is even more pronounced with multiple regression so it is worth repeating that the regressor variables X_i can be any powers of X.

CONFIDENCE LIMITS FOR THE ESTIMATES

As would be expected, the calculation of confidence limits for multiple regression coefficients and estimates of Y' and Y are more complicated than with simple regression. A fully worked example of regression with two regressor variables is provided at the end of the chapter. References are also given to enable the interested reader to pursue the subject when there are more than 2 regressor variables.

CORRELATION

The ratio of the regression sum of squares to the total sum of squares

$$\frac{\Sigma(Y'-\overline{Y})^2}{\Sigma(Y-\overline{Y})^2}$$

leads us, in the case of simple regression, to a measure of the strength of association between the two variables. This measure is provided by the square root of the ratio, which is denoted by r. Thus

$$r = \sqrt{\left\{\frac{\Sigma(Y'-\overline{Y})^2}{\Sigma(Y-\overline{Y})^2}\right\}}.$$

r is known as the *Pearson product-moment correlation coefficient*, or 'correlation coefficient' for short, named after Karl Pearson, one of the first great statisticians. It is called a 'product-moment' coefficient because in its most common form its formula is

$$r = \frac{\Sigma(X-\overline{X})(Y-\overline{Y})}{ns_X s_Y}$$

where s_X and s_Y are the standard deviations of X and Y respectively and the term $\frac{1}{n}\Sigma(X-\overline{X})(Y-\overline{Y})$ is known as the *first product-moment* or *covariance* of X and Y. The two expressions for r above are algebraically equal, the second is derived from the first by substituting for Y' and simplifying the resulting expression.

In the case of multiple regression, the ratio of the regression sum of squares to the total sum of squares is known as the *coefficient of multiple correlation* and is denoted by R^2. R^2 measures the extent to which the observed values of the dependent variable are fitted by the regression line; it is essentially a positive measure, unlike r which as we shall see can take negative values. Its square root R is not considered.

For the remainder of this chapter we shall concern ourselves only with the association between *two* variables. We return to a consideration of r. Since r^2, the ratio of the regression sum of squares to the total sum of squares, can only take values between 0 and 1, r clearly must lie in the range -1 to $+1$. r has several additional properties:

1. Complete lack of association between X and Y is represented by the value 0.

2. Perfect association is represented by +1 for positive association and −1 for negative association.
3. The strength of the association is independent of the choice of
origin for the variables and the units in which they are measured,
e.g. in Fig. 9.5 the strength of association is exactly the same

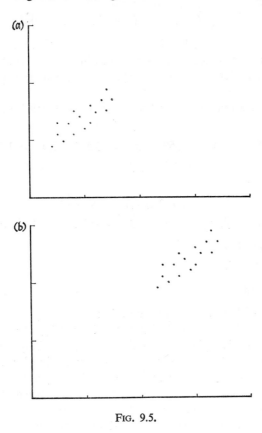

FIG. 9.5.

for (a) and (b); in Table 9.1 it does not matter whether the
values are expressed in £'000 or £million or pence.

We now proceed to illustrate the computation of r from the raw
data of Table 9.1. We shall not, however, apply the formula as it
stands, i.e.

$$r = \frac{\Sigma(X - \overline{X})(Y - \overline{Y})}{n s_X s_Y}$$

because it is a rather tedious expression to evaluate. Instead we shall manipulate the expression to obtain an alternative version which, although giving exactly the same answer, is considerably easier to compute. The alternative version is:

$$r = \frac{n\Sigma XY - \Sigma X\Sigma Y}{\sqrt{\{(n\Sigma X^2 - (\Sigma X)^2)(n\Sigma Y^2 - (\Sigma Y)^2)\}}}.$$

The reader might complain that this looks even worse than the first expression. So it does, but the factors in the new expression are very much easier to compute, especially with a desk calculating machine.

The calculation of r from the data of Table 9.1 proceeds as follows:

$$\Sigma X = 875, \quad \Sigma Y = 7\ 150 \quad \Sigma X^2 = 23\ 541, \quad \Sigma Y^2 = 1\ 372\ 500$$

$$\Sigma XY = 171\ 380, \quad (\Sigma X)^2 = 765\ 625, \quad (\Sigma Y)^2 = 51\ 122\ 500$$

$$n = 40$$

$$r = \frac{40(171\ 380) - (875)(7\ 150)}{\sqrt{\{(40(23\ 541) - 765\ 625)(40(1\ 372\ 500) - 51\ 122\ 500)\}}}$$

$$= \frac{598\ 950}{815\ 382}$$

$$= 0\cdot73$$

which is a fairly strong positive correlation.

Since $r = 0\cdot73$, $r^2 = 0\cdot54$, therefore 0·54 or 54% of the variation in sales revenue is explained by variations in advertising expenditure. This information is of more value to us than a value for r; for most purposes the square of the correlation coefficient is a better measure of the relationship between X and Y than the coefficient itself.

It is instructive to refer to Fig. 9.1 now that we know that $r = 0\cdot73$. An experienced statistician could perhaps have guessed at a value of about 0·7 from the appearance of the graph. The correlation is clearly positive since high values of advertising expenditure tend to accompany high values of sales revenue and the cluster of points moves up to the right. The actual value of r depends on the extent to which the points cluster around a straight line and the cluster in Fig. 9.1 would probably suggest a value of about 0·7 to the experienced statistician. He would not, of course, be content with guessing.

Some typical scatter charts are set out in Fig. 9.6, illustrating various degrees of correlation. Fig. 9.6(*a*) shows the shapeless pattern of points typical of a very low value of *r*. Figs 9.6(*b*), (*c*)

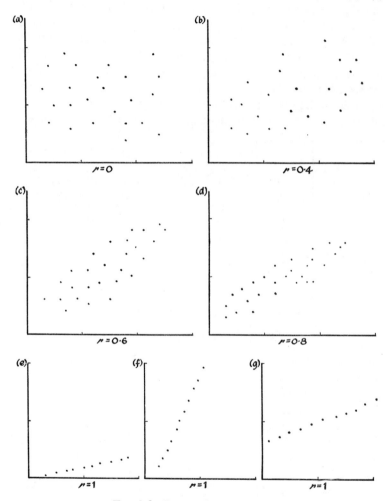

FIG. 9.6. Some typical scatter charts.

and (*d*) show how with increasing positive correlation the points approximate more and more to an upward sloping straight line until finally in Figs 9.6(*e*), (*f*) and (*g*) where *r* = 1 the points actually fall on a straight line. Thus we see that the absolute value of *r* depends on the extent to which the values of X and Y cluster around a straight line on the scatter chart (if this line slopes down to the right the correlation is negative of course). For this reason, we sometimes meet the expression '*linear correlation*' but this is misleading, since correlation is a dimensionless measure.

Figs 9.6(*e*), (*f*) and (*g*) have another point of interest. All three show perfect positive correlation, therefore the scatter chart in all three is a straight line. The slope of the line varies considerably, however, between the three scatter charts.   This fact has no effect whatever on the value of r but when several scatter charts are being compared it is very easy to assume automatically that the chart

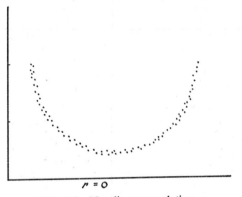

FIG. 9.7. Non-linear correlation.

showing the steepest relationship between the variables is the one with the highest value of r. We emphasise again, therefore, that the value of r depends on how closely the points on the scatter chart approximate to a straight line and *not* on the steepness of the straight line to which they approximate.

Finally in our brief look at typical scatter charts, we turn to Fig. 9.7 which shows an example of non-linear correlation. The two variables in Fig. 9.7 are quite clearly related but $r = 0$, illustrating the fact that r can only give a useful measure of the association between two variables when the variables are linearly related.

THE RELATIONSHIP BETWEEN THE REGRESSION COEFFICIENTS AND THE CORRELATION COEFFICIENT

The two regression coefficients and the correlation coefficient r are related in an interesting way. If we multiply b by b' the resulting expression is that for r^2. Thus the correlation coefficient is the geometric mean of the two regression coefficients.

EXAMPLE. From Table 9.1 we have $b = 3\cdot40$, $b' = 0\cdot16$ and $r = 0\cdot73$.

Therefore $\qquad bb' = (3\cdot40)(0\cdot16) = 0\cdot54$

and $\qquad\qquad \sqrt{0\cdot54} = 0\cdot73$ (to 2 decimal places)

$$= r.$$

THE INTERPRETATION OF r

We have just seen that a value of r is meaningless unless the two variables are linearly related. This is one of the two major pitfalls in interpreting values of r; the other one is the reading of cause and effect into an observed value of r. In terms of our example one might be tempted to conclude that a high level of advertising expenditure is a main contributary cause of high sales revenue. The correlation coefficient, however, is entirely neutral with regard to cause and effect. It measures the strength of association not the effect of one variable on another. In our example we might, for instance, argue that the level of advertising expenditure is determined by the level of sales, i.e. high sales lead to a high level of advertising. Such an argument might be based on the theory that advertising is something that a firm can 'afford' when things are going well and that its advertising budget is determined in the light of current and expected sales. (The reaction of some firms to a decline in sales, which is to cut the promotion budget, lends support to this theory.) The formula for r has, however, nothing whatever to contribute to this argument.

Finally to drive our point home we consider some of the many 'nonsense' correlations which can be constructed from any set of official statistics. There is a high positive correlation between the following pairs of variables:

1. Salaries of school teachers and consumption of alcoholic drink.
2. Number of Roman Catholics in Great Britain and number of divorces per year.
3. Number of television receiving licences and number of admissions to mental hospitals.

Many more can be found without difficulty.

Can we conclude that any of the above relationships are causal? Clearly not—at least not on the basis of a correlation coefficient.

The explanation of these and other nonsense correlations is that
there is some other factor or factors causing both variables to move
in the way they do, i.e. instead of cause and effect both variables in
the above examples are effects. For instance in the first example,
the increase in the general standard of living and the rise in real
incomes is responsible for both the increase in teachers' salaries and
the consumption of alcoholic drink, and similar broad underlying
causes readily suggest themselves for the other examples.

REGRESSION AND CORRELATION WITH GROUPED DATA

We often wish to calculate *b* or *r* from a grouped frequency distribution rather than the individual observations (only a *simple* regression

TABLE 9.3. *Correlation and Regression With Grouped Data*
Advertising Expenditure and Sales Revenue of 40 Retailers

Sales revenue (£'000)	Advertising expenditure (£'000)				
	5<15	15<25	25<35	35<45	Total
75<125	4	1	...	...	5
125<175	7	6	2	1	16
175<225	1	3	4	2	10
225<275	1	1	3	4	9
Total	13	11	9	7	40

coefficient can be calculated in this way). Such a need arises for
example when the number of observations is very large, or when
the information is presented to us in grouped form leaving us no
choice in the matter. In such cases we *estimate b* and *r* from the
distribution just as we *estimate x̄* from a frequency distribution.
We illustrate the method with the aid of our advertising and sales
example.

Table 9.3 presents the data of Table 9.1 in grouped form; this is
an example of a *bivariate distribution*. The number and size of the
classes is of course optional and depends on the number and range
of the observations. There is, however, one additional rule which
applies when *b* or *r* is to be calculated from the distribution. This
rule is that for each variable, class intervals must be equal. Thus,
for the *X* variable (advertising expenditure) the class interval is

£10,000 for all four classes; for the Y variable (sales revenue) the class interval is £50,000 for all classes. This requirement rules out the customary open-ended intervals at one or both ends of a frequency distribution.

We shall not work with the values in Table 9.3, i.e. with the mid-points of the class intervals. Instead we shall make use of working units, in the same way that we used working units in Chapter 2 when calculating means and variances. This is particularly labour-saving in calculating r from grouped data because it makes no difference to the value of r if we add or subtract a constant to the value of X and Y, or if we multiply or divide by a constant. We therefore get the same result whether we work with the class midpoints or whether we use working units. The practical effect of this is that having obtained our working units we do not have to turn them back into original units at the end of the calculation. In calculating b, however, we do have to convert our working units back into original units.

The data of Table 9.3 are expressed in terms of working units in Table 9.4. The calculations proceed as follows:

(1) b. We compute

$$b = \frac{n\Sigma uv - \Sigma u\Sigma v}{n\Sigma u^2 - (\Sigma u)^2},$$

i.e. the usual expression for b but with the working units u and v replacing X and Y.

From Table 9.4 we obtain:

$$\Sigma u = (-1 \times 13) + (0 \times 11) + (1 \times 9) + (2 \times 7) = 10$$
$$\Sigma u^2 = (-1^2 \times 13) + (0^2 \times 11) + (1^2 \times 9) + (2^2 \times 7) = 50$$
$$\Sigma v = (-1 \times 5) + (0 \times 16) + (1 \times 10) + (2 \times 9) = 23.$$

Σuv is slightly more complicated. It is obtained by summing all pairs of cross products uv multiplied by the corresponding frequency, e.g. the top left-hand corner term gives us $(-1 \times -1 \times 4)$. A convenient way of setting out this calculation is shown below:

u	-1	0	1	2	Total
Σv	-1	4	10	10	23
$u\Sigma v$	1	0	10	20	31

The u values in the table above are self-explanatory. Σv for a given value of u is obtained by summing the products of each value of v

K

TABLE 9.4. *Calculation of b and r in Working Units*

$$u = \frac{X-20}{10}; \quad v = \frac{Y-150}{50}$$

v	-1	0	1	2	Total
-1	4	1	...	...	5
0	7	6	2	1	16
1	1	3	4	2	10
2	1	1	3	4	9
Total	13	11	9	7	40

(column group header: u)

and the corresponding frequency, e.g. for $u = -1$, $\Sigma v = (-1 \times 4)$ $+(0 \times 7)+(1 \times 1)+(2 \times 1) = -1$, which is entered in the table above. The bottom row is simply the product of u and Σv. We stress that all we are doing in this rather unfamiliar calculation is to work out cross products uv and sum them over the whole sample of 40 observations.

We obtain a value for Σuv of 31.

Applying these values to the formula for b we have:

$$b = \frac{(40)(31)-(10)(23)}{(40)(50)-(10)}$$

$$= \frac{1010}{1900}.$$

But these are working units and if we left them in this form we would get a value for b of 0·53 which is obviously absurd. In order to regain our original scales of values we must multiply by the scaling factor in our working units. Now, the numerator of the formula for b involves cross products uv and the denominator involves u^2. The scaling factors for u and v are 10 and 50 respectively (see Table 9.4) so we multiply the numerator by (10) (50) and the denominator by 10^2. Then, in our original units we have

$$b = \frac{1010 \times 500}{1900 \times 100}$$

$$= 2·66$$

which compares with 3·40 obtained from the ungrouped data and illustrates the loss of information incurred in grouping the data.

(The reader who has forgotten the mechanics of working units is referred back to Chapter 2, in particular Table 2.1.)

Expressions for confidence limits for b and predicted values can also be obtained using similar methods, but they should be regarded as minimum limits on account of the loss of precision arising from the use of grouped data.

(2) r. In calculating r we do not need to convert working units back into original units because the scaling factors are self cancelling. This can be seen immediately from the formula for r. In working units we have

$$r = \frac{n\Sigma uv - \Sigma u \Sigma v}{\sqrt{\{(n\Sigma u^2 - (\Sigma u)^2)(n\Sigma v^2 - (\Sigma v)^2)\}}}$$

and in order to convert to original units we would multiply the numerator by (10)(50) and the denominator by $\sqrt{\{(10^2)(50^2)\}}$.

The numerator of r is in fact exactly the same as the numerator of b so we can write down its value at once: 1010. In order to evaluate the denominator we need one additional term Σv^2.

From Table 9.4.

$$\Sigma v^2 = (-1^2 \times 5) + (0^2 \times 16) + (1^2 \times 10) + (2^2 \times 9) = 51$$

therefore

$$r = \frac{1010}{\sqrt{\{((40)(50) - (10)^2)((40)(51) - (23)^2)\}}}$$

$$= \frac{1010}{\sqrt{\{(1900)(1511)\}}}$$

$$= \frac{1010}{1694}$$

$$= 0\cdot60$$

which compares with 0·73 obtained from the ungrouped data and again illustrates the loss of information incurred in grouping the data.

RANK CORRELATION

Up to now we have been interested in the correlation between pairs of measurements, i.e. actual values of our variables. It may be, however, that we are interested not so much in the actual measurements as in their *ranking* or it may be that we are dealing with some attribute which cannot be accurately measured but where observations can be *ranked*. A simple and well-known example of this is the kind of beauty contest where contestants are ranked by each one of a number of judges. Change the setting to that of an interview board assessing candidates for a particular job and you have a similar (if less attractive) problem in a common business situation.

We might wish to measure the correlation between the rankings of two judges. This can be done very simply by taking the two rankings, regarding the ranks as values of the variables x and y and calculating the familiar product moment correlation coefficient which is now, however, denoted by ρ (the Greek letter 'rho') to indicate that it refers to ranks.

We will illustrate the calculation of ρ with a very simple numerical example and then apply the method to our advertising expenditure and sales revenue data.

EXAMPLE.   Six prospective computer programmers are given an aptitude test before training and a proficiency test after six months post-training experience. The rankings of the candidates are set out below:

Candidate	Rank in aptitude test	Rank in proficiency test
A	3	5
B	4	3
C	1	2
D	6	6
E	2	1
F	5	4

Does this suggest that the aptitude test is a reliable predictor of programming ability?

If we apply the formula for r to the above data, with $X =$ rank in aptitude test and $Y =$ rank in proficiency test we arrive at a value of $+0.78$. We do not, however, apply the usual formula for r when dealing with ranked data. Instead, we use the following formula:

$$\rho = 1 - \frac{6(\Sigma d_i^2)}{n(n^2 - 1)}$$

where $d_i = X_i - Y_i$, i.e. d_i is the *difference* between the rankings for the *i*th observation. *n* is, of course, the number of observations.

The above formula looks totally unlike the formula for *r*, but in fact is a simplification of that formula. The simplification arises from the fact that with ranked data the values of *X* and *Y* consist of the numbers 1 to *n*. The derivation of the formula for ρ from that for *r* is set out in the Appendix to this chapter. It is not necessary to follow the derivation in order to calculate ρ; the important point to remember is that the value of ρ obtained from the simplified formula is exactly the same value as would be obtained from the straightforward application of the formula for *r*.

Returning to our example we have:

X	Y	d_i	d_i^2
3	5	-2	4
4	3	1	1
1	2	-1	1
6	6	0	0
2	1	1	1
5	4	1	1
		0	8

Therefore

$$\rho = 1 - \frac{(6)(8)}{6(36-1)}$$

$$= 1 - \frac{48}{215}$$

$$= 0.78$$

which suggests that the aptitude test is quite a good guide to future performance in the job.

We now apply the above methods to our advertising/sales example. For this purpose we recast the data of Table 9.1 in the form shown in Table 9.5. We re-arrange the order of the data so that the rankings of the *X* variable are set out in descending order.

The first thing we notice from Table 9.1 is that some of the values of *X*, and of *Y*, are equal, therefore the ranks of these values are also equal. The simplest method of dealing with this problem is to give each of the equal values the average of the ranks which they jointly occupy. For example, in Table 9.5, the 9th and 10th values of X are equal; we therefore give each value the rank $\dfrac{9+10}{2}$. We proceed similarly with the 14th and 15th, the 21st and 22nd values, etc.

The presence of tied ranks in the data raises a problem in using the simplified formula for ρ. This formula is strictly applicable only to untied ranks (as can be seen from the Appendix). When some ranks

TABLE 9.5. *Ranked Advertising Expenditure and Sales Revenue for 40 Retailers*

Adv. exp. (X)	Sales rev. (Y)	$X-Y$ (d)	Adv. exp. (X)	Sales rev. (Y)	$x-y$ (d)	Adv. exp. (X)	Sales rev. (Y)	$x-y$ (d)
1	1	0	14·5	13	1·5	26	36·5	−10·5
2	10·5	−8·5	14·5	29	−14·5	28·5	5·5	23·0
3	5·5	−2·5	16	19	−3·0	28·5	25·5	3·0
4	21·5	−17·5	17	8	9·0	30·5	17	13·5
5	2	3·0	18	29	−11·0	30·5	38·5	−8·0
6	8	−2·0	19	17	2·0	32	33·5	−1·5
7	13	−6·0	20	33·5	−13·5	33	25·5	7·5
8	3·5	4·5	21·5	13	8·5	34	38·5	−4·5
9·5	10·5	−1·0	21·5	25·5	−4·0	35·5	21·5	14·0
9·5	21·5	−12·0	23	25·5	−2·5	35·5	33·5	2·0
11	3·5	7·5	24	15	9·0	37·5	29	8·5
12	17	−5·0	26	21·5	4·5	37·5	36·5	1·0
13	8	5·0	26	31	−5·0	39	40	−1·0
						40	33·5	6·5

are tied the simplified formula will give a value of ρ different from that which would be obtained by the straightforward application of the product-moment formula; the extent of this difference depending mainly on the number of tied ranks. This is made clear in the following calculations.

From Table 9.5 we have

$$\Sigma d_i^2 = 2825\cdot50$$

therefore

$$\rho = 1 - \frac{6(2825\cdot50)}{40(1600-1)}$$

$$= 1 - \frac{16\ 953}{63\ 960}$$

$$= 0 \cdot 74.$$

If we calculate the product-moment correlation coefficient we have

$$\Sigma X = 820,\ \Sigma Y = 820,\ \Sigma X^2 = 22\ 134 \cdot 50,\ \Sigma XY = 20\ 333 \cdot 30$$

$$(\Sigma X)^2 = 672\ 400,\ (\Sigma Y)^2 = 672\ 400,\ n = 40.$$

Then

$$r = \frac{40(20\ 333 \cdot 30) - (820)(820)}{\sqrt{\{40(22\ 134 \cdot 50) - 672\ 400\}\{40(22\ 114 \cdot 50) - 672\ 400\}}}$$

$$= \frac{140\ 932}{212\ 580}$$

$$= 0 \cdot 66.$$

The difference between the two values reflects the errors arising from the application of the simplified formula of rankings which include ties. In the preceding example we had rather a lot of ties which resulted in an overestimate of ρ. When few ties occur, however, the simplified formula for ρ is acceptably accurate.

The simplified formula, and the symbol ρ, were introduced by a psychologist named Spearman and for this reason the coefficient is usually known as *Spearman's coefficient of rank correlation* or more briefly 'Spearman's ρ'. Other measures of rank correlation have since been developed, some of which have theoretical advantages over Spearman's ρ. The latter, however, remains the best known measure.

THE SIGNIFICANCE OF THE CORRELATION COEFFICIENT

The sampling distribution of r (and therefore of ρ) is rather complicated but, under the null hypothesis of zero correlation and for n reasonably large and X and Y Normally distributed, it approximates to the Normal distribution with mean 0 and standard deviation

$$\frac{1}{\sqrt{(n-1)}}.$$

We can therefore test an observed value of r against the null hypothesis of zero correlation by means of the following criteria:

(*i*) *At the* 0·05 *level of significance*
Accept null hypothesis if

$$\frac{-1\cdot96}{\sqrt{(n-1)}} \leqq r \leqq \frac{1\cdot96}{\sqrt{(n-1)}}.$$

Reject null hypothesis if r is outside above limits.

(*ii*) *At the* 0·01 *level of significance*
Accept null hypothesis if

$$\frac{-2\cdot58}{\sqrt{(n-1)}} \leqq r \leqq \frac{2\cdot58}{\sqrt{(n-1)}}.$$

Reject null hypothesis if r is outside above limits, etc.

EXAMPLE. From Table 9.1 we calculated $r = 0\cdot73$, $n = 40$. Therefore at the 0·05 level of significance we test against

$$\frac{1\cdot96}{\sqrt{39}} = \frac{1\cdot96}{6\cdot24} = 0\cdot31$$

r exceeds this value therefore r is significant at the 0·05 level.

APPENDIX: THE DERIVATION OF THE SIMPLIFIED FORMULA FOR SPEARMAN'S ρ

We start from the basic formula for the product moment correlation coefficient r. We have

$$r = \frac{\Sigma(X-\bar{X})(Y-\bar{Y})}{ns_X s_Y}$$

which, on expanding the denominator, can be written as

$$r = \frac{\Sigma(X-\bar{X})(Y-\bar{Y})}{\sqrt{\{\Sigma(X-\bar{X})^2 \Sigma(Y-\bar{Y})^2\}}}.$$

Since the variables are ranked, each variable takes the values 1, 2, ..., n. It is this fact which enables us to simplify the formula for r, since from elementary algebra we know that:

$$\Sigma(1+2+\ldots+n) = \tfrac{1}{2}n(n+1)$$

and $\qquad \Sigma(1^2+2^2+\ldots n^2) = \tfrac{1}{6}n(n+1)(2n+1).$

Therefore $\quad\quad\quad\quad \Sigma X = \Sigma Y = \frac{1}{2}n(n+1)$

and $\quad\quad\quad\quad\quad \overline{X} = \overline{Y} = \frac{1}{2}(n+1)$.

The next step is to simplify the numerator and denominator separately.

Numerator:

$$\Sigma(X-\overline{X})(Y-\overline{Y}) = \Sigma(XY-\overline{X}Y-X\overline{Y}+\overline{X}\overline{Y})$$
$$= \Sigma XY - n\overline{X}\overline{Y}.$$

But

$$\overline{X} = \overline{Y} = \frac{n+1}{2}.$$

Therefore

$$\Sigma(X-\overline{X})(Y-\overline{Y}) = \Sigma XY - \frac{n(n+1)^2}{4}.$$

To find ΣXY we make use of another result from elementary algebra:
$$\Sigma(X-Y)^2 = \Sigma X^2 + \Sigma Y^2 - 2\Sigma XY,$$
therefore

$$\Sigma XY = \frac{1}{2}(\Sigma X^2 + \Sigma Y^2 - \Sigma(X-Y)^2).$$

But

$$\Sigma X^2 = \Sigma Y^2 = \frac{1}{6}n(n+1)(2n+1),$$

therefore

$$\Sigma XY = \frac{1}{2}(\tfrac{1}{6}n(n+1)(2n+1) + \tfrac{1}{6}n(n+1)(2n+1) - \Sigma(X-Y)^2).$$

We denoted the term
$$\Sigma(X-Y)^2 \text{ by } \Sigma d^2,$$
therefore

$$\Sigma XY = \tfrac{1}{6}n(n+1)(2n+1) - \tfrac{1}{2}\Sigma d^2,$$

therefore

$$\Sigma(X-\overline{X})(Y-\overline{Y}) = \tfrac{1}{6}n(n+1)(2n+1) - \tfrac{1}{4}n(n+1)^2 - \tfrac{1}{2}\Sigma d^2$$
$$= \tfrac{1}{12}\{2n(n+1)(2n+1) - 3n(n+1)^2\} - 6\Sigma d^2$$

which reduces to $\quad\quad\quad \tfrac{1}{12}\{(n^3-n) - 6\Sigma d^2\}$.

Denominator:

Since $\Sigma(X-\overline{X})^2 = \Sigma(Y-\overline{Y})^2$ we can rewrite the denominator as $\Sigma(X-\overline{X})^2$.

We then have $\quad \Sigma(X-\overline{X})^2 = \Sigma(X^2 - 2X\overline{X} + \overline{X}^2)$
$$= \Sigma X^2 - 2n\overline{X}^2 + n\overline{X}^2$$
$$= \Sigma X^2 - n\overline{X}^2.$$

Using our results from elementary algebra we have

$$\Sigma(X-\bar{X})^2 = \tfrac{1}{6}n(n+1)(2n+1)-\frac{n(n+1)^2}{4}$$

which reduces to $\qquad \tfrac{1}{12}(n^3-n).$

Thus we have

$$\rho = \frac{\tfrac{1}{12}\{(n^3-n)-6\Sigma d^2\}}{\tfrac{1}{12}(n^3-n)}$$

$$= 1-\frac{6\Sigma d^2}{n^3-n}$$

or $\qquad\qquad \rho = 1-\frac{6\Sigma d^2}{n(n^2-1)}.$

Prove your prowess [1]

9.1. The figures below are the results of a series of observations made at a factory during the loading of gas cylinders on to lorries.

Time taken to load (minutes)	Number of cylinders in load
8	4
11	19
12	14
12	11
18	20
19	32
20	15
24	29
29	34
30	39

Express the relationship between the time taken to load and the number of cylinders in the load in the form of a regression equation.

(I.C.W.A., Part I)

9.2. Fit 95% confidence limits to b.

9.3. Estimate the time required to load 45 cylinders and fit 95% confidence limits to your estimate.

9.4. Calculate the regression line for the number of cylinders loaded against time taken to load.

[1] An asterisk against the exercise number indicates that the solution is given in detail.

9.5.* Fit a multiple regression line, relating crude steel output to activity in the vehicles and construction industries, to the following data. Calculate confidence limits for the regression co-efficients.

	Crude steel production (million tons)	Vehicles index	Construction index
		(1958 = 100)	
1958	19·6	100	100
1959	20·2	109	106
1960	24·3	118	111
1961	22·1	110	120
1962	20·5	112	121
1963	22·5	121	121
1964	26·2	126	135
1965	27·0	129	138
1966	24·3	131	139
1967	23·9	121	145

9.6.* Ignoring the fact that b_2 is not significant, estimate crude steel production for 1970 given that the forecast values for the two regressor variables are: Vehicles: 130; Construction: 150.

9.7. Show that the regression sum of squares in simple regression, $\Sigma(Y'-\bar{Y})^2$ can be written as $b\Sigma(X-\bar{X})(Y-\bar{Y})$.

HINT. Make things easier for yourself by assuming that $\bar{Y}$ and $\bar{X}$ are zero. Then $Y' = bX$.

9.8. Calculate the ratio of the regression issue of squares to the total sum of squares for the data of 9.1. Hence calculate r. Check that this answer agrees with that obtained by applying the more common direct formula for r.

9.9. For the data of 9.1 obtain r from the two regression coefficients. Test r for significance.

9.10. From the data of 9.5 calculate R^2.

HINT. In multiple regression, the regression sum of squares is simply

$$\Sigma_i \{b_i\Sigma(X_i-\bar{X})(Y-\bar{Y})\}.$$

9.11. An office contains 10 clerks. The longer-serving clerks feel that they should have a 'seniority increment' based on length of service built into their salary structure. An assessment of their efficiency, by their departmental manager and the personnel department produces a ranking of efficiency. This is shown below together with a ranking of their length of service. Do the data support the clerks' claim for a seniority increment?

Length of service	Efficiency
1	2
2	5
3	3
4	10
5	6
6	4
7	8
8	9
9	7
10	1

Note. For length of service 1 represents the shortest and 10 the longest service; for efficiency 1 represents the most and 10 the least efficient.

REFERENCES

DAVIES, O. L. (Ed.). 1961. *Statistical Methods in Research and Production*, 3rd Edn. Revised. Edinburgh, Oliver and Boyd for I.C.I. Ltd.
STEEL, R. G. D., and TORRIE, J. H. 1960. *Principles and Procedures of Statistics*. New York, McGraw-Hill.

Both of the above books contain an example of a computational routine for use with two or more regressor variables.
An excellent introduction to the theory of regression, which also covers the necessary matrix algebra is:

JOHNSTON, J. 1963. *Econometric Methods*. New York, McGraw-Hill.

10: THEIR SHADOWS BEFORE THEM
Forecasting

FORECASTING is using the knowledge we have at one moment of time to estimate what will happen at some future moment in time. The manager's need to forecast is based on the time lag between decisions or actions taken now and their outcomes. For example, a steel firm's decision to increase its selling capacity by installing a new rolling mill will take about five years to implement; it is therefore necessary to estimate what level of capacity will be required five years ahead. Similar considerations apply to all other investment situations; down to the basic question of whether there will be any demand at all for the firm's present products in five, ten, twenty years' time.

In a completely different context, industrial process control depends on monitoring the output of the system and taking the appropriate action whenever this is called for. The appropriate action is usually the adjustment of some physical constant, e.g. temperature, acidity, length, thickness. Because of the time lag between the need for action being spotted and the resulting action taking effect, it is necessary to forecast the state of the process a short time ahead. Solving a control problem requires the solution of two smaller problems. One is the problem of forecasting the state of affairs if no control is applied and the other is knowing how the system will respond to the controlling action applied. We shall not be concerned with the second problem.

An analogy is often made between driving a car and running a business. The car driver, it is said, should not commit himself further ahead than he can see. Thus, the good driver slows down at night, when approaching a bend, when approaching the brow of a hill, etc. Similarly, the good businessman does not commit his company's resources further ahead than he can forecast. If the analogy is accepted it implies two main duties for the forecaster:

1. To try to extend the range of vision into the future, by producing longer range forecasts.
2. To make quite clear how far into the future his vision really

does extend, by clearly stating the assumptions and possible margins of error in his forecasts.

After all, it is easy to drive a short distance through a ground mist on an unfamiliar road at very high speed, on the assumption that the road is broad, straight and clear.   Similarly, it is easy to run a business by making substantial financial commitments on equivalent assumptions, namely that there is a sellers' market, investment is cheap and labour is easy to get.   In both cases the ride is likely to get rather rough after a while.

A great amount of confusion and ambiguity seem to have grown up in the use of the words 'forecast', 'prediction' and 'projection'. One also encounters expressions such as 'statistical forecasting'. Before proceeding any further, therefore, we will define the terms to be used in this chapter.

A *prediction* is an estimate based solely on past data of the series under investigation.   It is a purely mechanical extrapolation.   All estimates arising from time series analysis, based on a single series of interest, e.g. the British Railways Freight Traffic example of Chapter 4 (Table 4) are *predictions*.

A *projection* is a prediction where the extrapolated values are subject to certain numerical assumptions.   The clearest examples arise in the area of population statistics where projections of numbers of births, deaths, marriages, etc., involve assumptions about changes in the birth rate, death rate, marriage rate, etc.

A *forecast* is an estimate which relates the series in which we are interested to external factors, e.g. sales revenue was related to advertising expenditure (and vice versa) in Chapter 9.   Forecasts are made by estimating future values of the external factors ('regressor variables' in the language of regression analysis) by means of prediction, projection or forecast and, from these values, calculating the estimate of the dependent variable.

The word forecast is also used for estimates which are based, or partly based, on subjective estimates arising from the judgement and experience of the forecasters.   For example, in a regression equation the future values of the independent variables might be estimated subjectively.   In this meaning of the word, forecasting involves using all our knowledge, from whatever source, about the situation.

Predictions and projections are sometimes referred to as 'naïve' methods of forecasting. This is not just a slur on their respectability or usefulness; it simply implies that no explicit attention is paid to any other information than that supplied by past data of the series concerned. The simplest forms of 'naïve' forecasting are very simple indeed, as we shall soon see, but it is often extraordinarily difficult to make much improvement in the accuracy of these predictions and certainly a disproportionate effort has to be made to secure a marginal improvement.

In this chapter we shall be concerned with the statistical techniques of prediction and forecasting. We shall make use of the results derived in the chapters on Time Series Analysis and Regression. We shall also introduce some further ideas developed from the ideas in these two chapters.

Before proceeding to our discussion of the available techniques there is one general aspect of forecasting which must be mentioned. This is the *range* of the forecast. The range of a forecast is the length of time between the *base point*, i.e. the date of the latest available data on which the forecast (or prediction, or projection) is based and the *forecast point*, i.e. the period we are forecasting, be it one second, one week or twenty years ahead. One commonly hears a distinction drawn between 'short-term' and 'long-term' forecasting or even 'short-term forecasting techniques' and 'long-term forecasting techniques'. (One also hears, but rather less frequently of 'medium-term' forecasts.) The implication seems to be that there is something basically different in the problems. This is not so. However, although the problems of long-term and short-term forecasts are not really different, the techniques used usually differ in practice. The main reason for this is that a number of very simple techniques, of the type christened 'quick and dirty methods' by an eminent American statistician some years ago, are in widespread use for short-term forecasting. These techniques involve simple prediction of a single series and are therefore rather shaky as a basis for long-term forecasting, nevertheless they are often adequate for short-term forecasting and their ease of computation is a strong (perhaps the main) point in their favour. We shall meet these techniques in the next section of this chapter. The important thing is that the forecasting method should be consistent with the data and with the range of the forecast.

(We mention in passing that frequently one is faced with a longer range of forecasting than is really necessary. For example we might, at the beginning of June, wish to estimate the June and July sales of a certain product. The latest data, however, might refer to March. Thus for June and July we have to make a 3-month and a 4-month forecast. This sort of situation is very common and in many cases it would be much more practical to spend money on speeding up the reporting of data than to spend it on more complicated forecasting systems.)

We are now ready to begin our discussion of statistical techniques of prediction and forecasting but we now face a problem of presentation. A logical progression would start with the simplest naïve techniques, proceed to more complex naïve techniques and then to the more sophisticated analytical techniques. This would involve considering moving averages, then simple regression, then back to variations on moving averages (and possibly some extremely complex time series analysis, except that this is outside the scope of this book), finally turning back to regression methods. We prefer not to jump from one technique to another and back again, so we have divided the techniques into two main sections: Time Series Methods and Regression Methods. Our first regression method however properly belongs to time series analysis.

TIME SERIES METHODS

The simplest method of predicting from a time series is to plot the data and apply the time-honoured technique of looking at it. Indeed, this should be the first stage in any serious time series study, unless the number of series is so large that it is not practicable. As a valid technique on its own, however, it would not usually get us very far and numerical manipulation of the data is called for.

The first stage in predicting a time series is usually a prediction of the trend. In Chapter 4, two methods of predicting trend values were discussed, the least squares line and the freehand extrapolation of the moving average. The advantages of the least squares line were obvious, but there are a number of pitfalls in its use. The first of these is that over a long forecasting range the trend, whether it be linear or non-linear is likely to change. Extrapolating a least squares line, as was mentioned in Chapter 9, is like using a hose: a very small movement of the wrist causes the jet to considerably over- or under-shoot the target, and the further away the target the bigger the error.

The second pitfall is that a given set of data can sometimes be fitted equally well by a straight line or a higher-order curve (see Table 4.1). In this situation which curve do we fit? We can test the data for its goodness of fit to each of the curves and select whichever curve gives the best fit, but still there may not be much to choose between two or more curves. A good rule of thumb is to choose the lowest order curve from among the alternatives; this is particularly good advice if we are predicting a long way ahead since the extrapolation of second and higher order curves frequently exaggerates the shape of our data.

EXAMPLE 1. Fig. 4.1 (p. 113) illustrates the problem. In this case, the linear trend appears to be more reliable as a predictor than the second order curve which is already turning upwards and which if extrapolated would result in ever-increasing predictions.

We see from the above example that the choice of curve has a profound effect on the predicted values, especially the more distant ones. In some practical situations, in fact, a curve might be chosen on the basis of what predictions it gives rather than on its closeness of fit to the data; in other words, the predictions determine the technique to be used rather than vice versa.

We have talked of predicting the trend by least squares methods. Where there is no trend, i.e. where the 'trend' line is horizontal, our extrapolated trend line is simply the mean of our observations ('a' in the equation $y = a+bx$). We must not forget, however, that though we have no trend we might have a cyclical effect plus, if our data are quarterly, monthly, weekly, daily, etc., a seasonal effect. The cyclical effect is a very difficult one to predict satisfactorily using simple methods unless the cycles are regular and have a strong influence on the series. The seasonal effect, however, can easily be measured; an example is given in an Appendix to this chapter.

MOVING AVERAGES

With many series, changes in the short-run, i.e. from one month to the next, are due more to the random component than to any other. For such series, if we are only interested in predicting one or two months ahead, some form of average, which is updated after every new observation becomes available, can be used as a prediction for next month's and subsequent months' values. We have already met this idea in our example of a 'zero-trend' least squares prediction, where the prediction was simply the mean of all our observations. The implication, in a least squares context, was that the prediction would remain unchanged until such time as the least squares analysis was repeated.

We are now suggesting a rather different approach. Given that we have a series, with no trend, and given that we are mainly interested in predicting just a few periods ahead, we can simply calculate the average value of our observations and use this as our prediction. Let us take monthly data as our example. We calculate the average monthly value of our data and use this to predict the next few values, say the next three values, of the series. When next month's value becomes available we recalculate the average and this now becomes our prediction. We are thus continually updating our prediction. The reader will immediately point out that, using this system, we will increase the span of our data by one every time we receive a new observation and that, apart from accumulating an ever-longer series, the addition of one extra value will make hardly any difference to our average, since this new observation will be averaged over a long run of data. This is perfectly true, so we abandon this simple idea without further ado and look for a better one. However, the argument has been essential and has served a purpose.

Since this section is headed 'Moving Averages' our next step is obvious. Instead of a simple arithmetic averaging process, applied to an ever-lengthening series, we use a moving average. We choose the number of observations in our average according to the methods outlined in Chapter 4; our object is to obtain a smoothed series, but one which reacts to a change in average level of the series more rapidly than the simple arithmetic average discussed above.

If our series has a significant seasonal component, the above method still applies, although now we adjust our predictions to allow for the seasonal effects.

Finally, we can abandon the restriction that our series must be trend-free, and consider how to allow for trend. We could of course go right back to estimating the trend by least squares, but if we did so we would probably use this, plus seasonal factors if seasonality was present, to produce our predictions. A simple way to estimate trend would be to take the latest value of the moving average and the value of say six months previously, take the difference between the two and divide this by six. Or we could take the latest value and the value nine months previously and divide the difference by 9, or twelve months previously and divide the difference by 12, etc. We could also with very little trouble devise some more complicated ways of getting at the same result, namely the trend expressed as an average monthly increase or decrease. Whatever method we choose, we add the trend to our prediction. Thus our prediction for one month ahead becomes our updated average plus the trend value just calculated; that for two months ahead our average plus 2 × trend and so on. These predictions are further adjusted for the seasonal effect, where this exists.

We now have a feasible prediction system, although it is not based on any powerful theoretical considerations, and clearly becomes more and more shaky as the range of the forecast increases. The system as it stands, however, is not commonly used, which is why we have not illustrated the above steps with examples. The variation of it which is in widespread use is based on a different form of the moving average. The moving averages used up to now have been simple moving averages; once the number of terms in the average has been decided every term has the same weight. But we are not restricted to this weighting system. We could construct a weighted average such that the most recent term has the largest weight, with the weights decreasing as the terms in the average become 'older'. There is nothing whatever to prevent us adjusting the weights in this way provided that we make the weights sum to unity. This new form of the weighted average will be more sensitive to changes in the data and will also be more up to date than, say, a seven-month moving average because the latter is centred on the fourth term and is therefore four months old as soon as it is calculated.

We will now investigate the properties of our new form of moving average and compare it with the simple type. This investigation will require the use of some simple but rather long-winded algebra, but by now the reader will be expecting this and will take it in his stride. We begin by defining our terms. Assume that we wish to predict monthly sales of a product, one month, two months, three months ahead, etc.

Let d_t = demand in month t and m_t = mean demand, calculated at month t.

294 MANAGING WITH STATISTICS

Consider first the simple unweighted moving average. This can be written, using the above symbols as:

$$m_t = \frac{1}{n}(d_t + d_{t-1} + d_{t-2} + ... + d_{t-(n-1)})$$

where n is the number of terms in the average.

This could also be written as:

$$m_t = a_0 d_t + a_1 d_{t-1} + a_2 d_{t-2} + ... + a_{n-1} d_{t-(n-1)}$$

where

$$a_0 + a_1 + ... + a_{n-1} = 1$$

and

$$a_0 = a_1 = a_2 = ... = a_{n-1}.$$

If more weight is to be given to the more recent terms, we need a systematic way of choosing the weights. This way must ensure that the weights add to unity. An easy way of meeting this requirement is to make the coefficients a_0, a_1 etc. the terms of a geometrical progression.

The reader will recall that the general form of a geometrical progression is:

$$a, ar, ar^2, ..., ar^n, ...$$

where a is the first term and r the common ratio (e.g. with $a = 1$ and $r = 2$ we have 1, 2, 4, 8, 16, etc.).

Now, in our case $r < 1$, since each successive term is smaller than the preceding one. From our knowledge of the geometrical progression we know that with $r < 1$, the sum to infinity is given by the expression $\frac{a}{1-r}$ (the series is infinite, i.e. never ending, but the terms gradually become infinitesimally small and the sum never exceeds the limit $\frac{a}{1-r}$).

Thus, our requirement that the weights should sum to unity can be expressed as

$$\frac{a}{1-r} = 1.$$

(We note at this point that we are implying that our moving average is of infinite length. This contrasts with the simple moving average where the number of terms is fixed. The new form of the moving average is close to our original idea, page 292, of a continually updated averaging process. We shall see in a moment that although we are keeping all the terms in our series the arithmetic is spectacularly simple.)

Since $\dfrac{a}{1-r} = 1$, given either a or r, the other is immediately fixed, so in order to assign numerical weights to our average we can choose either a value for a or a value for r. In practice we choose a value for r, and we do this by deciding at what rate the weights should decrease. Thus if we decided that each weight should be 80% of the preceding weight we choose $r = 0.8$ and we immediately have $a = 0.2$. Similarly if we choose $r = 0.75$, a is immediately 0.25. r is sometimes referred to as the *discount factor*.

The value we choose for r depends on how much weight we wish to give to the more recent as opposed to the more ancient observations in our series. If we choose $r = 0.99$ then the weights decrease very slowly indeed; if we choose $r = 0.01$ then the weights decrease extremely rapidly and only the one or two most recent observations have any effect on our average. In practice a value for r of 0.8 seems to be a good starting point; other values close to 0.8 can be tried if it seems worth it and the value which gives the best results by trial and error selected.

If we take $r = 0.8$ as suitable for our purpose, the geometric progression of weights is as follows:

$$a \qquad ar \qquad ar^2 \qquad ar^3 \qquad \text{etc.}$$

$$0.2 \quad 0.16 \quad 0.128 \quad 0.1024$$

Our weighted moving average then becomes:

$$m_t = 0.2d_t + 0.16d_{t-1} + 0.128d_{t-2} + 0.1024d_{t-3} + \dots.$$

This expression would be impossibly tedious to calculate every month, but fortunately the arithmetic can be reduced to the level of a triviality. To derive the appropriate expression we write the weighted average as

$$m_t = ad_t + ard_{t-1} + ar^2d_{t-2} + \dots$$

or $$m_t = a(d_t + rd_{t-1} + r^2d_{t-2} + \dots$$

Now, the average calculated for the previous month can be written:

$$m_{t-1} = a(d_{t-1} + rd_{t-2} + r^2d_{t-3} + \dots).$$

Therefore, multiplying both sides by r:

$$rm_{t-1} = a(rd_{t-1} + r^2 d_{t-2} + r^3 d_{t-3} + \dots).$$

Therefore,

$$m_t - rm_{t-1} = a(d_t + rd_{t-1} + r^2 d_{t-2} + \dots) - a(rd_{t-1} + r^2 d_{t-2} + \dots)$$

$$= ad_t.$$

Therefore

$$m_t = ad_t + rm_{t-1}.$$

But $r = 1 - a$.

Therefore $$m_t \doteq ad_t + (1-a)m_{t-1} \tag{1}$$

i.e. the current average is simply the weighted average of the latest observation (d_t) and last month's mean (m_{t-1}), with weights equal to a and $(1-a)$. Thus the infinitely long averaging process is reduced to a very easy weighted average. The factor a is often referred to as the *smoothing factor*.

We note that a must lie in the range

$$0 \leqq a \leqq 1.$$

Hence if $$a = 0, m_t = m_{t-1},$$

and if $$a = 1, m_t = d_t.$$

In general terms, the higher the value of a, the more weight is given to the most recent observations.

Another illuminating way of looking at the smoothing equation is to re-arrange the terms to give:

$$m_t = m_{t-1} + a(d_t - m_{t-1}). \tag{2}$$

This shows that the latest estimate of m_t is equal to the previous estimate plus a proportion of the difference between the latest observation and last month's estimate of the mean.

In the absence of any trend or seasonal effect, m_t is used to predict m_{t+1}, m_{t+2}, etc. Where seasonal effects are present, the predicted values are adjusted by the appropriate seasonal factor. Where a trend is present, however, the situation differs from the previous example. If we denote the trend by b, where b represents the latest estimate of the monthly trend, we can rewrite our basic equation, equation (1), as

$$m_t = ad_t + (1-a)(m_{t-1} + b_{t-1}). \tag{3}$$

If the trend itself is thought to be gradually changing, we can derive b_t from the previous value b_{t-1} by repeating our weighting operation.

The formula we use is

$$b_t = (1-h)b_{t-1} + h(d_t - m_{t-1}). \qquad (4)$$

Where h is another smoothing factor, but is much smaller than a. Experience suggests a value for h of about 0·02.

Our predictor is now $m_t + b_t$ for one month ahead, $m_t + 2b_t$ for two months ahead and so on.

We can put a final flourish on our formulae by once again re-arranging terms, this time in order to make the practical computation even easier.

If we write

$$e_t = d_t - (m_{t-1} + b_{t-1})$$

i.e. e_t = latest observation − last month's prediction

we can rewrite our formulae as:

$$m_t = m_{t-1} + b_{t-1} + ae_t, \qquad (5)$$

and $$b_t = b_{t-1} + he_t. \qquad (6)$$

Our prediction for k months ahead, p_{t+k}, is written:

$$p_{t+k} = m_t + kb_t \qquad (7)$$

or $$p_{t+k} = (m_t + kb_t)s_{t+k} \qquad (8)$$

where s_{t+k} is the seasonal factor for month $(t+k)$.

When we come to apply these formulae, as we will in a moment, we run up against the problem that we need starting values for m, and b. These can be approximated fairly roughly since their values are not critical. m can be estimated from the average of the past twelve months, if rates are fairly stable. Alternatively, a centred twelve-month moving average can be calculated and extended back freehand to estimate m.

For b, a method similar to that described when dealing with simple moving averages can be applied.

It is important to have a run of past data over which to 'run-in' the system before using it for predicting future values. Not only does the running-in period correct any errors in the selection of the starting values but it also enables us to experiment, if necessary, with different values to see which give the best fit to the data.

The system we have just studied is the basic form of the so-called *exponential smoothing*. The particular version we have presented is known as 'Holt's Method' after Professor C. C. Holt, who introduced it. A number of variations have been produced but the underlying method is basically the same. One of the variations is to use only one parameter, instead of two. Formulae (5) and (6) then become:

$$m_t = m_{t-1} + b_{t-1} + (1 - \beta^2)e_t \qquad (9)$$

$$b_t = b_{t-1} + (1 - \beta)^2 e_t. \qquad (10)$$

Here we have the single parameter β instead of the two parameters a and h. If we choose $\beta = 0.9$ we have the equivalent of our previous version with $a = 0.81$ and $h = 0.01$. A single-parameter version is rather easier to experiment with since we need only specify one set of values.

We illustrate the method of exponential smoothing with a fully worked example using this single-parameter version. The example is set out in an Appendix to this chapter.

Another variation which has become very well known is the *Box-Jenkins* method, named after its originators. This is a single-equation extension of exponential smoothing which takes into account not only the familiar error term e_t but also the difference between the last two errors:

$$D(e_t) = e_t - e_{t-1}$$

and the sum of all the errors to date

$$S(e_t) = e_t + e_{t-1} + e_{t-2} + \dots$$

The prediction equation for m_{t+1} is:

$$m_{t+1} = m_t + \gamma_{-1}D(e_t) + \gamma_0 e_t + \gamma_1 S(e_t). \qquad (11)$$

The terms in the equation are related to the control engineer's proportional control ($\gamma_0 e_t$), differential control ($D(e_t)$), and integral control ($S(e_t)$).

The parameters γ_{-1}, γ_0 and γ_1 are estimated by trial and error using a computer. The criterion used for a good prediction is the sum of squares of the errors, i.e. Σe_t^2. With this criterion, and with a range of trial values for the parameters, the computer is programmed to select the set of parameters which minimises Σe_t^2.

The term $\gamma_{-1}D(e_t)$ can be dropped from the prediction equation without affecting the accuracy of the prediction appreciably, although this term turns up in the full control problem. Furthermore, Σe_t^2 is not very sensitive to the precise choice of γ_0 and γ_1.

The Box-Jenkins method was devised for the purposes of industrial process control but is widely used in applications such as stock control, and is built into a number of computer packages.

Exponential smoothing is widely used today, especially in the field of stock control and production scheduling where a particular firm may have hundreds of series to predict every month. It will be apparent to the reader that as a basis for long-term forecasting, exponential smoothing is very shaky. For short-term predictions, however, it is often very difficult to make much improvement on exponential predictions. The outstanding advantage of the system is the extreme ease of computation and the very small quantity of data which has to be carried. This makes the system very easy to use with a computer: commercial type computers are usually only equipped to carry out simple arithmetical operations while the scientific type often have very limited storage space. When a company has hundreds of series to predict this consideration is a weighty one. It must be said, however, that the dazzling perfection of the arithmetic tends to divert attention from the imperfections of the underlying theory (or lack of it).

OTHER TIME SERIES METHODS

The time series methods dealt with in the foregoing sections were very simple, involving nothing more than moving averages, trend estimation and seasonal factors. There are far more complex methods of time series analysis available but these are outside the scope of this book. One important one is worth mentioning here, however. This is the use of a single series of data in a regression equation. In such an equation the independent variables are, in fact, lagged values of the dependent variable. For example, if we denote the dependent variable by y_t the regression equation might be

$$y_t = ay_{t-1} + a_2 y_{t-2} + a_3 y_{t-3} + \dots$$

where $a_1, a_2, a_3 \dots$ are coefficients estimated from the data.

There is no necessity for the dependent variables to be the immediately preceding values of the dependent variable, nor need they be in unbroken sequence as they are in the example above. If the lag between the dependent variable and the most recent lagged value is greater than one period, the equation can obviously be used for prediction purposes. For example, if the dependent variable y_t

depended on y_{t-3} and earlier values in the series, we could predict up to three periods ahead of current data.

There are a number of pitfalls in using regression analysis in this way. The main problem arises from the fact that in most economic time series the terms are correlated. The effect of this will be discussed in more detail in the next section.

REGRESSION METHODS

All the time series methods considered in the previous section were naïve in that they used no information other than that contained in the series under investigation. Such methods enable us to make predictions about future values of the series, but such predictions, especially if we merely fit a least squares curve to the data and extrapolate, are often of dubious value since they assume no important changes in the mechanism which is creating the series in question, e.g. a prediction of steel consumption based solely on past data would assume that the relative importance of substitute products, such as plastics, would remain unchanged. More sophisticated predictions can, therefore, be made by relating the series in which we are interested to other series. Thus if we wish to predict the demand for cars we might relate car production (or new car registrations in the U.K.) to factors such as consumer expenditure, business income, Bank Rate and H.P. controls. The technique of multiple regression enables us to relate car registrations to as many other variables as we wish. The general form of the relationship can be illustrated with the example of car registrations.

Let y = new registrations

$\quad x_1$ = consumer expenditure

$\quad x_2$ = business income

$\quad x_3$ = Bank Rate

$\quad x_4$ = H.P. controls.

We then have an equation of the form

$$y = b_0 + b_1 x_1 + b_2 x_2 + b_3 x_3 + b_4 x_4 + u$$

where b_0 is the constant term, $b_1, b_2, \ldots$ are the regression coefficients and u is the error term, assumed to be Normally distributed with zero mean. The above equation happens to be linear, but there is no reason why it should not be curvilinear if the data suggest this.

Multiple linear regression, however, accounts for the overwhelming majority of multiple regression analysis. Equations such as the one above are frequently termed *regression models* for reasons which will become clear in a moment. They have at least three distinctive uses:

1. They are analytical, i.e. they identify the mechanism giving rise to the time series and indicate the relative importance of various factors.
2. They enable forecasts to be made. The dependent variable is forecast by first predicting or forecasting the regressor variables and then using the equation to obtain a forecast of the dependent variable.
3. They enable the effects of various assumptions about the future levels of the regressor variables to be studied. This is what is meant by the term 'model'. The model can be manipulated and, where the firm, or perhaps the government, has control over some of the regressor variables, the likely outcomes of various courses of action can be estimated, and the most suitable course of action undertaken.

We are concerned mainly with the second of the three uses, in this chapter. We have already seen in the chapter on regression how the coefficients in the equation are derived by the least squares method. In this chapter we are more concerned with:

(*a*) How do we decide which variables to include in a regression model?

(*b*) How do we predict (or forecast) the regressor variables?

CHOICE OF VARIABLES

In the example of car registrations above, we presented the regressor variables without comment on their source. In practice, one selects those series which it is reasonable to think might be related to the dependent series. This selection might be based on economic theory, common sense, engineering knowledge, or just plain hunch. Often, several of the series selected as regressor series are rejected in the course of analysis because they are not statistically significant. This part of the forecasting exercise, the building and testing of the model, is very time consuming, because it proceeds by trial and error, and is very frustrating, because with all economic time series the precise form of the data required usually differs from that available.

There are a number of other problems concerning the application of regression techniques to economic data. These problems arise from the fact that regression analysis is based on a number of assumptions, most of which are broken by economic data. Some of the more important problems are described below:

Autocorrelation. Regression theory assumes that the error terms (u in the equation above) are uncorrelated. This is rarely the case with economic data, where successive observations are often highly correlated. The effect of applying straightforward regression methods to autocorrelated series is to obtain estimates of the coefficients, which are not only inaccurate but which are likely to be accepted as statistically significant when in fact they are not.

Multi-collinearity. If two or more of the regressor variables are highly correlated it becomes very difficult, if not impossible, to disentangle their separate effects. In other words, we gain very little information from the additional 'independent' series, e.g. a regression model which included both Gross National Product and Net Disposable Personal Incomes, over a long period, as regressor variables would not gain much by having both of these series present, and would make the estimation of the two regression coefficients very difficult.

Errors in variables. Most economic statistics contain errors of measurement arising from difficulties in collecting the data or difficulties in the concepts themselves, e.g. the concept of 'the general level of retail prices' is rather vague and there are many practical difficulties in collecting price data, so an index of retail prices is certainly subject to errors. Standard regression analysis assumes that all regressor variables are measured without error. If this assumption does not hold, estimates of the parameters in the equation will be biased.

The problems outlined above can be tackled using methods which are beyond the scope of this book. The study of such problems is part of the subject-matter of *Econometrics*; we mention them here merely to point out the danger of applying straightforward regression techniques to economic series. The danger is particularly acute in these days of computers. Every computer manufacturer provides a standard multiple regression 'package' which acts as a sausage machine, receiving any data that anyone cares to insert, mincing it up thoroughly, and producing a string of regression coefficients at the other end. The misuse of these packages, by their application to invalid data, is probably the most common crime committed in the name of statistics today.

FORECASTING THE REGRESSOR VARIABLES

When we have completed the testing and revision of our model[1] we can begin to use it for forecasting purposes. We will have an equation of the general form

$$y = b_0 + b_1 x_1 + b_2 x_2 + b_3 x_3 + b_4 x_4 + \ldots$$

only by now we will have numerical values for the b_0, b_1, ...

If we wish to forecast the value of y at time t, we have to estimate the values of x_1, x_2, ... at time t, insert these values into the equation and calculate y. The estimation of future values of x_1, x_2, ... is carried out as best we can. Some of the values might be predicted from past data, some (especially population size and structure, which occur in many forecasting exercises) might be projected, some are based on judgement (or guess-work) and some might be taken from existing official and other forecasts. This last point is quite important. One of the practical problems in using regression models is that a variety of economic series might be included in the model. If the statistician tries to become expert in all the fields of economic activity which give rise to these series, so that he can make valid forecasts of the independent variables, he might never get to the stage of forecasting his own firm's product. Nor can the problem be solved by employing a few assistants: the number required would certainly create an empire for him but would almost certainly be vetoed by top management. (We are not speaking of the top few industrial giants here but of the great majority of medium-to-large firms.)

The most common solution to the problem is for the statistician to familiarise himself with the sources of forecasts for the series in which he is interested and perhaps to make contacts among the individuals who produce these forecasts. He can then check, to

---

[1] Many models consist of more than one equation, with some variables appearing in more than one equation. We shall restrict ourselves to the single equation model of the kind shown above.

some extent, the confidence which his contacts have in their forecasts and thus decide whether to use them as they stand, or modify them if they appear to be too optimistic or pessimistic, or reject them altogether.

There are of course situations in which straightforward predictions of the regressor series yield reliable forecasts of the dependent variable. This situation could arise where, say, sales of a particular product depend on the combined effect of a number of external factors, each of which gives rise to a stable, well-behaved series for which a simple prediction is an adequate means of forecasting. Such situations are rare, however, and all too often the forecaster, with the aid of a multiple regression package, transforms his problem from that of predicting a single series (his own product) to that of predicting half a dozen other series and then using these predictions in a regression model.

It is now time to illustrate the ideas of this section by means of an example. We are unable to give much general advice on how to choose the regressor variables: this depends on a deep knowledge of the subject-matter, the availability of data and good luck. Nor can we illustrate the calculations without going into a great deal of detail which is beyond the scope of this book. In fact, when there are more than two regressor variables the calculations become unbearably tedious on a desk machine and resort is made to a computer package.

What we shall do, therefore, is to quote a real-life example showing the regressor variables used in the equation, and the numerical values of the coefficients.

The example is taken from a paper by O'Herlihy entitled 'Demand for Cars in Great Britain', published in *Applied Statistics*, 1965, **14**, Nos. 2 and 3.[1] The regressor variables used in the final equation were:

r = real personal disposable income per head at 1954 prices,

p_0 = price index of new cars relative to the price of all other goods and services,

I = index of supply, the ratio of the used car price index to the new car price index,

t = 'taste' factor, represented by a time trend,

[1] Although not published until 1965, this paper was written in 1962.

$H =$ hire purchase terms, represented by the average contract period,

$B =$ real business income at 1954 prices,

$q-1 =$ new registrations in the previous year.

All the above were in the form of annual figures.

A further basic assumption, which does not appear explicitly in the equation is that the depreciation rate can be approximated by a 50% reducing balance scheme, i.e. at the end of a given year the car is worth 50% of its value at the beginning of the year, at constant prices.

The description of the regressor variables makes it obvious that as much, or more, work needs to be put into the construction of the data as into the mathematics of the model. Only rarely can one turn to a set of ready-made data and simply copy them out of a statistical return.

In fact, the regressor variables above were used in a rather more complicated manner than might be expected. Instead of taking the values as defined above, the difference between each value and the corresponding value for the previous year was added to the actual values. If we denote the difference by the symbol Δ the independent variables were used in the form:

$$(1+\Delta)r, (1+\Delta)p_0, \text{ etc.}$$

The point of mentioning all this is to emphasise that a knowledge of multiple regression techniques is often less than half the battle in building a realistic model. The reader is recommended to look at the original paper for a fuller discussion of these points.

The final equation fitted to the data was

$$q = -2\cdot484+1\cdot120r-0\cdot401p_0+0452I+0\cdot066t+0\cdot113H$$

$$+1\cdot496B-0\cdot372(q-1)$$

where q, the dependent variable represents new car registrations in year q.

The graph below shows the results of the exercise. The model was fitted to the data of 1948-61. The broken line labelled 'Estimated' was obtained from the equation above. Forecasts of the regressor variables were then made and fed into the equation, producing projected figures of car registrations for the years 1962–1966.

The original paper includes a section on the forecasts of the regressor variables.

Since the model was constructed, data for the projected years

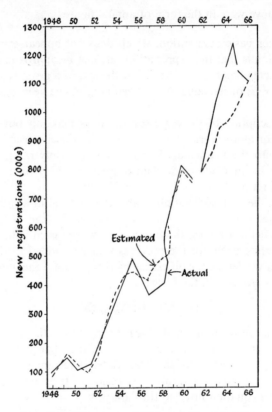

FIG. 10.1. New car registrations in Great Britain.

1962-66 have become available, and have been inserted in Mr O'Herlihy's chart. The immediate conclusion from an inspection of the chart is that the model was of limited use in predicting future values. A comparison of the forecast figures with the actual outturn has been made by Mr O'Herlihy and published in the *National Institute Economic Review* for May 1967. The reader is strongly recommended to study this comparison.

APPENDIX: SHORT-TERM FORECASTING

WORKED EXAMPLE

Data

Net Home Deliveries from Home Production of
Steel Sheets to the Motor Industry

| | Weekly averages—thousand tons | | | | | |
	1961	1962	1963	1964	1965	1966
January	...	13·51	14·20	13·66	20·37	18·43
February	...	15·02	18·57	21·55	21·40	20·40
March	...	17·54	19·24	21·50	16·26	18·46
April	...	14·42	16·43	18·97	15·84	17·79
May	...	19·12	17·28	18·25	21·50	18·29
June	...	15·57	17·93	16·35	18·70	17·39
July	11·89	10·06	11·34	9·77	11·26	9·83
August	10·47	13·58	16·72	15·85	18·30	15·90
September	13·05	16·41	18·33	16·90	16·35	13·45
October	12·05	13·70	19·55	18·96	17·73	11·89
November	10·83	13·83	18·14	18·42	17·03	11·10
December	9·84	10·95	15·25	15·23	13·82	9·97

Source: *Monthly Statistics*, B.I.S.F.

Formulae

Notation:

d_t = actual deliveries in tons for month t,

m_t = average monthly deliveries (exponentially weighted) calculated at month t,

b_t = trend value calculated at month t,

$F_k(t)$ = forecast for k months ahead made at month t,

$e_t = d_t - F_1(t-1)$, i.e. error in forecast for month t made in previous month,

s_j = seasonal factor for month j. $j = 1 - - 12$,

s_{j+k} = seasonal factor for k months ahead,

β = smoothing constant.

Average Level of Deliveries:

$$m_t = m_{t-1} + b_{t-1} + (1-\beta^2)e_t,$$
$$b_t = b_{t-1} + (1-\beta)^2 e_t.$$

L

Forecast k Months Ahead:

$$F_k(t) = (m_t + kb_t)s_{j+k}.$$

Note. For a series with no significant seasonality,

$$s_j = 1 \text{ for all } j \text{ and therefore } s_{j+k} = 1 \text{ also.}$$

Seasonal Adjustment Factors (based on data from July 1961 to June 1965)

Step 1. Estimate trend values by means of a 12-month centred moving average. This is identical to a 13-month weighted moving average with weights 1, 2, 2, ..., 2, 2, 1. The latter is easier to compute.

Estimation of Trend Values

	1962		1963		1964	
	13-month total	Weighted mean	13-month Total	Weighted mean	13-month Total	Weighted Mean
January	324·8	13·5	365·6	15·2	417·6	17·4
February	326·1	13·6	370·1	15·4	415·2	17·3
March	332·5	13·9	375·1	15·6	412·9	17·2
April	337·6	14·1	382·9	16·0	410·9	17·1
May	342·2	14·3	393·1	16·4	410·6	17·1
June	346·3	14·4	401·7	16·7	410·8	17·1
July	348·1	14·5	405·4	16·9	417·5	17·4
August	352·4	14·7	407·9	17·0	424·1	17·7
September	357·6	14·9	413·1	17·2	418·7	17·4
October	361·3	15·1	417·9	17·4	410·3	17·1
November	361·5	15·1	421·4	17·6	410·4	17·1
December	362·0	15·1	420·8	17·5	416·1	17·3

Step 2. Deduct trend values from their corresponding actual values and form a table of these 'residuals'.

Step 3. Calculate the average residual value for each of the twelve months and the overall average.

Step 4. Deduct the overall mean from each of the twelve monthly residual means. The twelve values should now sum to zero.

Calculate the mean of the original values from July 1961 to June 1965.

Steps 2 *and* 3

Residual Values

	1962	1963	1964	Totals	Means
January	...	−1·0	−3·7	−4·7	−1·567
February	1·4	3·2	4·3	8·9	2·967
March	3·6	3·6	4·3	11·5	3·833
April	0·3	0·4	1·9	2·6	0·867
May	4·8	0·9	1·2	6·9	2·300
June	1·2	1·2	−0·7	1·7	0·567
July	−4·4	−5·6	−7·6	−17·6	−5·867
August	−1·1	−0·3	−1·8	−3·2	−1·067
September	1·5	1·1	−0·5	2·1	0·700
October	−1·4	2·2	1·9	2·7	0·900
November	−1·3	0·5	1·3	0·5	0·167
December	−4·1	−2·3	−2·1	−8·5	−2·833
				2·9	0·081

Step 4

	Residual monthly averages corrected for their mean (1)	Col. 1 +15·923* (2)	Col. $2 \times \dfrac{100}{15\cdot923}$ = seasonal adjustment factor (3)
January	−1·648	14·275	89·7
February	2·886	18·809	118·1
March	3·752	19·675	123·6
April	0·786	16·709	104·9
May	2·219	18·142	113·9
June	0·486	16·409	103·1
July	−5·948	9·975	62·6
August	−1·148	14·775	92·8
September	0·619	16·542	103·9
October	0·819	16·742	105·1
November	0·086	16·009	100·5
December	−2·914	13·009	81·7
	0		1199·9

* 15·923 is the average value of the original values used in obtaining seasonal adjustment factors, i.e. original data from July 1961 to June 1965.

Add this to each of the twelve monthly means.

Adjust these twelve new values to sum to 1200. This is done by multiplying each monthly mean by 100/mean of original values. The resulting values are the twelve seasonal adjustment factors.

Note. There are a number of other methods for estimating seasonal adjustment factors. This one is relatively easy to follow and to use on a desk machine.

Starting Values

The formulae for m_t and b_t are recursive, i.e. their values in a given month are based on values obtained in the previous month. Starting values are therefore required in order to launch the system.

Denoting the initial values by the suffix o we need m_o, b_o and $F_1(o)$. No elaborate calculations are necessary, we can choose values which are near enough for our purpose by reference to the graph of the data. Let

$$m_o = 14 \cdot 00$$

$$b_o = 0 \cdot 20,$$

therefore $F_1(o) = 8 \cdot 89$, i.e. $(14 \cdot 00 + 0 \cdot 20)\ 62 \cdot 6$

Note. In practice a 'running-in' period for the system is allowed. This is why too much trouble need not be taken over the starting values.

Let $\beta = 0 \cdot 9$. This value is obtained from an empirical rule. In practice, especially with a computer, several values of β could be tested and the one giving the best results selected.

Notes on Computation

The steps in the computation proceed as follows:

1. Choose a value for β, calculate $(1-\beta)^2$ and $(1-\beta^2)$ and enter these three figures at the top of the work sheet.

2. Set out the columns of the work sheet. It will be found helpful to include a column $(m_t + kb_t)$ for all the values of k to be used, e.g. we are forecasting 1, 3, 6 and 9 months ahead so we have columns

$$(m_t + b_t),\ (m_t + 3b_t),\ (m_t + 6b_t),\ (m_t + 9b_t).$$

3. Enter in the work sheet the starting values of m_t, b_t and $F_1(t)$, also the original data d_t and the seasonal factors s_j.

$$\beta = 0\cdot9; \quad (1-\beta)^2 = 0\cdot01; \quad (1-\beta^2) = 0\cdot19$$

	d_t	m_t	b_t	m_t+b_t	m_t+3b_t	m_t+6b_t	m_t+9b_t	s_j	e_t	$(1-\beta)^2 e_t$	$(1-\beta^2)e_t$	$F_1(t)$	$F_3(t)$	$F_6(t)$	$F_9(t)$
		14·00	0·20									8·89			
1961															
July	11·89	14·77	0·23	15·00	15·46	16·15	16·84	62·6	3·00	0·03	0·57	13·92	16·25	14·49	17·67
August	10·47	14·34	0·20	14·54	14·94	15·54	16·14	92·8	−3·45	−0·03	−0·66	15·11	15·01	18·35	18·39
September	13·05	14·15	0·18	14·33	14·69	15·23	15·77	103·9	−2·06	−0·02	−0·39	15·06	12·00	18·82	16·26
October	12·05	13·76	0·15	13·91	14·21	14·66	15·11	105·1	−3·01	−0·03	−0·57	13·98	12·75	15·38	9·46
November	10·83	13·31	0·12	13·43	13·67	14·03	14·39	100·5	−3·15	−0·03	−0·60	10·97	16·14	15·98	13·35
December	9·84	13·22	0·11	13·33	13·66	13·99	14·32	81·7	−1·13	−0·01	−0·21	11·96	16·88	14·42	14·88
1962															
January	13·51	13·62	0·13	13·75	14·01	14·40	14·79	89·7	1·55	0·02	0·29	16·24	14·70	9·01	15·54
February	15·02	13·52	0·12	13·64	13·88	14·24	14·60	118·1	−1·22	−0·01	−0·23	16·86	15·81	13·21	14·67
March	17·54	13·77	0·13	13·90	14·16	14·55	14·94	123·6	−0·68	0·01	−0·13	14·58	14·60	15·12	12·21
April	14·42	13·87	0·13	14·00	14·26	14·65	15·04	104·9	−0·16		−0·03	15·95	5·66	15·40	13·49
May	19·12	14·60	0·16	14·76	15·08	15·56	16·04	113·9	3·17	0·03	0·60	15·22	13·99	15·64	18·94
June	15·57	14·83	0·16	14·99	15·31	15·79	16·27	103·1	0·35		0·07	9·38	15·91	12·90	20·11
July	10·06	15·12	0·17	15·29	15·63	16·14	16·65	62·6	−0·68	0·01	−0·13	14·19	16·43	14·48	17·47
August	13·58	15·17	0·16	15·33	15·65	16·13	16·61	92·8	−0·61	−0·01	−0·12	15·93	15·73	19·05	18·92
September	16·41	15·42	0·16	15·58	15·90	16·38	16·86	103·9	0·48	−0·01	−0·09	16·37	12·99	20·25	17·38
October	13·70	15·07	0·13	15·20	15·46	15·85	16·24	105·1	−2·67	−0·03	−0·51	15·28	13·87	16·63	10·17
November	13·83	14·92	0·12	15·04	15·28	15·64	16·00	100·5	−1·45	−0·01	−0·28	12·29	18·05	17·81	14·85
December	10·95	14·79	0·11	14·90	15·12	15·45	15·78	81·7	−1·34	−0·01	−0·25	13·37	18·69	15·93	16·40

$$\beta = 0.9; \quad (1-\beta)^2 = 0.01; \quad (1-\beta^2) = 0.19 \quad (continued)$$

	d_t	m_t	b_t	m_t+b_t	m_t+3b_t	m_t+6b_t	m_t+9b_t	s_j	e_t	$(1-\beta)^2 e_t$	$(1-\beta^2)e_t$	$F_1(t)$	$F_3(t)$	$F_6(t)$	$F_9(t)$
1963															
January	14.20	15.06	0.12	15.18	15.42	15.78	16.14	89.7	0.83	0.01	0.16	17.92	16.18	9.88	16.96
February	18.57	15.30	0.13	15.43	15.69	16.08	16.47	118.1	0.65	0.01	0.12	19.07	17.87	14.92	16.55
March	19.24	15.46	0.13	15.59	15.85	16.24	16.63	123.6	0.17	—	0.03	16.35	16.34	16.87	13.59
April	16.43	15.61	0.13	15.74	16.00	16.39	16.78	104.9	0.09	—	0.02	17.93	10.02	17.23	15.05
May	17.28	15.62	0.12	15.74	15.98	16.34	16.70	113.9	−0.65	−0.01	−0.12	16.23	14.83	16.42	19.72
June	17.93	16.06	0.14	16.20	16.48	16.90	17.32	103.1	1.70	0.02	0.32	10.14	17.12	13.81	21.41
July	11.34	16.43	0.15	16.58	16.88	17.33	17.78	62.6	1.20	0.01	0.23	15.39	17.74	15.55	18.65
August	16.72	16.83	0.16	16.99	17.31	17.79	18.27	92.8	1.33	0.01	0.25	17.65	17.40	21.01	20.81
September	18.33	17.12	0.17	17.29	17.63	18.14	18.65	103.9	0.68	0.01	0.13	18.17	14.40	22.42	19.23
October	19.55	17.55	0.18	17.73	18.09	18.63	19.17	105.1	1.38	0.01	0.26	17.82	16.23	19.54	12.00
November	18.14	17.79	0.18	17.97	18.33	18.87	19.41	100.5	0.32	—	0.06	14.67	21.65	21.49	18.01
December	15.25	17.85	0.19	18.04	18.42	18.99	19.56	81.7	0.58	0.01	0.11	16.18	22.77	19.58	20.32
1964															
January	13.66	17.56	0.16	17.72	18.04	18.52	19.00	89.7	−2.52	−0.03	−0.48	20.93	18.92	11.59	19.97
February	21.55	17.84	0.17	18.01	18.35	18.86	19.37	118.1	0.62	0.01	0.12	22.26	20.90	17.50	19.47
March	21.50	17.87	0.16	18.03	18.35	18.83	19.31	123.6	−0.76	−0.01	−0.14	18.91	18.92	19.56	15.78
April	18.97	18.04	0.16	18.20	18.52	19.00	19.48	104.9	0.06	—	−0.01	20.73	11.59	19.96	17.47
May	18.25	17.73	0.14	17.87	18.15	18.57	18.99	113.9	−2.48	−0.02	−0.47	18.42	16.84	18.66	22.43
June	16.35	17.48	0.12	17.60	17.84	18.20	18.56	103.1	−2.07	−0.02	−0.39	11.02	18.54	14.87	22.94
July	9.77	17.36	0.11	17.47	17.69	18.02	18.35	62.6	−1.25	−0.01	−0.24	16.21	18.59	16.16	19.25
August	15.85	17.40	0.11	17.51	17.73	18.06	18.39	92.8	−0.36	—	−0.07	18.19	17.82	21.33	20.95
September	16.90	17.26	0.10	17.36	17.56	17.86	18.16	103.9	−1.29	−0.01	−0.25	18.24	14.35	22.07	18.72
October	18.96	17.50	0.11	17.61	17.83	18.16	18.49	105.1	0.72	0.01	0.14	17.70	15.99	19.05	11.57
November	18.42	17.75	0.12	17.87	18.11	18.47	18.83	100.5	0.72	0.01	0.14	14.60	21.39	21.04	17.47
December	15.23	17.99	0.13	18.12	18.38	18.77	19.16	81.7	0.63	0.01	0.12	16.25	22.72	19.75	19.91

Table header: $\beta = 0.9$; $(1-\beta)^2 = 0.01$; $(1-\beta^2) = 0.19$ (continued)

	d_t	m_t	b_t	m_t+b_t	m_t+3b_t	m_t+6b_t	m_t+9b_t	s_j	e_t	$(1-\beta)^2 e_t$	$(1-\beta^2)e_t$	$F_1(t)$	$F_3(t)$	$F_6(t)$	$F_9(t)$
1965															
January	20·37	18·90	0·17	19·07	19·41	19·92	20·43	89·7	4·12	0·04	0·78	22·52	20·36	12·47	21·47
February	21·40	18·86	0·16	19·02	19·34	19·82	20·30	118·1	−1·12	−0·01	−0·21	23·50	22·03	18·39	20·40
March	16·26	17·64	0·09	17·73	17·91	18·18	18·45	123·6	−7·24	−0·07	−1·38	18·60	18·47	18·89	15·07
April	15·84	17·21	0·06	17·27	17·39	17·57	17·75	104·9	−2·76	−0·03	−0·52	19·67	10·89	18·47	15·92
May	21·50	17·62	0·08	17·70	17·86	18·10	18·34	113·9	1·83	−0·02	0·35	18·25	16·57	18·19	21·66
June	18·70	17·79	0·08	17·87	18·03	18·27	18·51	103·1	0·45	—	0·09	11·19	18·73	14·93	22·88
July	11·26	17·88	0·08	17·96	18·12	18·36	18·50	62·6	0·07	—	0·01	16·67	19·04	16·47	19·41
August	18·30	18·27	0·10	18·37	18·57	18·87	19·17	92·8	1·63	0·02	0·31	19·09	18·66	22·29	21·83
September	16·35	17·85	0·07	17·92	18·06	18·27	18·48	103·9	−2·74	−0·03	−0·52	18·83	14·76	22·58	19·05
October	17·73	17·71	0·06	17·77	17·89	18·07	18·25	105·1	−1·10	−0·01	−0·21	17·86	16·05	18·96	11·42
November	17·03	17·61	0·05	17·66	17·76	17·91	18·06	100·5	−0·83	−0·01	−0·16	14·43	20·97	20·40	16·76
December	13·82	17·54	0·04	17·58	17·66	17·78	17·90	81·7	−0·61	−0·01	−0·12	15·77	21·83	18·33	18·60
1966															
January	18·43	18·09	0·07	18·16	18·30	18·51	18·72	89·7	2·66	0·03	0·51	21·45	19·20	11·59	19·67
February	20·40	17·96	0·06	18·02	18·14	18·32	18·50	118·1	−1·05	−0·01	−0·20	22·27	20·66	17·00	18·59
March	18·46	17·30	0·02	17·32	17·36	17·42	17·48	123·6	−3·81	−0·04	−0·72	18·17	17·90	18·10	14·28
April	17·79	17·25	0·02	17·31	17·31	17·37	17·43	104·9	−0·38	—	−0·07	19·62	10·84	17·13	15·63
May	18·29	16·98	0·01	16·99	17·01	17·04	17·07	113·9	−1·33	−0·01	−0·25	17·52	15·79	13·91	20·16
June	17·39	16·97	0·01	16·98	17·00	17·03	17·06	103·1	−0·13	—	−0·02	10·63	17·66	15·10	21·09
July	9·83	16·83	—	16·83	16·83	16·83	16·83	62·6	−0·80	−0·01	−0·15	15·62	17·69	19·94	17·65
August	15·90	16·88	—	16·88	16·88	16·88	16·88	92·8	0·28	—	0·05	16·88	16·96	19·60	19·23
September	13·45	16·10	−0·04	16·06	15·98	15·86	15·74	103·9	−4·09	−0·04	−0·78	16·88	13·06	15·28	16·23
October	11·89	15·11	−0·09	15·02	14·84	14·57	14·30	105·1	−4·99	−0·05	−0·95	15·10	13·31	15·35	8·95
November	11·10	14·26	−0·13	14·13	13·87	13·48	13·09	100·5	−4·00	−0·04	−0·76	11·54	16·38	13·33	12·15
December	9·97	13·83	−0·15	13·68	13·38	12·93	12·48	81·7	−1·57	−0·02	−0·30	12·27	16·54		12·97

4. Update the values of e_t, b_t and m_t, in that order, using the formulae of page 307. For July 1961 this gives:

(i) $e_t = 11\cdot89 - 8\cdot89 = 3\cdot00$
    therefore $(1-\beta)^2 e_t = 0\cdot03$ and $(1-\beta^2)e_t = 0\cdot57$,
(ii) $b_t = 0\cdot20 + 0\cdot03 = 0\cdot23$,
(iii) $m_t = 14\cdot00 + 0\cdot20 + 0\cdot57 = 14\cdot77$.

5. The $(m_t + kb_t)$ columns can now be filled in, e.g. for July 1961
$$m_t + b_t = 14\cdot77 + 0\cdot23 = 15\cdot00, \text{ etc.}$$

6. Compute $F_1(t)$ for $k = 1$ and any other required values of k. $F_1(t)$ *must* be computed because it comes into the computation of e_t. This is not so for other values of k and in practice it will be found quicker and easier to leave the other forecasts until all values of m_t, b_t, etc., have been calculated. Once the system is running, of course, all updating and all forecasts will be made as each new piece of data becomes available.

For July 1961, $F_1(t) = (15\cdot00)(92\cdot8)^* = 13\cdot92$. Note that this forecast (which is for August) is entered in the *July* row of the table. Similarly $F_3(t)$, i.e. the forecast made in July for October is given by $(15\cdot46)(105\cdot1) = 16\cdot25$.

In words, to make a forecast say six months ahead, you take the current value of the average (m_t) and the current value of the trend (b_t) and assume that the average level in six months time will be equal to current average plus six times the current trend (which simply says that m_t is increasing or decreasing by b_t per month). In order to estimate the *actual* value six months ahead you take the estimated average level and apply to this the seasonal factor for the month you are forecasting.

7. As a further example, take the month of August 1961. We have:

(i) $e_t = 10\cdot47 - 13\cdot92 = -3\cdot45$,
       therefore $(1-\beta)^2 e_t = -0\cdot03$ and $(1-\beta^2)e_t = -0\cdot66$.
(ii) $b_t = 0\cdot23 - 0\cdot03 = 0\cdot20$.
(iii) $m_t = 15\cdot00 - 0\cdot66 = 14\cdot34$.
(iv) $F_1(t) = (14\cdot54)(103\cdot9) = 15\cdot11$,
       $F_3(t) = (14\cdot94)(100\cdot5) = 15\cdot01$,
       $F_6(t) = (15\cdot54)(118\cdot1) = 18\cdot35$,
       $F_9(t) = (16\cdot14)(113\cdot9) = 18\cdot38$.

* All seasonal factors are in the form of percentages.

Note

(*a*) The slightly tricky 'diagonal' adding, e.g. in obtaining e_t, where values from one line in the table are added (or subtracted) to values in the preceding line, becomes much simpler once the system is running since from then on each line in the table builds up from a new value of d_t and only one line per month is dealt with.

(*b*) The figures in the example are not guaranteed accurate to the second decimal place although nearly all of them are that accurate. On a computer, more decimal places would be used since e_t, b_t and therefore m_t are quite sensitive to the accuracy in the second and subsequent decimal places.

(*c*) A short run-through the computations will demonstrate the advantages of having a computer available.

Graphs of Results

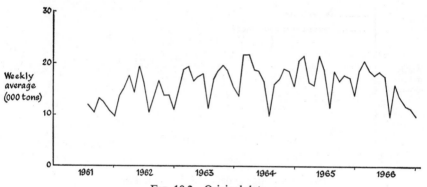

Fig. 10.2   Original data.

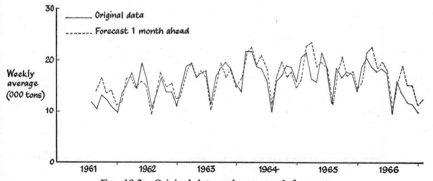

Fig. 10.3.   Original data and one-month forecast.

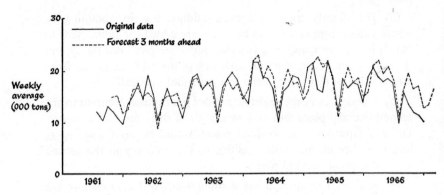

Fig. 10.4. Original data and three-month forecast.

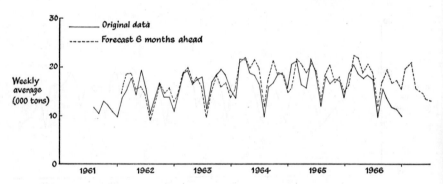

Fig. 10.5. Original data and six-month forecast.

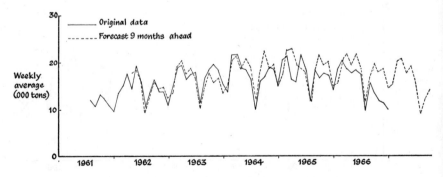

Fig. 10.6. Original data and nine-month forecast.

The Accuracy of the Forecasts

A simple measure of the accuracy of a series of forecasts is obtained by taking the average absolute percentage error of the forecasts, i.e. ignoring the signs of the percentage errors. Refinements can be made by considering the quartiles, etc. of the absolute errors. The table below gives the original data, the forecasts and the percentage error.

	d_t	$F_1(t)$	% error	$F_3(t)$	% error	$F_6(t)$	% error	$F_9(t)$	% error
1961									
Aug.	10·47	13·92	+33·0						
Sep.	13·05	15·11	+15·8						
Oct.	12·05	15·06	+25·0	16·25	+34·8				
Nov.	10·83	13·98	+29·1	15·01	+38·6				
Dec.	9·84	10·97	+11·5	12·00	+22·0				
1962									
Jan.	13·51	11·96	−11·5	12·75	−5·6	14·49	+7·2		
Feb.	15·02	16·24	+8·1	16·14	+7·4	18·35	+22·2		
Mar.	17·54	16·86	−3·9	16·88	−3·8	18·82	+7·3		
Apr.	14·42	14·58	+1·1	14·70	+1·9	15·38	+6·6	17·67	+22·5
May	19·12	15·95	−16·6	15·81	−17·3	15·98	−16·4	18·39	−3·8
June	15·57	15·22	−2·2	14·60	−6·2	14·42	−7·4	16·26	+4·4
July	10·06	9·38	−6·8	5·66	−43·7	9·01	−10·4	9·46	−6·0
Aug.	13·58	14·19	+4·5	13·99	+3·0	13·21	−2·7	13·35	−1·7
Sep.	16·41	15·93	−2·9	15·91	−3·0	15·12	−7·9	14·88	−9·3
Oct.	13·70	16·37	+19·5	16·43	+19·9	15·40	+12·4	15·54	+13·4
Nov.	13·83	15·28	+10·5	15·73	+13·7	15·64	+13·1	14·67	+6·1
Dec.	10·95	12·29	+12·2	12·99	+18·6	12·90	+17·8	12·21	+11·5
1963									
Jan.	14·20	13·37	−5·8	13·87	−2·3	14·48	+2·0	13·49	−5·0
Feb.	18·57	17·92	−3·5	18·05	−2·8	19·05	+2·6	18·94	+2·0
Mar.	19·24	19·07	−0·9	18·69	−2·9	20·25	+5·2	20·11	+4·5
Apr.	16·43	16·35	−0·5	16·18	−1·5	16·63	+1·2	17·47	+6·3
May	17·28	17·93	+3·8	17·87	+3·4	17·81	+3·1	18·92	+9·5
June	17·93	16·23	−9·5	16·34	−8·9	15·93	−11·2	17·38	−3·1
July	11·34	10·14	−10·6	10·02	−11·6	9·88	−12·9	10·17	−10·3
Aug.	16·72	15·39	−8·0	14·83	−11·3	14·92	−10·8	14·85	−11·2
Sep.	18·33	17·65	−3·7	17·12	−6·6	16·87	−7·9	16·40	−10·5
Oct.	19·55	18·17	−7·1	17·74	−9·3	17·23	−7·1	16·96	−13·2
Nov.	18·14	17·82	−1·8	17·40	−4·1	16·42	−9·5	16·55	−8·8
Dec.	15·25	14·67	−3·8	14·40	−5·6	13·81	−9·4	13·59	−10·9

	d_t	$F_1(t)$	% error	$F_3(t)$	% error	$F_6(t)$	% error	$F_9(t)$	% error
1964									
Jan.	13·66	16·18	+18·4	16·23	+18·9	15·55	+13·8	15·05	+10·2
Feb.	21·55	20·93	−2·9	21·65	+0·5	21·01	−2·5	19·72	−8·5
Mar.	21·50	22·26	+3·5	22·77	+5·9	22·42	+4·3	21·41	−0·4
Apr.	18·97	18·91	−0·3	18·92	−0·3	19·54	+3·0	18·65	−1·7
May	18·25	20·73	+13·6	20·90	+14·5	21·49	+17·8	20·81	+14·0
June	16·35	18·42	+12·7	18·92	+15·7	19·58	+19·8	19·23	+17·6
July	9·77	11·02	+12·8	11·59	+18·6	11·59	+18·6	12·00	+22·8
Aug.	15·85	16·21	+2·3	16·84	+6·2	17·50	+10·4	18·01	+13·6
Sep.	16·90	18·19	+7·6	18·54	+9·7	19·56	+15·7	20·32	+20·2
Oct.	18·96	18·24	−3·8	18·59	−2·0	19·96	+5·2	19·97	+5·3
Nov.	18·42	17·70	−3·1	17·82	−3·3	18·66	+1·3	19·47	+5·7
Dec.	15·23	14·60	−4·1	14·35	−5·8	14·87	−2·4	15·78	+3·6
1965									
Jan.	20·37	16·25	−20·2	15·99	−21·5	16·16	−20·7	17·47	−14·2
Feb.	21·40	22·52	+5·2	21·39	−	21·33	−0·3	22·43	+4·8
Mar.	16·26	23·50	+44·5	22·72	+39·7	22·07	+35·7	22·94	+41·1
Apr.	15·84	18·60	+17·4	20·36	+28·5	19·05	+20·3	19·25	+21·5
May	21·50	19·67	−8·5	22·03	+2·5	21·04	−2·1	20·95	−2·6
June	18·70	18·25	−2·4	18·47	−1·2	19·75	+5·6	18·72	+0·1
July	11·26	11·19	−0·6	10·89	−3·3	12·47	+10·7	11·57	+2·8
Aug.	18·30	16·67	−8·9	16·57	−9·5	18·39	+0·5	17·47	−4·5
Sep.	16·35	19·09	+16·8	18·73	+14·6	18·89	+15·5	19·91	+21·8
Oct.	17·73	18·83	+6·2	19·04	+7·4	18·47	+4·2	21·47	+21·1
Nov.	17·03	17·86	+4·9	18·66	+9·6	18·19	+6·8	20·40	+19·8
Dec.	13·82	14·43	+4·4	14·76	+6·8	14·93	+8·0	15·07	+9·0
1966									
Jan.	18·43	15·77	−14·4	16·05	−12·9	16·47	−10·6	15·92	−13·6
Feb.	20·40	21·45	+5·1	20·97	+2·8	22·29	+9·3	21·66	+6·2
Mar.	18·46	22·27	+20·6	21·83	+18·3	22·58	+22·3	22·88	+23·9
Apr.	17·79	18·17	+2·1	19·20	+7·9	18·96	+6·6	19·41	+9·1
May	18·29	19·62	+7·3	20·66	+13·0	20·40	+11·5	21·83	+19·4
June	17·39	17·52	+0·7	17·90	+2·9	18·33	+5·4	19·05	+9·5
July	9·83	10·63	+8·1	10·84	+10·3	11·59	+17·9	11·42	+16·2
Aug.	15·90	15·62	−1·8	15·79	−0·7	17·00	+6·9	16·76	+5·4
Sep.	13·45	17·54	+30·4	17·66	+31·3	18·10	+34·6	18·60	+38·3
Oct.	11·89	16·88	+42·0	17·69	+48·8	18·26	+53·6	19·67	+65·4
Nov.	11·10	15·10	+36·0	16·96	+52·8	17·13	+54·3	18·59	+67·5
Dec.	9·97	11·54	+15·7	13·06	+31·0	13·91	+39·5	14·28	+43·2

Average Absolute Percentage Errors (*Unweighted*) :

$$F_1(t) = 10\cdot5$$
$$F_3(t) = 12\cdot5$$
$$F_6(t) = 12\cdot0$$
$$F_9(t) = 13\cdot7$$

Upper Quartiles of Absolute Percentage Errors (i.e. values such that one-quarter of the observed absolute values lie above the given value):

$$F_1(t) = 15\cdot7$$
$$F_3(t) = 18\cdot3$$
$$F_6(t) = 17\cdot8$$
$$F_9(t) = 17\cdot6$$

Prove your prowess

10.1 Apply Holt's Method (ignoring trend and seasonality) to the following data:

January	20·37
February	21·40
March	16·26.

For each month calculate m_t, $F_1(t)$, $F_3(t)$, $F_6(t)$.

For starting values take $m_0 = 17\cdot96$, $F_1(o) = 16\cdot20$, $a = 0\cdot2$.

10.2. Repeat 10.1 introducing a trend term. Take $b_0 = 0\cdot10$, $h = 0\cdot02$.

10.3. Repeat 10.2 assuming the seasonal pattern derived in the worked example in the Appendix.

10.4. A worked example of a regression model is given in Examples 9.5 and 9.6.

REFERENCES

BROWN, R. G. 1963. *Smoothing, Forecasting and Prediction of Discrete Time Series*. New York, Prentice-Hall.

COUTIE, G. A., DAVIES, O. L., HOSSELL, C. H., MILLAR, D. G. W. P., and MORRELL, A. J. H. 1964. *Short Term Forecasting*, I.C.I. Monograph No. 2.

GREGG, J. V., HOSSELL, C. H., and RICHARDSON. 1964. *Mathematical Trend Curves*, I.C.I. Monograph No. 1.

Published for I.C.I. Ltd. by Oliver and Boyd, Edinburgh.

HADLEY, G. 1968. *Introduction to Business Statistics.* New York, Holden-Day.

JOHNSTON, J. 1963. *Econometric Methods.* New York, McGraw-Hill.

11: SO MUCH FROM SO FEW
Sample Surveys

THE concepts of sampling are fundamental to statistical theory and its application. In nearly all of the preceding chapters we were dealing with samples—often very small samples—and making inferences about the populations from which they were drawn. In these earlier chapters our samples were all of the type known as Simple Random Samples. The methods used to calculate confidence limits and to test hypotheses, etc., in these chapters are only applicable to this form of sampling. However, in practice, especially when sampling from large human populations, simple random sampling is hardly ever used because other methods giving a much greater return for the time and money spent on them are available.

In this chapter we consider some of these more complicated types of sampling and, since most of the developments in sampling have come from statisticians interested in sampling from human populations, we also consider the broader aspects of sample surveys.

The selection of a sample of humans is usually undertaken for the purpose of investigating some aspect of human behaviour, e.g. purchasing behaviour, medical experience, social characteristics, voting intentions. In addition, individuals in their capacity as business or professional men are frequently the subject of a survey. In the execution of such surveys sampling is only one of many kinds of expertise necessary to achieve successful results but it is the one which has attracted most attention; from the theorists because it is the aspect of survey work with the strongest theoretical background, and from the layman because it seems incredible to him that valid inferences about populations can be made on the basis of samples which are relatively small. We shall, therefore, devote the larger part of this chapter to a discussion of sampling techniques and spend very little time on other important aspects of survey work like questionnaire design and non-response. Most of our discussion will be in terms of human populations, but most of the principles of survey design apply equally to the sampling of business records, or farms, or forests, or fish, or engineering components, etc.

THE DESIGN OF SAMPLE SURVEYS

Surveys, whether based on samples or on complete coverage of the defined population, are carried out for a variety of reasons, but whatever the purpose of a survey, the stages of planning and execution come under the same general headings. All the stages, except for the second, are equally relevant to a sample survey or a complete enumeration, i.e. a census.

1. The objectives of the survey and the population to be covered must be defined. At the same time the resources available, i.e. time, money and staff, need to be considered.

2. The question of whether to attempt a complete enumeration of the defined population or to take a sample must be settled. If a sample is decided on, the type and size of sample has to be settled.

3. The precise form of data to be collected and the method of collecting it have to be decided. This is rarely as straightforward as it might sound and in certain types of survey, e.g. motivation research, can become very complicated indeed.

4. The questionnaire, or research brief, must be designed and tested. The questionnaire is perhaps the weakest link in survey methodology and a great deal of time and effort is often needed to produce a workable one. The testing stage is absolutely essential to check for vague, offensive, impossible and other kinds of bad question.

5. The fieldwork must be organised and executed. Most of the organisational problems in sample surveys occur in the administration of the field staff. This is a very large subject in itself, and one in which a good deal of research has been carried out. We will have to omit any discussion of it, for reasons of space; the interested reader will find references to the relevant literature at the end of the chapter.

6. The data from the questionnaires or from the observations must be analysed and summarised. This again is a major subject in itself.

7. The survey results must be presented to the survey sponsors. This is undoubtedly the most neglected part of survey method-

ology. It is frequently forgotten that however carefully all the preceding stages are carried out, the effort is wasted if the results are presented in such a way that the sponsors cannot understand them, or are discouraged by bad presentation from trying to. Much worse, of course, is the presentation which by its omissions, obscurities and ambiguities leads to clearcut but wrong conclusions being drawn by the reader. Regrettably, such situations are not unknown.

We emphasise the point that all the above stages, with the possible exception of Stage 2, are equally relevant to a sample survey or a census.

The above steps are not, of course, sequential but iterative. At any stage, the findings or decisions might affect decisions already reached in earlier stages. For example, a consideration of the precise form of data to be collected (Stage 3) might reveal that certain information vital to the initial objectives of the survey is simply not available, or is impossible to collect. In this situation the survey objectives must be reconsidered and amended. If necessary, the survey must be abandoned.

From the steps outlined above, it is obvious that the skill of the statistician is only one of a number of skills required to carry out a survey. Other fields in which expertise is or might be required include:

(*i*) The subject-matter of the survey.

(*ii*) Psychology, for the framing of the questions.

(*iii*) Fieldwork supervision. The fieldwork supervisor (and, of course, the interviewers) exert a major influence on the quality and punctuality of the results.

(*iv*) Data processing. As research methods become more sophisticated and survey objectives more ambitious, more complex methods of data processing are required. Analysis must become speedier and yet more detailed. More and more survey analysis is being carried out by computers these days, partly because of the increasing complexity of the required analysis and partly because the computer manufacturers and the independent bureaux are now offering standard survey analysis programs.

Sample surveys attract a fair amount of criticism or scepticism from the press and public. Most of this is totally unwarranted. Lay criticism frequently concentrates on the fact that a sample, probably representing only a very small proportion of the population concerned, is used as a basis for conclusions about the population. This criticism is occasionally valid, where for example a sample taken in one part of the country only, is used to draw conclusions about the country as a whole, but the other more valid criticisms such as the form of questions, the interviewing technique, and the proportion of refusals are rarely if ever made, and these factors not only give rise to errors greater than the errors due to sampling but present even more of a problem in a census than in a sample survey. We shall see the reason for this in the next section when we begin to consider the sampling process.

SAMPLING

The general idea of sampling is, of course, familiar to everyone. Sampling essentially consists of obtaining information from a portion—often a very small portion—of a larger group. Sampling has a number of advantages over the census method of collecting information. Sampling is:

1. Cheaper—although more expensive per unit sampled.

2. Labour saving—because a smaller staff is needed for the field-work and tabulation.

3. Quicker—for the reasons above. It must be remembered that even with unlimited resources it still takes time to produce results, e.g. the Census of Population results appear over a number of years. Information which is required for management decision is usually required quickly.

The above advantages, however, are rather negative. Some positive ones are:

4. More detailed information can be collected from each informant because of the better quality and better trained staff who can be used and because of the longer time which can be spent with each informant.

5. A higher degree of accuracy can be achieved from a sample because work at all stages of the survey can be of higher quality. Staff can be given more training than in a census, fieldwork

checks become economic, greater care can be taken in editing the completed questionnaires, and so on. Non-sampling errors (i.e. errors due to the misunderstanding of the questions by the informant, incorrect recording by the interviewer, reluctance of informants to give any answer at all, etc.), which are present whether the survey is based on a sample or census, are usually much more serious than sampling errors. They can be controlled much more closely in a sample survey than in a census.

In no sense therefore can we regard a sample survey as an inferior substitute for a census.

The reader, no doubt, is tempted to pursue the above arguments to their logical conclusion and ask 'If sample surveys are so good, why should anyone ever bother to carry out a census?' Why indeed? One obvious reason, of course, is that the application of sampling methods is still relatively new and there are still pockets of resistance to the new methods. Another reason, arising from the first, is that the existence of non-sampling errors is an even more recent discovery. An attitude of 'If it comes from a census it must be correct' can still be found among businessmen, politicians, administrators and, of course, the general public, who have all the wrong ideas about statistics.

A third, and more valid, reason for using the census approach for certain enquiries is that if accurate figures, in extreme detail, for numerous small sections of the population are required, then a census might be the only way of obtaining these. If, at the same time, an accurate 'benchmark' is required, against which to measure variations in subsequent periods a census might be better for the purpose than a sample survey. The decennial U.K. Census of Population is a case in point. In such a situation, the problem of non-sampling errors must be kept in mind throughout the census and resources devoted to the training of enumerators, and the checking of results.

The subject-matter of the survey also has a strong bearing on the practicability of a census. In the case of the Census of Population, the questionnaires, although fairly complicated, are factual. In addition, the Registrar-General, who is responsible for the Census, has statutory powers to compel the return of completed schedules— a useful sanction not possessed by the majority of survey practitioners.

Two final situations in which a census might be preferable to a sample survey occur for technical reasons, and will be illustrated by examples. The first of these concerns an investigation into a group of docks, where the revenue accounting was based on shipping notes which moved with the goods. At any given moment, hundreds of thousands of these notes were in existence somewhere in the docks and the dock authorities wanted a system to indicate at the end of each month how much revenue was outstanding, i.e. what was the value of all the shipping notes outstanding. It was suggested that a sample of shipping notes would give information of sufficient reliability for management purposes. The problem encountered in trying to select a sample of shipping notes, however, was that it was impossible to select a sample without first identifying and locating all the units, i.e. shipping notes, in the population. An unknown number of shipping notes was spread around the docks and these needed to be counted before any sample could be drawn. The piece of information to be obtained from each shipping note was simply the amount due to the docks—a single figure. Thus, it would take virtually no longer to locate a shipping note and record the information than it would to just locate it. In this situation a census was less trouble, and just as good with regard to non-sampling errors, as a sample survey. As we shall see in the next section, the sampling problem in the docks was the lack of a convenient *sampling frame*, i.e. a list of all the units in the defined population from which the sample can be selected.

The second example concerns the carrying of freight on the railways. The organisation concerned based much of its commercial and operating information on data contained in the drivers' and guards' work tickets and waybills. It was thought that the burden of collecting and analysing the thousands of documents involved could be lightened, without much sacrifice of accuracy, by introducing sampling techniques. Accordingly, an investigation was started into the use actually made by managers of the information obtained from the documents. Much of the information related to only part of the organisation, e.g. a particular type of traffic, or a particular form of train operating such as single-line working. Thus, any given piece of management information was likely to be based on relatively few returns. As a result it was found that for an acceptable level of sampling error a given piece of information required a sample consisting of something between 60% and 80% of the returns containing that piece of information. Such a situation required an overall sampling fraction of about 75%. The selection of such a sample, i.e. the rejection of the 25% of returns not to be included in the sample, was more complicated and more expensive than an analysis of all the returns. The recommendation was, therefore, to retain census coverage and concentrate on streamlining the clerical procedures.

Thus there are still situations where a census is desirable or necessary. Since most surveys are based on samples, however, we now turn to consider some of the basic ideas underlying sampling techniques.

BASIC IDEAS OF SAMPLING

The reader is already familiar with the ideas of *population, sample* and *sampling distribution* of a statistic; these were discussed at length in Chapter 8. Three more terms need defining before we take a closer look at the fundamental concepts of sampling: *sampling frame, sampling fraction* and *raising factor*. (The last two of these were mentioned very briefly in Chapter 8.)

Sampling frame. A *sampling frame* is some kind of list or other description of the units comprising the population in which we are interested. The sampling frame is vitally important in sample design since it governs the selection procedures open to the statistician. In its simplest form the sampling frame is literally a list; a widely used example being the U.K. Electoral Register. A sample of adults can be selected from the Electoral Register in a variety of ways; the important thing is that there must be something to select *from* if we wish to apply random sampling techniques. This problem does not necessarily arise with non-random samples, although very often it does. We shall see the reasons for this later in the chapter.

Where a list is not available, some other type of frame can often be constructed. For example, an area on a map can be divided into sub-areas by superimposing a grid on the map. A random sample can then be drawn by selecting a number of sub-areas at random. All individuals in the selected sub-areas might be included in the sample, alternatively, a further stage of sampling might be carried out. We shall return to this idea later.

Where no frame exists, the researcher is faced with three main alternatives. He can set about constructing a frame; he can decide to use a non-random method of selection; or he can abandon the proposed research. Constructing a frame is usually a formidable task and one which is rarely feasible for an *ad hoc* survey. The docks example in the previous section is a good example of a situation in which the relevant sampling frame would have to be constructed before each sample could be selected.

In assessing a frame, or in constructing one, the task varies from the trivial to the impossible. A sampling frame for selecting a sample of schoolchildren from a certain area would be based on school registers and other readily available records; a sampling frame for farms would run into serious problems of definition, e.g. when does a small-holding become a farm; a sampling frame for firms in a particular industry who use our product and/or our competitors' products would present all kinds of problems.

Most sampling frames fall short of perfection. Most are at least slightly out of date, or include certain units which do not fall within the defined population, or exclude certain units which do fall into the defined population. In most cases the researcher has to accept minor shortcomings and assume that his results will not be significantly affected.

Sampling fraction. The *sampling fraction* is simply the number of units in the sample expressed as a fraction of the number of units in the population. Thus if the defined population consisted of 1000 firms and a sample of 100 firms was selected, the sampling fraction would be $\dfrac{100}{1000} = \frac{1}{10}$. We shall denote the sampling fraction by f, in this chapter.

Raising Factor. The raising factor is simply the reciprocal of the sampling fraction. If the sampling fraction is $f = \frac{1}{10}$, then the raising factor is 10. We denote the raising factor by g. For example, if a sample of firms use £1 000 000 of a certain product per annum, and if $f = \frac{1}{10}$ then the estimate for the population is

$$£1\,000\,000 \times 10 = £10\,000\,000.$$

Armed with the above definitions we can now consider the fundamental concepts of sampling. There are three: *accuracy, precision* and *bias*.

Accuracy. The accuracy of a particular sample estimate is simply how close it comes to the population value. Since the population value is very rarely known (if it were there would be no need to carry out a sample survey), the accuracy of a sample estimate is not usually known. Nevertheless the concept is a useful one.

In certain cases the accuracy of a particular sample estimate can be checked and, if satisfactory, provide a rough indication of the reliability of the sample. For example, the sales of a particular firm's products during a certain period will be known to a high degree of accuracy within the firm. If a sample survey of customers was carried out to discover, say, their attitude to the firm's products *vis-à-vis* those of its competitors, the customers might be asked to give their purchases from the sponsor company for the period in question. If the sample estimate is very close to the population figure (derived from sales statistics), then confidence in the other results from the sample is increased.

Precision. Precision is a concept based on the idea of a sampling distribution which we first met in Chapter 8. The precision of a sampling scheme is measured by the variance of the sampling distribution of the statistic concerned.

Bias. If the mean of the sampling distribution of our statistic is the 'true' population value, the sampling scheme is said to be *unbiased.* If it is not, the sampling scheme is said to be *biased.* If, for example, certain large firms consistently refuse to co-operate in surveys then no matter how many samples are taken the sampling distribution will not be centred on the population value. The amount of bias is equal to the difference between the population value and the value about which the sampling distribution is centred. The latter is referred to as the *Expected Value*, an idea which was first met in Chapter 6.

Precision and bias, unlike accuracy, are essentially 'long-term' concepts, based on the idea of repeated sampling. Their importance lies in the fact that we can never be sure of the accuracy of a particular set of sample results. If we wish to examine their validity, we can only examine the validity of the sampling scheme which gave rise to them.

The three concepts—accuracy, precision and bias—can be explained using an analogy. Consider a rifle clamped in a vice aimed at a target. The object is to hit the bull's-eye. A number of shots are fired at the target producing a scatter of hits.

Accuracy is the distance between any given shot and the bull's-eye.

Precision is the degree of concentration of the shots.

Bias is the extent to which the group of shots is centred on or off the bull. A new rifle, aimed accurately at the bull, will deliver the shots in a tight cluster centred on the bull, but if the aim is off-centre the shots, although clustered as tightly as before, will centre on a point off the bull, i.e. will be biased although precise.

An old rifle, on the other hand, might spray bullets in all directions. The cluster will be very wide, i.e. precision will be low, but the spread will still be centred on the bull, i.e. be unbiased. If the aim is off centre, the shots will be both imprecise and biased.

One final point can be made before we look at different types of sampling scheme. Any sample estimate is the result of two separate procedures: the *selection procedure* and the *estimating procedure*. Given an acceptable sampling frame there are a number of sample designs to choose from in the selection of the sample. Having selected the sample, we then have available a number of alternative estimating procedures. For example, if we wish to estimate the population mean we can take the sample mean, or the sample median, or half the sample range, or any other statistic which we might think gives an estimate of the population mean. We know, from Chapter 8, that the sample mean gives the best estimate because its sampling distribution has the smallest variance. In other words the sample mean gives the most *precise* estimate of the population mean from a given sample. Thus, when we talk about the precision or bias of a sample estimate we must remember that these depend not only on what kind of sample we select but also on how we use the results from the selected sample.

We can now investigate various kinds of sample design.

SAMPLE DESIGN

In this section we consider, very briefly, the main types of sample design available to the survey practitioner. A major division is between *random* and *non-random* sampling. Most of our discussion will concern random sampling; the only form of non-random sampling which is widely used is *Quota Sampling* which we shall consider after our discussion of random sampling.

Random sampling. In Chapter 8 we defined a random sample as one in which every unit in the population had a known and non-zero probability of selection. Nothing in the general definition of random sampling implies that the selection probabilities should be equal. The simplest kind of random sampling, however, is based on equal probabilities of selection. It is known, appropriately, as *simple random sampling*.

Simple random sampling. A simple random sample (abbreviated as '*s.r.s.*') is one in which:

(*i*) Every unit in the population has an equal chance of selection.

(*ii*) Every combination of units has an equal chance of selection.

Thus if an *s.r.s.* of 100 individuals from a population of 10 000 was selected, every individual would have a probability of $\frac{100}{10\,000}$ or 1 in 100 of being selected and every combination of 100 individuals would have an equal chance of selection. (The reader is left to calculate this probability as an exercise. *Hint*: The material of Chapter 6 provides the method.)

Simple random sampling is the type on which the examples of Chapter 8 were based. We can, therefore, quote from the results of that chapter for our expressions for sampling error. The standard error of the sample mean was given in Chapter 8 as:

$$\sigma_{\bar{x}} = \sqrt{(s^2/n)}$$

with a qualification, later in the chapter, that when the population size was small, an additional factor known as the finite population correction factor (f.p.c.) entered into the calculation.

Introducing this factor (without deriving it theoretically) we write the expression for the standard error of the sample mean as

$$\sigma_{\bar{x}} = \sqrt{\{(s^2/n)(1-n/N)\}}$$

where n is the sample size and N is the population size. When n is very small compared to N, $(1-n/N)$ approximates to 1, and the factor can be ignored. In practice, we do this where n/N (which, of course, is the sampling fraction) is less than $\frac{1}{10}$. Thus where we are dealing with very large populations we can nearly always ignore the f.p.c. We can also arrange for our population to be infinitely large by sampling *with replacement*.

With this system every unit is returned to the population (after its value has been recorded) before the next unit is selected. This allows the expressions for sampling error to be simplified by dropping the f.p.c., but it also introduces the possibility that a unit will be selected more than once. If this happens, that unit must be included as many times as it appears. Such a situation did occur, in fact, in one of the Ministry of Labour enquiries into Household Expenditure, some years ago. An area in Scotland was selected twice and included both times in calculating the statistics.

Simple random sampling illustrates the two most important attributes of random sampling:

1. It is possible from the one sample actually selected to calculate not only estimates of the population values but also estimates of the limits between which each population value lies, at any given level of probability.

2. The proportion of the population included in the sample is almost completely irrelevant in determining the precision of the sample estimates. What matters is the absolute size of the sample and the variability of units in the population.

The expression for the standard error of the sample mean also gives a valuable clue to one way in which precision can be increased. Ignoring the f.p.c., the standard error depends on two factors, the variability of the data, i.e. s^2, and the size of the sample n. Nothing can be done about s^2 but n can be increased in order to reduce s^2/n. We should note however that since the standard error involves the square root of s^2/n, precision increases only as the square root of n. If we wish to double the precision, i.e. 'tighten' the cluster of statistic values, we must increase n by $2^2 = 4$.

Systematic, Quasi-random, or List sampling. A *systematic sample* is one in which every kth item in a list is included in the sample. For example if the sampling frame consisted of a list of 1000 firms and if a sample of 100 firms was required, every 10th firm would be selected. A starting point would be chosen by selecting a random number between 1 and 10. If this number were 4, say, the sample would consist of the 4th, 14th, 24th, ... firm on the list up to the 994th. (Selection of the starting number would be made with the aid of a table of random numbers, the use of which will be explained in a later section.)

Systematic sampling is often confused with simple random sampling because in both types of sample every unit in the population has an equal chance of selection. But in systematic sampling, the second requirement of simple random sampling—equal probabilities for all conbinations of units—is not met. Once the first unit is chosen, all the subsequent units are determined. Only if the list itself is arranged in a random order are systematic and simple random sampling equivalent. Most lists are in an order which is far from random, e.g. Personnel lists (arranged alphabetically within sections and/or departments), the Electoral Register (streets within polling districts within electoral wards within constituencies).

List sampling has a number of potential advantages over simple random sampling. To begin with it is usually easier and cheaper to draw the sample, and this can be done in the field, e.g. an interviewer can be given an address and be instructed to interview at every kth house, with instructions for turning left or right at intersections and for dealing with cul-de-sacs, etc. This method of sampling is referred to as a *Random Walk* procedure. Another common application of list sampling is where the sampling frame consists of a card index. It is often sufficient to measure the overall depth of the cards in the index, rather than count them, and select a systematic sample by making the sampling interval equal to a certain depth, e.g. the card occurring every $1\frac{1}{2}$ inches might be selected.

In many situations systematic sampling gives more precise estimates than a simple random sample of equal size because some of the extreme samples, e.g. the first n units, are ruled out and the sample is spread more evenly over the population. Oddly enough, however, there is no theoretically sound expression for the standard error of an estimate based on a systematic sample. The sampling theory of simple random sampling cannot properly be applied; although, since it gives a conservative estimate of the true standard error, it sometimes is applied.

Systematic sampling is very widely used and where there appears to be no possible relationship between the subject of the survey and the order within the list the resulting sample is often treated as a simple random sample.

Stratified random sampling. Simple random sampling and systematic sampling are the basic forms of random sampling but they are not used in their 'pure' form very often. One reason is that sample designs which give more precise estimates for a given size of sample (or given expenditure) exist; the other main reason is that the administrative problems and expense of handling a simple random or systematic sample of, say, 3000 adults drawn from the Electoral Register would be prohibitive. We now consider the first of these two reasons and introduce the idea of *stratified sampling*.

The basis of the idea is that if we already know something about the population, e.g. its age and sex distribution, obtained from official statistics or company records, etc., we can use this information to increase the precision of our sample design. To take a concrete but hypothetical example, suppose we are interested in estimating, by means of a random sample, the cigarette consumption of a group of 1000 workers in a particular factory. We select a sample of 100 say, calculate from this the average number of cigarettes smoked per worker per week, and multiply this average by the total number of workers (1000) to obtain our required estimate.

Now suppose that this group of 1000 workers consists of 600 men and 400 women. A simple random sample of 100 might contain all men, all women, or any combination of men and women which adds up to 100. Instead of selecting one overall simple random sample of 100, however, we can use the information about sex distribution and divide our sample between men and women in the same proportions. We then select a sample consisting of 60 men and 40 women and we do this by selecting a *s.r.s.* of 60 men from the 600 men on the list and 40 women from the 400 on the list.

This latter type of selection is an example of *stratified random sampling*. Before any sampling takes place, the population is divided into *strata* on the basis of external information, in our case sex, from company records. Then a separate sample is selected from each stratum, usually with equal probabilities of selection. If the sample is stratified proportionately to the population (as it was in the example above) the probabilities of selection for all units in the sample are also equal, but this is not necessary. For example, we might have had good reasons for splitting our sample into 50 men and 50 women, instead of 60 and 40. In this case the probability of selection for a man would be $\frac{50}{600}$, i.e. 1 in 12 while for women it would be $\frac{50}{400}$, i.e. 1 in 8. There is nothing wrong in doing this, indeed it frequently gives more precise estimates provided that the overall mean is based on a weighted average of the stratum means. This is the fundamental rule of estimating from stratified samples. All overall estimates must be based on weighted estimates. 'Stratification' and 'weighting' always go together. We will illustrate the rule by inserting some hypothetical figures in our example.

Suppose that the following results were obtained:

Men: $\bar{x}_M = 100$ cigarettes per week

Women: $\bar{x}_W = 80$ cigarettes per week

The overall mean, denoted by $\bar{x}_{st}$ (where 'st' implies 'stratified') is given by:

$$\bar{x}_{st} = \frac{600\bar{x}_M + 400\bar{x}_W}{1000}$$

$$= 92 \text{ cigarettes per week.}$$

The system of allocating the number of observations in each stratum in proportion to the corresponding population numbers is known, for obvious reasons as proportional allocation. If, for some reason, we had chosen a sample containing 50 men and 50 women, and if $\bar{x}_M$ and $\bar{x}_W$ were 100 and 80 respectively, as before, the stratum means *would still be weighted in the proportions* 600 : 400. The weights are always based on the *population* proportions.

The standard error of an estimate from a stratified sample requires, as one would expect, a more complicated calculation. We show below a typical formula for the standard error of $\bar{x}_{st}$. Denoting this standard error by s.e. $(\bar{x}_{st})$ we have:

$$\text{s.e. } (\bar{x}_{st}) = \sqrt{\left\{ \sum_{h=1}^{k} (W_h^2 S_h^2 / n_h)(1 - f_h) \right\}}$$

where: $h = 1, 2, ..., k$ gives the number of strata,

W_h = the weight of the hth stratum,

S_h^2 = the sample variance in the hth stratum,

n_h = the number of observations in the hth stratum,

f_h = the sampling fraction in the hth stratum.

This expression looks more complicated than it really is. All we have to do is keep our data from the various strata separate and for each stratum calculate the stratum variance (from the usual expression

$$S^2 = \frac{1}{n-1} \sum_{i=1}^{n} (x_i - \bar{x})^2 \Bigg)$$

and the sampling fraction (which can be ignored if it is less than 0·1). We then calculate for each stratum the expression

$$(S_h^2 / n_h)(1 - f_h)$$

and combine these values into a weighted average using the stratum sizes as weights. In our cigarette example we had

$$W_M = \frac{600}{1000} = 0{\cdot}6; \; W_W = \frac{400}{1000} = 0{\cdot}4,$$

$$n_M = 60; \; n_W = 40,$$

$$f_M = \frac{60}{600} = 0{\cdot}1; \; f_W = \frac{40}{400} = 0{\cdot}1,$$

and we can denote the sample variances by S_M^2 and S_W^2.

The standard error of $\bar{x}_{st}$ is, therefore:

$$\text{s.e.} \, (\bar{x}_{st}) = \sqrt{\{(0{\cdot}6^2 S_M^2/60)(1-0{\cdot}1)+(0{\cdot}4^2 S_W^2/40)(1-0{\cdot}1)\}}.$$

Note: Since the value of f_h is on the border line we have retained the term $(1-f_h)$ in the above calculation.

What are the advantages of stratified random sampling over simple random sampling? The first one is that for the overall sample estimates, e.g. the sample mean as an estimate of the population mean, *stratifying almost always gives more precise estimates.* The reason for this is rather technical but can be outlined as follows: Given a population which is fairly heterogeneous, and in which different groups have different values for the characteristic in which we are interested, e.g. men in general are heavier smokers than women, the sampling error from a simple random sample arises from two causes:

(*i*) The straightforward fact that, since not all the population have been included, any sample results are likely to differ from the population values.

(*ii*) Within the sample actually selected, the proportions falling into the different strata will not necessarily be equal to the corresponding population proportions.

Now, by stratifying we can eliminate the second cause of sampling error, thus estimates based on stratified random samples are almost always more precise than those based on simple random samples of equivalent size. It is assumed, in saying this, that the particular stratifying factors used are correlated with the subject of the enquiry. The higher the correlation, the greater the increase in precision brought about by stratification. In fact, the best possible stratifying factor is the very characteristic we are investigating, e.g. in our cigarette example, if we could have stratified men and women according to their average weekly consumption of cigarettes we would have achieved the ideal stratification, but of course if we were able to do this we would not need to carry out the survey in the first place.

If our stratifying factors bear little or no relationship to the characteristics we are investigating, no harm is done. Only in extremely rare cases can one find a stratifying factor which actually results in less precise estimates than from simple random sampling. The general objective in stratifying should be to divide the population into groups which are as homogeneous as possible within each group, but which differ between groups as much as possible.

The second major advantage of stratified random sampling is that if results are required for a number of strata separately, which is usually the case in sample surveys, then stratification will ensure that each stratum is represented in the sample. Furthermore, if data of specified precision is required for certain subdivisions of the population the sample size necessary to yield estimates of the required precision can be calculated in advance for each of the subdivisions concerned. The total sample size is then obtained by summing the component sample sizes.

A third advantage is that the administration of the survey may be made easier by stratification. One obvious way in which this could come about is by using geographical areas as a stratifying factor. The survey organisation might well have a system of area offices each of which can supervise one section of the survey. Another way in which administration might be eased is by including in one stratum those units in the population which share similar sampling problems. For example, in a survey affecting the whole population, people living in hotels, institutions, hospitals, prisons, etc. are often placed in a separate stratum because the sampling poses different problems. Or in sampling firms, the few very large firms in a particular industry might all be included. This is equivalent to specifying a stratum of large firms and sampling with a sampling fraction of 1.

Optimal allocation. We have seen that the simple procedure of stratifying with proportional allocation gives us more precise estimates of population values. Can precision be still further increased by means of some other system of allocation? The answer is 'Yes'. We can increase precision by varying our sampling fraction according to the variability of data in the stratum. Expressed algebraically, for a given overall sample size of n we have:

$$n_h = n \frac{N_h S_h}{\Sigma N_h S_h}$$

where N_h is the population size of the hth stratum.

We would attempt to estimate the relative variability in the different strata either from past experience (our own or somebody else's) or by carrying out a pilot survey, of which more later. In simple terms, what we are saying is that the more variable a particular stratum, the larger the proportion of total sample units it should contain.

If, in addition to variability within strata, the costs of collecting data also vary, two additional general principles must be observed. For a given stratum we take a larger sample if:

(*i*) the stratum is larger, or

(*ii*) sampling is cheaper in that stratum.

The idea of the cost of collecting data entering into our statistical formulae may be strange to the reader, but sampling is a severely practical area of statistics. The objectives of sampling can be expressed in two ways, which are, in effect, the two sides of the same coin. Sampling seeks either to:

(*a*) produce information of a specified precision for the lowest possible cost; or

(*b*) produce information of the highest possible precision for a given outlay.

Stratification after selection. We have made the tacit assumption up to now that when stratifying we not only have the necessary information with which to set stratification factors but can also recognise from the sampling frame into which stratum a particular unit falls. This is not always the case. To illustrate this point we can return to our cigarette smoking example and make the unwarranted assumption that the company records, which form our sampling frame, show only the surname and initials of the worker. In other words, we cannot tell from the list whether a given individual is male or female, although we do know that 600 are male and 400 female. We can still use a stratified sampling scheme, however. All that is required is that we should be able to establish to which stratum a sample unit belongs when that unit is interviewed. We have two alternatives. We can either allocate informants to their respective strata as we collect the information, until both strata have at least the required number of observations, or, when one stratum is full, we can ignore further informants from that stratum and question only those informants who fall into the stratum which has still to be filled. Either way we shall have to select more than our nominal 100 informants in order to fill both strata. Stratification after selection is, therefore, rather less efficient than stratification before selection, but we may have no choice in the matter.

An important use of stratification after selection is that in a given survey it may become apparent as sampling proceeds that some relevant stratification factor, which was unsuspected before the fieldwork began, exists. If the necessary population totals are available this factor can now be incorporated into the sampling scheme.

This concludes our discussion of stratified random sampling. We have spent a long time on it because stratification provides a highly efficient means of increasing precision without increasing sample size. Virtually all major sample surveys are based on stratified sampling. There is, however, another technique, known as *multi-stage sampling*, which is also very widely used, usually in conjunction with stratified sampling (the typical real-life random sample is a *stratified multi-stage sample*). We now turn to consider this form of sampling.

Multi-stage sampling. We begin our discussion by introducing the idea of a *cluster sample*. Sampling frames are often laid out in sections or 'clusters', for example the Electoral Register for urban areas is arranged by streets within polling districts within electoral wards within constituencies. A polling district can be regarded as a cluster of streets, an electoral ward as a cluster of polling districts, and so on. Now, if we wish to select a sample of electors, instead of taking a simple random sample or a systematic sample from the entire list, we can select one or more streets and interview all electors living in these streets. Alternatively we could select one or more polling districts and interview all the electors in the selected polling districts.

M

The advantages of this method of sampling are in the administration of the survey. Fieldwork costs can be kept to a minimum by concentrating the interviewing into a small number of areas. Quality checks of the interviewers' work are also much more economical. There is a price to pay for this, of course. The price is the inherent loss of precision in a cluster sample compared with a simple random sample of the same size. This loss of precision occurs because populations are rarely homogeneous as a whole but occur in homogeneous clusters. Since the object of cluster sampling is to select a limited number of clusters, important types of unit might be missed altogether. If the population is thoroughly mixed, cluster sampling is equivalent to simple random sampling but populations, especially human ones, are never thoroughly mixed. The effect of this is that for any characteristic in which we are interested, units within a particular cluster are likely to show positive correlation. For example, a family can be regarded as a cluster of individuals. Assuming an average of 3 individuals per family in our defined population, consider the question of a sample survey into religious beliefs and churchgoing, and consider the merits of simple random sampling versus cluster sampling. If we require a sample of 300 individuals we can obtain these either by selecting 300 individuals from our frame or selecting (at random) 100 families and interviewing all the individuals in the selected families. In the latter case we would expect to find high correlation between individuals in a given family and the sample will not yield as much information as the simple random sample of 300. This correlation between members of a cluster is known in the statistical jargon as *intra-class correlation* and the formula for the standard error of an estimate based on cluster sampling takes this into account.

We can illustrate these ideas further by returning to our cigarette smoking example. Let us now make the highly artificial assumption that the 1000 workers in the company were divided into twenty sections, each containing exactly 50 workers. A sample of 100 can now be selected by picking two sections at random from the twenty, and collecting information from all the workers in the two selected sections. Now, there will probably be a number of sections which are all male, or all female, e.g. the typing pool; similarly, some sections will be engaged on heavy manual work while others are occupied with more sedentary work. Differences such as these are quite likely to be correlated with cigarette smoking—at least we would be very foolish to assume that they were not. Selecting two sections out of twenty therefore will probably lead to very imprecise estimates of average cigarette consumption per worker, compared with a simple random sample of 100 individuals. The resulting cluster sample might, for example, exclude all women, or all men.

It would seem, therefore, that cluster sampling has little to offer, and in the above example this is certainly true. However, we must remember the two sides of the sampling coin. We are interested in getting the most for our money and we can usually get so many additional interviews for a given sum of money by using cluster sampling that the net effect is a more precise sample than we could afford using simple random sampling. In addition there is the point, which could be vital, that a sampling frame is only required in terms of clusters, e.g. a list of addresses rather than a list of individuals.

A widely used version of cluster sampling is found in the United States, where there is no list of electors comparable to the U.K. Electoral Register. Samples are selected using the grid system, by superimposing a rectangular grid on a large-scale map of an area, selecting a number of squares from the grid at random, and including all units of the population found in the selected squares.

With the idea of cluster sampling clear in our minds we can proceed to consider *multi-stage sampling*, which is an extension of cluster sampling. As the name implies, sampling proceeds by stages. At the first stage a number of clusters is selected from the population. At the second stage, a further sample is selected from each first-stage cluster. At the third and any subsequent stages, further samples are selected from those clusters left after the preceding stage of sampling. For example, a four-stage sample selected from the Electoral Register would appear as follows:

Stage 1: Sample of *constituencies*.

Stage 2: Sample of *electoral wards* drawn from each constituency selected at Stage 1.

Stage 3: Sample of *polling districts* drawn from each electoral ward selected at Stage 2.

Stage 4: Sample of *individuals* drawn from each polling district selected at stage 3.

A typical national survey based on 2000 interviews might include 50 constituencies selected at Stage 1, 4 electoral wards selected from each of the 50 constituencies, 2 polling districts selected from each of the 200 electoral wards and 5 individuals selected from each of the 400 polling districts.

Clearly the burden of administration and the time and cost involved in achieving the required interviews is far less than would be the case if 2000 individuals were selected from the complete electoral register. In fact, the arguments for and against cluster sampling apply with greater force to multi-stage sampling. The art of designing a multi-stage sample lies in knowing the extent to which it is safe to concentrate interviews into a few final sampling units. The most important part of this decision is deciding how many first-stage units to select. In our example of multi-stage sampling we ended up by selecting a sample of 5 individuals from each polling district. There is nothing to prevent us including all units at any given stage of sampling, e.g. we could have included *all* individuals in the 400 polling districts (although this would give us an unnecessarily large and expensive sample). In such a situation we would refer to '100% sampling' at Stage 3. In this sense, cluster sampling can be regarded as two-stage sampling with 100% sampling at Stage 2.

The reader who has followed the example carefully will probably be wondering how, with multi-stage sampling, the probabilities of selection are controlled. Constituencies, electoral wards, etc., do not contain equal numbers of individuals. How then can we achieve a sample of individuals with, say, equal chances of selection? This in fact is very easy to achieve; we make use of two compensating factors:

(*i*) By selecting clusters with *probability proportional to size* we ensure that at all but the last stage of selection, the bigger the cluster the greater its chance of inclusion in the sample.

(*ii*) At the final stage of selection we select a fixed number of individuals. Thus the larger the final cluster, the smaller the probability of selection for any individual in that cluster.

The interaction of these two effects ensures equal probabilities of selection. We will demonstrate the truth of this statement with a simple example of a two-stage sampling scheme.

Consider a population of 10 000 individuals distributed as follows:

TABLE 11.1

Cluster no.	Size of cluster	Cumulative number of individuals
1	800	800
2	500	1 300
3	1 400	2 700
4	2 000	4 700
5	600	5 300
6	100	5 400
7	1 000	6 400
8	900	7 300
9	2 500	9 800
10	200	10 000
Total	10 000	

Suppose we wish to select a sample of 200 individuals and we decide to use a two-stage sampling scheme, selecting two clusters at the first stage and 100 individuals from each selected cluster at the second stage.

We select our two first-stage units with probability proportional to size (p.p.s.). This we do by listing the clusters and cumulating the total number of individuals. We then select from a table of random numbers two numbers between 1 and 10 000. If one of the numbers lies in the range 1-800, Cluster 1 is selected; if one of the numbers lies in the range 801-1300, Cluster 2 is selected, and so on. Thus the two random numbers will identify two clusters; furthermore the probability of selection for any particular cluster is in direct proportion to its size i.e. the range of numbers which covers that cluster. Cluster 7, for example, has 10 times the probability of selection that Cluster 6 has.

Having selected our two first-stage units in this manner, all that remains is to select a simple random sample of 100 individuals from each of them. If we now calculate the overall probabilities of selection we find that every individual in the population has an equal chance of selection. Table 11.2 demonstrates this.

The question now arises of what happens if both our random numbers fall in the range for one particular cluster? This is a contingency for which we should be prepared before any selection is carried out. We have to decide whether we are sampling *with replacement* or *without replacement*. We touched on this point in our earlier discussion of the f.p.c. In sampling with replacement we are in effect drawing a unit out of the hat and throwing it back again before selecting the next unit. This ensures that the probabilities of selection remain constant from one drawing to the next. In sampling without replacement we are in effect drawing a unit out of the hat, then drawing the next without replacing the first. This means that each successive drawing is from a population one unit smaller than at the previous drawing. Clearly, if we are sampling with replace-

TABLE 11.2

Cluster no.	Probability of selection	Probability of selection for an individual	Overall probability of selection for an individual
1	$\dfrac{800}{10\,000}$	$\dfrac{100}{800}$	$\dfrac{800}{10\,000} \times \dfrac{100}{800} = \dfrac{100}{10\,000}$
2	$\dfrac{500}{10\,000}$	$\dfrac{100}{500}$	$\dfrac{500}{10\,000} \times \dfrac{100}{500} = \dfrac{100}{10\,000}$
3	$\dfrac{1400}{10\,000}$	$\dfrac{100}{1400}$	$\dfrac{1400}{10\,000} \times \dfrac{100}{1400} = \dfrac{100}{10\,000}$
etc.			

ment we stand a chance of selecting a particular unit more than once. If the population is very large in relation to the sample size this chance is very small and the two methods of sampling are, for practical purposes, equivalent. In our simple example the two methods are clearly not equivalent and we have to decide which method we wish to use. Sampling without replacement is the obvious method since we do not wish to run the risk of obtaining a sample of two first-stage units which consists of the same unit twice over. Sampling without replacement, however, complicates the expression for the standard error. In practice, with relatively large populations the method of sampling with replacement is used and if one particular unit is selected more than once then it is included in the analysis more than once, as in the Household Budget survey quoted earlier.

We said earlier on that the typical real-life sample survey is based on a stratified multi-stage sample. Having considered in some detail the two types of sample design separately, the reader might be rather uncertain about how to combine the two methods. The process is fairly straightforward. For example, if area is a stratifying factor, first stage units (e.g. constituencies) are grouped into their respective strata before any selection takes place. In a similar fashion, at the other end of the sampling process, the final sample of individuals might be stratified by age, sex, social class, etc. A considerable amount of work is often necessary, before any selection takes place, to define the cluster/strata combinations and to fix the number of observations to be made in each combination, and obviously the information necessary for stratification must be available in the detail required. It may also be necessary to rely on stratification after selection. The main effect of this detailed breakdown of the sample design is that the expressions for sampling errors in stratified multi-stage sampling become very complicated, especially when optimal allocation with varying costs of collection are involved. In fact, expressions for the sampling error in multi-stage sampling generally are so complex compared with the corresponding expressions for simple random sampling, that they are excluded from our consideration.

Multi-phase sampling. Many sample surveys seek information about a number of related topics, e.g. a market research survey into purchases of branded goods might seek to establish

(a) the size of the market and market shares;

(b) the proportion of housewives who recognise and correctly identify the trade mark of a particular brand.

In this situation the sample size required for a given level of precision might differ between (a) and (b), in fact acceptable answers to (b) might be obtained from about half the number of interviews required for (a). Putting the point the other way round, about one-half of the answers on (b) would be unnecessary. If the questions relating to (b) add appreciably to the total length of the interview, it will be worth cutting these questions from half of the questionnaires.

Thus some questions will appear on all questionnaires while others will appear on only a proportion. The allocation of 'long' questionnaires to respondents will, of course, be random and will conform to the general design of the sample. This kind of sampling is known as *multi-phase sampling*. In a complicated survey where interviews last up to an hour and occasionally longer, a carefully designed multi-phase sample might save sufficient interviewing time to allow an appreciably larger sample to be selected than would otherwise be the case.

The multi-phase technique is not confined to sample surveys. It can also be used in a census. For example, the 1961 Census of Population sought information about certain household amenities, e.g. bathrooms, from one in ten households. Every tenth schedule was a 'long' schedule (although to illustrate the power of positive thought the official explanation said that no less than nine out of ten schedules was a 'short' one).

Replicated or Interpenetrating sampling. This is another technique which can be applied to any basic sample design. In the simplest case the sample (whatever the design) is randomly split into two halves, one interviewer, or group of interviewers, taking each half. Each half is then analysed separately; the two sets of results being combined for the full analysis. This approach has several merits. First, an indication of some of the major results can be given quickly by analysing one half before the other. This could be useful if the questionnaires were long and complicated, requiring considerable editing and punching effort.

Secondly, it is possible to check for interviewer biases by considering the variation between the two sets of results.

Thirdly, it is possible to demonstrate to the lay sponsor (or the sceptic) that sampling is a reliable technique by comparing the two halves of the survey showing the small differences between them. (The combined results would also be given, of course.)

The simple case described above can be extended to cover more than two divisions of a sample. With more than two interviewers, for example, the variation between individual interviewers can be investigated by dividing the total sample into as many equal parts as there are interviewers. The practical disadvantage of this, however, is that since each division must be a random sub-sample of the whole, each interviewer will have to cover the whole sample area which will raise interviewing costs enormously.

A different use for this form of sampling is to divide the total sample into about ten random sub-samples, analyse these sub-samples separately and estimate sampling errors by considering the *range* of the various sample values, e.g. the ten values of, say, average family income will be combined to give the overall estimate, and the sampling error of this estimate will itself be estimated from the range of the ten values.

This method of sampling has a number of very strong adherents, especially in America, but it is not used widely in practice, possibly because computers now make it possible to calculate sampling errors using the straightforward formulae.

The selection procedure. In discussing random sampling we have so far dodged the question of how the selection is actually carried out. We have made passing references to 'drawing from a hat' and (with reference to systematic and p.p.s. sampling) the use of a table of *random numbers.*

A table of random numbers provides the most convenient means of selecting a random sample. An abbreviated version of such a table is shown in Table 11.3. Every digit in this table is selected from the range 0-9 with equal probability. Selection is made using some mechanical or electronic system; the principle is that of a roulette wheel. The digits generated from this process are set out in the table in sets of 5 merely to facilitate reading them.

Since every digit is random, every run of n digits is also random. Digits can be read horizontally, vertically, diagonally or any other way.

The use of the table will be illustrated with examples drawn from earlier sections of this chapter.

EXAMPLE 1

In the example of systematic sampling on page 343 we required a number between 1 and 10 to fix the starting-point for our sample. If we let 0 represent 10, we have 10 single digit numbers. For the sake of simplicity let us take the top left-hand digit on the first page of Table 11.3. This digit is 0, so we take the 10th firm on our list as the starting-point for our sample and proceed with the 20th, 30th, etc., firm up to the 1000th. This gives us our sample of 100 firms.

If we wished to devise a more complicated procedure for selecting our starting-point than merely taking the top left-hand digit we could easily do so. We have two pages of digits, each of which consists of 50 rows and 50 columns. We can begin by shutting our

Table 11.3. *Random Digits*

03 991	10 461	93 716	16 894	98 953	73 231	39 528	72 484	82 474	25 593
38 555	95 554	32 886	59 780	09 958	18 065	81 616	18 711	53 342	44 276
17 546	73 704	92 052	46 215	15 917	06 253	07 586	16 120	82 641	22 820
32 643	52 861	95 819	06 831	19 640	99 413	90 767	04 235	13 574	17 200
69 572	68 777	39 510	35 905	85 244	35 159	40 188	28 193	29 593	88 627
24 122	66 591	27 699	06 494	03 152	19 121	34 414	82 157	86 887	55 087
61 196	30 231	92 962	61 773	22 109	78 508	63 439	75 363	44 989	16 822
30 532	21 704	10 274	12 202	94 205	20 380	67 049	09 070	93 399	45 547
03 788	97 599	75 867	20 717	82 037	10 268	79 495	04 146	52 162	90 286
48 228	63 379	85 783	47 619	87 481	37 220	91 704	30 552	04 737	21 031
88 618	19 161	41 290	67 312	71 857	15 957	48 545	35 247	18 619	13 674
71 299	23 853	05 870	01 119	92 784	26 340	75 122	11 724	74 627	73 707
27 954	58 909	82 444	99 005	04 921	73 701	92 904	13 141	32 392	19 763
80 863	00 514	20 247	81 759	45 197	25 332	69 902	63 742	78 464	22 501
33 564	60 780	48 460	85 558	15 191	18 782	94 972	11 598	62 095	36 787
90 899	75 754	60 833	25 983	01 291	41 349	19 152	00 023	12 302	80 783
78 038	70 267	43 529	06 318	38 384	74 761	36 024	00 867	76 378	41 605
55 986	66 485	88 722	56 736	66 164	49 431	94 458	74 284	05 041	49 807
87 539	08 823	94 813	31 900	54 155	83 436	54 158	34 243	46 978	35 482
16 818	60 311	74 457	90 561	72 848	11 834	75 051	93 029	47 665	64 382
34 677	58 300	74 910	64 345	19 325	81 549	60 365	94 653	35 075	33 949
45 305	07 521	61 318	31 855	14 413	70 951	83 799	42 402	56 623	34 442
59 747	67 277	76 503	34 513	39 663	77 544	32 960	07 405	36 409	83 232
16 520	69 676	11 654	99 893	02 181	68 161	19 322	53 845	57 620	52 606
68 652	27 376	92 852	55 866	88 448	03 584	11 220	94 747	07 399	37 408
79 375	95 220	01 159	63 267	10 622	48 391	31 751	57 260	68 980	05 339
33 521	26 665	55 823	47 641	86 225	31 704	88 492	99 382	14 454	04 504
59 589	49 067	66 821	41 575	49 767	04 037	30 934	47 744	07 481	83 828
20 554	91 409	96 277	48 257	50 816	97 616	22 888	48 893	27 499	98 748
59 404	72 059	43 947	51 680	43 852	59 693	78 212	16 993	35 902	91 386
42 614	29 297	01 918	28 316	25 163	01 889	70 014	15 021	68 971	11 403
34 994	41 374	70 071	14 736	65 251	07 629	37 239	33 295	18 477	65 622
99 385	41 600	11 133	07 586	36 815	43 625	18 637	37 509	14 707	93 997
66 497	68 646	78 138	66 559	64 397	11 692	05 327	82 162	83 745	22 567
48 509	23 929	27 482	45 476	04 515	25 624	95 096	67 946	16 930	33 361
15 470	48 355	88 651	22 596	83 761	60 873	43 253	84 145	20 368	07 126
20 094	98 977	74 843	93 413	14 387	06 345	80 854	09 279	41 196	37 480
73 788	06 533	28 597	20 504	51 321	92 246	80 088	77 074	66 919	31 678
60 530	45 128	74 022	84 617	72 472	00 008	80 890	18 002	35 352	54 131
44 372	15 486	65 741	14 014	05 466	55 306	93 128	18 464	79 982	68 416
18 611	19 241	66 083	24 653	84 609	58 232	41 849	84 547	46 850	52 326
58 319	15 997	08 355	60 860	29 735	47 762	46 352	33 049	69 248	93 460
61 199	67 940	55 121	29 281	59 076	07 936	11 087	96 294	14 013	31 792
18 627	90 872	00 911	98 936	76 355	93 779	52 701	08 337	56 303	87 315
00 441	58 997	14 060	40 619	29 549	69 616	57 275	36 898	81 304	48 585
32 624	68 691	14 845	46 672	61 958	77 100	20 857	73 156	70 284	24 326
65 961	73 488	41 839	55 382	17 267	70 943	15 633	84 924	90 415	93 614
20 288	34 060	39 685	23 309	10 061	68 829	92 694	48 297	39 904	02 115
59 362	95 938	74 416	53 166	35 208	33 374	77 613	19 019	88 152	00 080
99 782	93 478	53 152	67 433	35 663	52 972	38 688	32 486	45 134	63 545

TABLE 11.3. *Random Digits (contd.)*

27 767	43 584	85 301	88 977	29 490	69 714	94 015	64 874	32 444	48 277
13 025	14 338	54 066	15 243	47 724	66 733	74 108	88 222	88 570	74 015
80 217	36 292	98 525	24 335	24 432	24 896	62 880	87 873	95 160	59 221
10 875	62 004	90 391	61 105	57 411	06 368	11 748	12 102	80 580	41 867
54 127	57 326	26 629	19 087	24 472	88 779	17 944	05 600	60 478	03 343
60 311	42 824	37 301	42 678	45 990	43 242	66 067	42 792	95 043	52 680
49 739	71 484	92 003	98 086	76 668	73 209	54 244	91 030	45 547	70 818
78 626	51 594	16 453	94 614	39 014	97 066	30 945	57 589	31 732	57 260
66 692	13 986	99 837	00 582	81 232	44 987	69 170	37 403	86 995	90 307
44 071	28 091	07 362	97 703	76 447	42 537	08 345	88 975	35 841	85 771
59 820	96 163	78 851	16 499	87 064	13 075	73 035	41 207	74 699	09 310
25 704	91 035	26 313	77 463	55 387	72 681	47 431	43 905	31 048	56 699
22 304	90 314	78 438	66 276	18 396	73 538	43 277	58 874	11 466	16 082
17 710	59 621	15 292	76 139	59 526	52 113	53 856	30 743	08 670	84 741
25 852	58 905	55 018	56 374	35 824	71 708	30 540	27 886	51 732	65 454
46 780	56 487	75 211	10 271	36 633	68 424	17 374	52 003	70 707	70 214
59 849	96 169	87 195	46 092	26 787	60 939	59 202	11 973	02 902	33 250
47 670	07 654	30 342	40 277	11 049	72 049	83 012	09 832	25 571	77 628
94 304	71 803	73 465	09 819	58 869	35 220	09 504	96 412	90 193	79 568
08 105	59 987	21 436	36 786	49 226	77 837	98 524	97 831	65 704	09 514
64 281	61 826	18 555	64 937	64 654	25 843	41 145	42 820	14 924	39 650
66 847	70 495	32 350	02 985	01 755	14 750	48 968	38 603	70 312	05 682
72 461	33 230	21 529	53 424	72 877	17 334	39 283	04 149	90 850	64 618
21 032	91 050	13 058	16 218	06 554	07 850	73 950	79 552	24 781	89 683
95 362	67 011	06 651	16 136	57 216	39 618	49 856	99 326	40 902	05 069
49 712	97 380	10 404	55 452	09 971	59 481	37 006	22 186	72 682	07 385
58 275	61 764	97 586	54 716	61 459	21 647	87 417	17 198	21 443	41 808
89 514	11 788	68 224	23 417	46 376	25 366	94 746	49 580	01 176	28 838
15 472	50 669	48 139	36 732	26 825	05 511	12 459	91 314	80 582	71 944
12 120	86 124	51 247	44 302	87 112	21 476	14 713	71 181	13 177	55 292
95 294	00 556	70 481	06 905	21 785	41 101	49 386	54 480	23 604	23 554
66 986	34 099	74 474	20 740	47 458	64 809	06 312	88 940	15 995	69 321
80 620	51 790	11 436	38 072	40 405	68 032	60 942	00 307	11 897	92 674
55 411	85 667	77 535	99 892	71 209	92 061	92 329	98 932	78 284	46 347
95 083	06 783	28 102	57 816	85 561	29 671	77 936	63 574	31 384	51 924
90 726	57 166	98 884	08 583	95 889	57 067	38 101	77 756	11 657	13 897
68 984	83 620	89 747	98 882	92 613	89 719	39 641	69 457	91 339	22 502
36 421	16 489	18 059	51 061	67 667	60 631	84 054	40 455	99 396	63 680
92 638	40 333	67 054	16 067	24 700	71 594	47 468	03 577	57 649	63 266
21 036	82 808	77 501	97 427	76 479	68 562	43 321	31 370	28 977	23 896
13 173	33 365	41 468	85 149	49 554	17 994	91 178	10 174	29 420	90 438
86 716	38 746	94 559	37 559	49 678	53 119	98 189	81 851	29 651	84 215
92 581	02 262	41 615	70 360	64 114	58 660	96 717	54 244	10 701	41 393
12 470	56 500	50 273	93 113	41 794	86 861	39 448	93 136	25 722	08 564
01 016	00 857	41 396	80 504	90 670	08 289	58 137	17 820	22 751	36 518
34 030	60 726	25 807	24 260	71 529	78 920	47 648	13 885	70 669	93 406
50 259	46 345	06 170	97 965	88 302	98 041	11 947	56 203	19 324	20 504
73 959	76 145	60 808	54 444	74 412	81 105	69 181	96 845	38 525	11 600
46 874	37 088	80 940	44 893	10 408	36 222	14 004	23 153	69 249	05 747
60 883	52 109	19 516	90 120	46 759	71 643	62 342	07 589	08 899	05 985

eyes and stabbing at the pages with a pin in the time-honoured fashion. If the digit we alight on is odd we take the first page, if it is even we take the second page. We now repeat the pin operation on the selected page and read off two-digit numbers until we have two between 01 and 50. The first of these can be taken as our column reference, the second as our row reference. Thus, if we select the first page of our table, and if our two references are 21 and 46, we find that the random digit at this point of intersection is 6, and this is our starting-point.

EXAMPLE 2

In our discussion of p.p.s. sampling (see Table 11.1, page 343) we required two random numbers between 1 and 10 000 (or 0000 and 9999, which is more convenient). We carry out the same exercise as before, but now we can read the numbers from Table 11.3 in sets of 5, in which form they are conveniently laid out, ignoring the first digit. (We are not compelled to do this, but it will make the example easier to follow.) Table 11.3 can now be regarded as containing 10 columns, each of 5 digits.

Suppose now that our preliminary stages of selection lead us to the second page of Table 11.3, column 8, row 33. The 5-digit number at this point of intersection is 00307. For our second number we can repeat the whole operation or take any number adjacent to 00307. Let us take the next number in column 8, i.e. 98932. Ignoring the first digit, we have 0307 and 8932. Then from Table 11.1 we have selected clusters 1 and 9.

The user of Table 11.3 is free to devise any method he likes for choosing which page, row, column, etc., of the table is to be used. There is no limit to the amount of complication which can be built-in to these procedures. The principle involved and the end-product remains the same.

Quota sampling. The sample designs considered so far in this chapter have all been based on random sampling. In this section we consider the best known and most widely used form of non-random sampling: the method of *quota sampling*.

The fundamental difference between quota sampling and random sampling is that in the former the ultimate selection of the person (or firm) to be interviewed is left to the interviewer (or research executive), while in the latter selection is based on rigorous selection from a frame.

In the typical survey based on quota sampling—say a market research survey into brand shares in a consumer-goods market—interviewers are issued with a schedule showing the numbers of respondents in various categories, such as sex, age and social class with whom they should obtain interviews. The schedule sometimes contains additional instructions relating to when and where interviewing should take place. The choice of which particular men and women to interview, however, is left to the interviewer. The characteristics specified in the interviewer's schedule are known as *quota controls*. Commonly used controls are sex, age, social class, marital status and family size. Quota controls may be *open* or *interlocking*. An example will illustrate the difference. Consider an assignment of 100 interviews and suppose that two controls are imposed: sex and age; the schedule might set out the requirements as follows:

TABLE 11.4

Sex	Total	Age	Total
Male	40	16–34	20
Female	60	35–54	50
		55 and over	30
	100		100

In this form the controls are open, i.e. they are independent of one another. There is nothing, therefore, to prevent the interviewer selecting all the 16–34 age group from male respondents, or all the males from the 35–54 age group, or any other combination of controls. This possibility can be obviated, and the coverage of the survey made more representative by arranging for the controls to be interlocking. With interlocking controls the 100 interviews might be specified in the form of a two-way table, as below (Table 11.5).

Obviously, interlocking controls impose much more stringent conditions on the interviewer. Equally obviously, these controls cannot be set unless the necessary population information is available. This tends to confuse the difference between quota and random sampling for a number of people. The above controls seem to result in a stratified sample. Indeed they do, but it is a *stratified quota sample*. In fact, most of the features of random sample design can be built into a quota sample; we can have *multi-stage stratified quota sampling* or *clustered quota sampling* or *multi-phase quota sampling*, etc. It is only at the final stage of selection—the selection of the specific individual to be interviewed—that the two methods differ.

Quota sampling is widely used in market research and other kinds of sample survey. It is usually much more economical than random sampling because, for example, there is no question of finding the respondent out and having to call back, perhaps several times. Quota sampling is also easier to administer and can often be carried out where no sampling frame exists. It is also useful where fieldwork has to be completed very quickly and in fact may be the only method possible in such circumstances. For example, B.B.C. Audience Research figures are based on interviews carried out on the day following the programme referred to. A large number of interviews

TABLE 11.5

Sex	Age 16–34	Age 35–54	55 and over	Total
Male	8	20	12	40
Female	12	30	18	60
Total	20	50	30	100

are conducted every day (about 5000) and it would be quite impossible to obtain 5000 interviews in one day with random sampling, even with unlimited resources.

Against these factors in favour of quota sampling must be set a number of disadvantages. The two most important of these are, first, that sampling errors cannot be calculated from quota samples because their calculation is not based on probability theory, and secondly that quota sampling often yields biased samples because no human being ever achieves perfectly random selection.

We have now met the main forms of sampling; both random and non-random. We move on to consider briefly a number of other aspects of survey design beginning with the all-important question of sample size.

SAMPLE SIZE

Almost the first question asked by the sponsor when planning or commissioning a survey is: 'How big should my sample be?' The answer in practice is usually 'How much money have you got?' or 'How much time do we have?' Aside from these considerations, however, a proper answer to the question cannot be given until a good deal is known about the aims of the survey, the population to be covered and the detail in which results are required. When this information is available, three main factors must be considered (in addition to the two already given):

1. The degree of precision required in the answers. At this stage the concept of precision might have to be explained to the sponsor. If all goes well, a specification of precision will be produced. A typical specification is that we require an error not greater than 5%, at the 95% significance level. In most surveys, of course, a number of questions are put to respondents. Precision requirements might vary from one question to another, in which case we either fix sample size according to the most stringent requirement or we design a multi-phase sample and calculate the sample size needed for different questions or groups of questions. (We are reminded in connection with the specification of precision, of the immortal words of the businessman talking to a survey statistician: 'You just give me the figures—I'll do my own doubting.')

2. The inherent variability in the data. If there was no variability with respect to the information we seek, i.e. if all the units in the population were identical in this respect, a sample of one would be sufficient. Obviously the more variable the population the larger the sample required for a given degree of precision.

3. The number of sub-samples for which separate analysis is required. In most surveys, sub-sample results and differences between sub-samples are as interesting as the overall results, often more so. A minimum number of observations for each separate sub-sample must be specified, based on the precision required and the variability in that particular sub-sample. The total sample size is then obtained by adding together the sub-sample sizes.

Given satisfactory information on the above three points we can proceed to estimate sample size. Information concerning items 1 and 3 will be obtained from the survey sponsor. Information about item 2 might be obtained from previous research (our own or some-one else's). In the absence of any such information an estimate of variability can be obtained from a pilot survey.

There is one other vital practical requirement if we are using random sampling. This is to allow, in fixing sample size, for the inevitable *non-response*. Thus, if the average response rate for the kind of survey we are conducting is 80%, we should increase our calculated sample size by 25% in an attempt to ensure enough observations in sub-samples for some sort of analysis to be made. We say 'some sort of' analysis. Non-response presents a number of problems, most of which cannot be solved simply by increasing sample size, as we shall see in the next section.

One of the problems which can be solved, however, is the problem of not having sufficient observations to allow any detailed analysis at all. The solution is simply to increase sample size.

NON-RESPONSE

Selecting a random sample of individuals or firms and obtaining information from all those selected are two different things. We will fail to get information from some members of the sample for any one of a number of reasons:

1. The sampling frame may be out of date and some individuals or firms may no longer exist at all, or may have moved out of the area.

2. Problems of definition and the availability of a suitable frame may mean that some of those included in the frame ought not to be there at all.

3. Some members of the sample will simply be unavailable at the time, e.g. away on holiday, or on a business trip, or in jail or hospital, or just out every time the interviewer calls.

4. Some members will not be able to provide the information we seek because they do not have it themselves.

5. If we use a postal questionnaire some will forget or not bother to reply.

6. Some members will simply refuse to co-operate.

Between them, the above factors might result in a *non-response rate* of around 30%. The non-response rate varies considerably between different surveys but anything below 30% can be considered very good.

Why does non-response matter? We have already noted that in calculating the required sample size an allowance should be made for non-response. Is this not sufficient to overcome the problem? The answer is 'No'. Numbers may still be inadequate in certain sub-samples because the non-response rate is unlikely to be uniform over all sub-samples. More important than this, however, is the fact that results based on 'volunteer' respondents will almost certainly be biased. The bias is often serious enough to invalidate the findings and therefore something must be done, first to minimise non-response, secondly to adjust the results, if possible, to make a better guess at the right answers.

Bias arises because frequently those individuals or firms which respond are not representative of the whole, i.e. are not a random sub-sample of the original sample. For example, a survey of family incomes might find higher refusal rates among the very rich and very poor respondents than among the remainder of the sample. Similarly, big firms might be less willing to co-operate than small firms, or vice versa, in a particular survey. The danger is that availability, ability or willingness to respond is often correlated with the subject of the survey. In a survey on cinema-going it would be most unwise to confine interviewing to the evening without insisting that the interviewer calls again if people are out.

With non-response, the effect is as if the members of the sample imposed their own additional stratification on the sample; the respondents and the non-respondents. An unbiased estimate from the sample requires a weighted average of these two new strata; if we merely increase sample size we obtain more interviews in the respondent stratum but have absolutely no effect on the non-respondent stratum. We must therefore make some attempt to obtain information from the non-respondents, in other words we must aim at increasing the response rate. The best way of doing this is probably polite persistence. When an individual is out, or claims to be busy, the interviewer should find out, if possible, when he is likely to be in and call back. Several call-backs, in fact, should be made before admitting defeat. Most survey organisations require their interviewers to make two or three call-backs. If a respondent refuses to co-operate, a further attempt by a more experienced interviewer— probably the supervisor—is quite often successful. With postal questionnaires, reminder letters should be sent at pre-determined intervals to the non-respondents. No reputable survey practitioner would dream of basing his results on data collected without follow-ups—unless the initial response was 100%.

Polite persistence is the main method of minimising non-response, but much can be done to help before any interviewing has taken place. Factors such as the sponsorship of the survey, the subject-matter, the type of question, and the length of the questionnaire all influence the response rate. One bad question, e.g. vague, impossible, offensive, can easily bring the interview to a halt or dispatch the postal questionnaire into the waste-paper basket. These things can and should be tested in the pilot survey.

If, in spite of all efforts, interviews are unobtainable it is sometimes possible to collect some basic information about the sample member, e.g. age, social class, family size and composition, size of firm, market segment. This information can be compared with corresponding information from respondents and can be used to check the structure of the non-response stratum against the response stratum. Agreement between the strata provides some assurance as to the validity of the sample results but, of course, it tells us nothing about the information we are really seeking.

Finally, a certain amount of juggling with the results, involving weighting the means, totals, etc., will give better estimates. The general idea is quite simple. Data from initial respondents are kept separate from data obtained after follow-ups. The two sets of data are then combined with weights proportional to the numbers in the response and non-response strata. This idea can be easily extended by separating data obtained after one call-back, two call-backs, three call-backs, etc., and combining these sets of data with the appropriate weights. The basic idea underlying this method is that respondents who only provide data after a number of call-backs are more typical of the non-respondents than are the initial respondents. The final open-ended category of call-back data therefore is taken to be representative of the remaining non-respondents and weighted accordingly.

Other methods have been proposed for estimating the effects of non-response. Some of these methods are fairly complex, and none of them is commonly used so we will not investigate them.

Our discussion of non-response has assumed that the sampling being carried out was random sampling. Non-response also occurs with quota sampling (and with censuses for that matter) but the fact is sometimes unrecognised and quota sampling is often justified on the grounds that it avoids non-response. It does not—all that happens is that the non-response is concealed by the fact that sampling continues until the required number of interviews has been obtained. The important thing, however, is the number of attempted interviews that were necessary to achieve the required sample size.

QUESTIONNAIRE DESIGN

Most sample surveys rely on some form of questionnaire as the means of collecting information. Other methods of collection are available, including observation, automatic data logging, and accounting procedures, but something of the order of 90% of all surveys use a questionnaire.

The design of the questionnaire is as important as the design of the sample, if not more so, because with a badly designed sample something can usually be salvaged but if the wrong questions are asked the whole survey is a failure.

Questionnaire design is undoubtedly the most difficult part of survey design. There is no compact body of theory and techniques to make use of as there is with sampling, for example. A good deal of research has been carried out into questionnaire design and much has been written about the subject. Unfortunately, most of the research has demonstrated the existence of problems without being able to provide much specific advice on how these problems can be solved. A few general principles have emerged, however, and current research offers hope of further advances.

A good questionnaire, like many other things, is much easier to recognise than it is to construct. A good questionnaire is as short as possible, it is interesting both to interviewer and to respondent, it contains no superfluous questions nor any that are impractical, neither does it omit any item of importance. The wording of the questions is simple and unambiguous, the questions themselves are arranged in the proper order and the various 'routes' through the questionnaire, i.e. the alternative questions which depend on the answer to a previous question, are easy to follow. Finally, the layout of the form makes the task of recording the answers easy and facilitates editing, coding and punching into cards or tape.

The authors have not yet met this ideal questionnaire, but it should be the aim of all survey designers. We will consider very briefly what can be done to point us in the right direction.

To begin with, the aims and objectives of the survey should be re-examined. There is a great temptation to cover too wide a field of enquiry. This is particularly true of social research where surveys are generally aimed at fields of enquiry rather than a specific topic. The effect on the questionnaire can easily be to swamp the really relevant questions with a number of 'fringe area' questions, with the result that the questionnaire becomes too long. In cutting back the length to a manageable size, it is not unknown for some of the vital questions to be thrown out with some of the fringe ones.

All questions which survive this first stage of screening should be vigorously tested by means of a *pre-test* followed by a pilot survey. A pre-test is carried out by simply trying out the first draft of the questionnaire on anyone who is available—colleagues, friends, relatives, etc. The aim is to detect any gross errors. A pilot survey is a kind of 'dress rehearsal' for the main survey and fulfils a number of functions, one of which is to test the questionnaire for length and practicability. It is no use asking about things of which the respondent has no knowledge, or which make unreasonable demands on his memory or tolerance.

Before we can test, however, we must draw up the questions to be tested. We can distinguish three main aspects of question design: *content*, i.e. subject-matter, *order* and *wording*.

Question content is concerned with the information we are seeking and the possible sources of that information. We have to remember that accurate information depends on both the *ability* of respondents to give an accurate answer and their *willingness* to do so. In many cases willingness is greater than ability; for example, it is not unknown for surveys to seek information about family behaviour and rely on the housewife to give such information as husband's earnings, children's leisure activities and so on. Such information is often highly unreliable.

In discussing question content, a basic distinction has to be drawn between *factual* and *opinion* questions. Factual questions usually present fewer problems than opinion questions because with most factual questions there is a right answer which does not require deep thinking. With opinion questions there is often no single 'correct' answer. For example, if someone is asked if he is in favour of Great Britain joining the Common Market his answer might be hedged with a number of qualifications and conditions. He may be in favour in principle but find the practical difficulties overwhelming;

he may be in favour subject to special protective measures for far-
mers; he may want a trial period, and so on. Possibly he knows
nothing at all about the subject, or may never have reached a decision
one way or the other. He may in fact be prompted by the interview
to think about the subject for the first time.

A further problem with opinion questions is that a straight 'Yes' or
'No' may conceal the intensity with which an opinion is held. If intensity of opinion is regarded as important in the survey, special techniques may be necessary in order to measure it, and additional questions
have to be asked. This kind of research comes under the heading of
Attitude Scaling, a topic which is outside the scope of this chapter.

Answers to opinion questions are much more sensitive to the
influence of the interviewer. It is very important, therefore, to try
to achieve uniformity of presentation by interviewers. Even more
important is the need to control the amount of *probing* allowed by
the interviewer. With factual questions any reasonable steps to
ensure that the questions are clearly understood and interpreted and
the answers correctly recorded, e.g. by repeating or re-phrasing
them, are allowed. A well-known method is the use of *aided-recall*
questions, in which the interview prompts the respondent with a list
of possible answers. For example, in television audience research,
instead of asking 'which programmes, if any, did you watch last
night?' copies of the *Radio Times* and the *TV Times* might be shown
and the question asked in the form 'Did you watch any of these
programmes?' A similar system is used in newspaper readership
surveys. A book containing the 'mastheads' of all the newspapers
is shown to the respondent, who is asked to indicate which of them
he has seen in the past week, or whatever period is under review.
In most opinion polls probing is either forbidden altogether or
strictly controlled. Uncontrolled probing is too risky; wording,
intonation, etc. could materially affect answers.

Question order is sometimes overlooked as a factor affecting survey
results, but a lot of evidence exists to show that it does. As with
question content, the effect is more serious with opinion questions
than with factual ones.

In some questionnaires it is possible to vary the order of the
questions for different sub-samples. The scope for this is limited,
however, because there is usually a logical sequence of questions
through which the respondent should be led. It is sometimes

possible to randomise the order of blocks of questions, however. For example, in a survey into the image of various textile fibres respondents were shown a number of samples of textiles, including carpets, suitings, and knitwear. Each particular end-use had its appropriate set of questions and the order in which the sets of samples were presented to respondents was varied. The aim of the randomisation was to overcome any fatigue or boredom effect as respondents went through the sets of materials.

Similarly, in the newspaper readership example above the order in which mastheads are presented to respondents is varied. If there is any doubt about the effects of varying the order of questions on a particular questionnaire, elaborate testing with perhaps several pilot stages may be necessary.

Question wording is the key issue in questionnaire design. Many books and articles have been written on the subject but, in spite of these, little specific advice can be given on how questions should be worded. We give below a number of pointers, most of which are so obvious that they might appear to insult the reader's intelligence. Yet when an example of a bad questionnaire is produced, one or more of these obvious requirements is usually found to be lacking.

Always use the *simplest* words that convey the exact meaning of the question. If the house caught fire we would shout 'Help—fire!' not 'Assistance pray—conflagration'. It is very easy, however, to fall into this kind of error when expressing questions in writing instead of speech. A particular danger comes from using words which have slightly different meanings in statistics or sociology, etc., from their everyday meaning. 'Population', 'sample' and 'random' are obvious examples from the statistician's jargon. Another example is the word 'book', which means magazine or journal in some parts of the country. 'Dinner' is also a very dangerous word, and numerous other examples are readily available.

Words which are *ambiguous*, words which are *vague* and words which are *insufficiently specific* should likewise be avoided at all costs. As an example, consider the following question taken from an insurance company's proposal form for the insurance of household contents, sent recently to one of the authors:

'Question 5: (*a*) Is the dwelling house at present occupied?

　　　　　　　(*b*) Will it be left unattended regularly during the daytime?

(c) For what period do you estimate it will normally be left unattended during any one year?'

It would be difficult to improve on this question as an example illustrating the points to be avoided. Apart from the faults mentioned at the beginning of this paragraph, the question illustrates in part (b) another very common error, namely, the confusion between *regularity* and *frequency*. Does the question mean 'regularly' or does it mean 'frequently'.

Leading questions, i.e. questions which lead the respondent towards the answer sought, by the questioner must also be avoided. Such questions are very common in the less reputable opinion polls and in general among surveys which are designed to provide 'evidence' in support of their sponsors' prejudices.

Hypothetical questions are quite common in badly designed questionnaires. The answers to such questions are usually not worth very much. For example, if in a survey into the market for a proposed new product, housewives were asked 'if this product was available at a price of X would you buy it?', many answers would be made in a mood of 'sure, I'll try anything once' and would provide a very shaky basis for deciding whether to launch the product or not. It is one of the problems of market and social research that one of the most unreliable ways of predicting people's future actions is to ask them what they are going to do.

Presuming questions are another type of question to be avoided. A presuming question presumes that a respondent behaves in a certain way without establishing the fact first. He is then questioned about this presumed behaviour. Thus, a questionnaire on tobacco consumption might begin by asking how many cigarettes the respondent smokes per day, week, month, etc., without first establishing whether he is a smoker or a non-smoker. With factual questions this might not matter too much—the respondent can always answer 'None'—but in opinion surveys it is often vital to establish whether the respondent has an opinion on the subject at issue before asking what that opinion is in order to avoid a biased answer.

Open and pre-coded questions. There is one major consideration in question design which we have not yet mentioned. This is the choice between *open* and *pre-coded* questions.

Open questions leave the length, nature and wording of the answer entirely to the respondent. The interviewer's task is to record as much of this answer as possible.

Pre-coded questions are of two kinds. In the first kind the respondent is given a limited number of alternatives to choose from, e.g. 'Do you use any of these products?', followed by a list of the products concerned. In the second kind the question is put as an open one but the interviewer codes the answer according to a predetermined list on the questionnaire.

The essential difference between open and pre-coded questions is the stage at which the answers are coded. Before the information collected on the questionnaire can be used some scheme of classification has to be devised. This is particularly important if the data processing is to be carried out mechanically or by computer. Most classification schemes involve numerical coding of the answers. With open questions, editing and coding can be carried out in the office, under uniform conditions (one editor can edit the same question or set of questions on all questionnaires). The actual words of the respondent are on the questionnaire so coding can be checked. With pre-coded questions, coding is done at the time of the interview and cannot be checked. There is also the danger that answers are forced into a category to which they do not belong. It is essential, therefore, to have an additional category of 'Other answers—please specify' into which answers which do not appear to fit into any of the pre-coded categories can be placed by the interviewer.

In general, the aim should be pre-coded questions where possible. In order to construct an efficient set of categories, however, we need to know something about the range of answers. This is something else to be investigated in the pilot survey.

Finally, we must face the fact that not all the information we would like to collect is amenable to collection by questionnaire. We have already seen that valid answers to questions depend on both the ability and willingness of the respondent. There are some situations in which respondents might give false answers deliberately. For example, there is the widely reported 'prestige effect' observed in television audience measurement where respondents often claim to have watched a serious 'highbrow' programme rather than the quiz game or the serial on the other station. Similarly, in surveys on such prickly problems as racial tolerance there will be a tendency for respondents to give stereotyped answers which fall into one of a number of standard categories when in fact their real opinions are more extreme.

Two further tips can be given. First, all questions should be spoken aloud as they are drafted. A number of questions which ought never to reach the pre-test stage get through because this simple step is omitted. A question which looks perfectly innocuous on paper might sound quite different. As an example, a proposed questionnaire on men's clothing included a question which was intended to check whether a particular suit was the individual's 'best' suit. (Previous questions would establish whether the respondent regarded any particular suit as 'best'.) The check question simply said 'Is this your best suit?' The reader is invited to say this in as many different ways as possible to savour its full beauty.

Secondly, as questions are drafted, skeleton tables showing how the information is to be used should also be drafted. We cannot emphasise too strongly that the time to start thinking about the analysis of the data, and the shape of the final report, is at the questionnaire stage. It is not at all uncommon to find that a question over which fierce arguments raged at the questionnaire stage is not used at all in the analysis simply because nobody thought about it from the analysis point of view. Failure to observe this requirement is the main reason for the depressingly low standard of questionnaires generally, and for the intolerable time lags between field work and final report, which are all too common. *Every* stage of a survey should be planned before field work commences.

POSTAL SURVEYS

Many sample surveys are carried out by means of questionnaires sent through the post. They are especially common in industrial market research, as many suffering managers will verify. In principle, postal surveys are no different from those based on interviews, but there are several practical limitations.

Postal surveys are apparently cheap to conduct, compared with interviewer surveys, but this cheapness is often illusory. Postal surveys usually suffer from a very high rate of non-response compared with interview surveys and postal questionnaires have to be shorter and simpler. The result of these two factors is that cost per unit of information can easily be as high or higher than in an interview survey.

Apart from questionnaire content, and non-response, the problems encountered in postal surveys are no different from those found with interview surveys. We can therefore proceed to study the two major differences.

The questionnaire length and content is all important in postal surveys. One difficult, vague, offensive, etc., question will be all that is necessary to consign the questionnaire to the waste paper basket. The same question in an interview survey can be glossed over by the interviewer and something retrieved from the interview. In a postal survey, however, all might be lost. For this reason, careful piloting of the questionnaire is essential. One way of doing this is to pilot the questionnaire by means of interviews. Alternatively, the pilot questionnaire can be sent through the post and collected by interviewers.

The other problem is non-response. Here the main action we can take is to send follow-up letters, enclosing further copies of the questionnaire to prod non-respondents into answering. Several follow-up letters should be sent to non-respondents before abandoning them. The techniques for adjusting results to allow for non-response mentioned earlier in connection with interview surveys can also be used.

Our discussion of postal surveys rather portrays them as a means of conducting a survey on the cheap. This is a bit unjustified because in certain circumstances postal questionnaires can be a more efficient information-collecting tool than interviews. In some surveys, for example, the information required may not be immediately available and the respondent, although willing, would be unable to give an immediate answer. He might, however, be prepared to look up the information and return it on a form. In such a situation the postal questionnaire could be more efficient. Similarly, a survey into medical history, provided it is conducted by a responsible body, might obtain more information from a postal questionnaire than an interview. Apart from having to check his own recollections, the respondent might be inhibited against revealing his various past diseases to an interviewer.

Thus in certain circumstances postal surveys might have advantages over interview surveys. The disadvantages, however, are formidable. In addition to those already mentioned, there is no guarantee that the individual to whom the questionnaire is addressed will be the one who actually fills it in. Even if he is, there is great scope for imagination, facetiousness and plain cussedness in filling the thing in and, human nature being what it is, advantage is no doubt often taken of this scope. Nevertheless, a number of survey practitioners report satisfactory results from postal surveys. One factor which undoubtedly influences the quality of the results and the response rate, is the subject-matter of the survey and its sponsorship. The ideal situation for a successful postal survey is a homogeneous population, e.g. a professional association, where

(*a*) the subject of the survey is of interest to the members and relevant to their work;

(*b*) they have the information necessary to answer the questions;

(*c*) the survey is sponsored by their association or professional body.

PILOT SURVEYS

Pilot surveys are essential ingredients in well-conducted sample surveys and are equally vital to census inquiries. A pilot survey should be regarded as a rehearsal of the main operation and more than one rehearsal may prove necessary before the main survey is launched. The functions of a pilot survey include:

1. A test of the questionnaire for length, clarity, feasibility, ease of administration, etc. If necessary, alternative versions of questions can be tested by means of random sub-samples. The average and maximum length of interview can also be estimated.

2. To provide an estimate of variability in the population for the purpose of calculating sample size, if this is necessary.

3. To test the adequacy of the sampling frame.

4. To estimate the final response rate and the likely breakdown of non-response into separate categories.

5. To test the suitability of the proposed data-collecting medium, e.g. the feasibility of a postal questionnaire.

6. To check the efficiency of the fieldwork organisation and the central office administration, e.g. in editing, coding, etc.

7. To determine the coding schemes for all pre-coded questions.

8. To provide estimates of the cost and duration of the fieldwork.

Pilot surveys are usually based on 100 or so interviews, although this will obviously vary with the amount of information required from the pilot. The sample may be random or quota, again depending largely on what items need to be established on the strength of the pilot. If it is necessary to estimate the population variance in order to determine sample size, then obviously the pilot sample must be random. If the proposed sample design is a complex one, then the pilot survey must mirror this design if it is required to estimate sample size. For complicated sample designs, therefore, the pilot survey itself becomes large and complex, and is unlikely to be worthwhile in small-scale surveys such as the run of the mill market research survey. In such cases information from previous research projects is likely to be of more use—and will certainly be cheaper—than information from a complicated pilot survey. The usual approach is to estimate sample size and cost, etc., from previous survey material and use the pilot survey mainly to test the questionnaire.

OMNIBUS SURVEYS

It will have become clear to the reader by now, we hope, that a sample survey—especially one carried out on a national scale—is not something to be undertaken lightly. A lot of time and money is involved. Now, it often happens that an answer is required to one specific question or to a very limited range of questions. Frequently these answers are required quickly. The time and money required to mount a full-scale survey for the sake of two or three questions is usually prohibitive; however, if several firms could share the costs, the price to each could be brought down to an economic level. This is the idea behind the *omnibus survey*. An omnibus survey is a sample survey in which several different sponsors combine their questions in one questionnaire. Omnibus surveys are offered as a regular service by several market research firms. Space is offered on the questionnaire to interested parties and the questions are put to a sample of specified size and design. The advantages of this system to the sponsors are that results are quicker and cheaper than would otherwise be the case. There are a number of possible disadvantages, however. In most cases a given sponsor does not know who will be sharing the questionnaire with him nor will he know whereabouts on the questionnaire his questions will come. Our earlier discussion of question content, wording and order will suggest to the reader the numerous pitfalls which might be lying in wait.

A slightly different version of omnibus sampling may be found in some market research firms. In this version, the questionnaire is drawn up for the market researcher's own purposes but a limited number of simple questions may be added at the end. This method is sometimes referred to as 'hitch-hiking'.

RATIO ESTIMATION

The production of good estimates from survey data depends on two distinct factors:

(a) the method used to collect the data, i.e. sample design and questionnaire design;

(b) the method of estimation applied to this data.

All our discussion up to now has concerned (a). In fact, very little attention has been paid in general to problems of estimating, compared with that paid to problems of sample design. One reason for this undoubtedly is that in most surveys a great many estimates have to be made, so the simplest methods, e.g. sample means, sample distributions, sample proportions, etc., are used. In this final section of the chapter we consider one alternative method of estimating population totals, which in the appropriate circumstances can be very useful. This is the method of *ratio estimation*. We will illustrate the idea with a hypothetical example.

Suppose we wanted to estimate the total population in a specified number of towns and cities in Great Britain in 1968 and suppose further that the only existing data in this field related to the 1961 Census of Population. We have insufficient resources to conduct another census so we are forced to take a sample of towns.

Let us assume that from our specified towns a simple random sample of ten towns is selected and yields the results shown in Table 11.6.

From the results we could estimate the required parameter by multiplying the sample total by the raising factor. We could also calculate confidence limits for this estimate. This is our orthodox way. We are now going to demonstrate an alternative method. This method depends on two factors:

(a) we can look up the 1961 Census Reports and extract the 1961 figure for each town selected in the sample;

(b) from the Census Reports we can determine the *total population* at 1961 in the whole set of towns in which we are interested.

TABLE 11.6

Town	Population in 1968
1	20 000
2	500 000
3	16 000
4	100 000
5	150 000
6	230 000
7	8 000
8	360 000
9	35 000
10	60 000
Total	1 479 000

Let us denote by y, x and X the three quantities: sample total in 1968, sample total in 1961, population total in 1961. We already have $y = 1\ 479\ 000$. Let us assume that $x = 1\ 063\ 000$ and $X = 12\ 187\ 000$. Our estimate of the 1968 total is then given by the expression

$$1968\ \text{total} = \frac{y}{x} X$$

$$= \frac{1\ 479\ 000}{1\ 063\ 000} \times 12\ 187\ 000$$

$$= 16\ 956\ 000.$$

The advantage of the ratio estimate over the more usual method is that the ratio estimate, in this kind of situation, is more precise. Ratio estimates can be used in many fields of application. An obvious business application is the estimation of inventory value. If a full-scale stocktaking exercise is carried out once a year, estimates at, say, the half-year can be made by selecting a sample of stock items, valuing these, extracting the corresponding values from the previous annual census and performing calculations similar to those above.

We have assumed simple random sampling for our example but ratio estimates are equally applicable in all kinds of sample, including quota. As usual, however, the arithmetic becomes heavier as the sample design becomes more complex.

CONCLUDING REMARKS

In this chapter we have outlined some of the most important considerations in the design of sample surveys. Several major topics, however, have been omitted or given the briefest of passing mentions. We have said nothing about the conduct of the fieldwork or about the analysis and presentation of the results, nor have we described the work of psychologists in devising questions aimed at revealing attitudes and opinions. We have omitted all discussion of non-sampling errors, e.g. those errors which arise from inefficient recording of answers, or biased interviewing, or deliberately false replies from respondents. A guide to these, and other topics, is given in the references to this chapter.

REFERENCES

KISH, L. 1966. *Survey Sampling.* New York, Wiley

MOSER, C. A. 1958. *Survey Methods in Social Investigation.* London, Heinemann.

PAYNE, S. L. 1951. *The Art of Asking Questions.* Princeton, University Press.

12: STATISTICAL SQUARE-BASHING
Analysis of Variance

INTRODUCTION

IN Chapter 2 we learned the meaning of the spread or dispersion of values about a population average and discovered ways of measuring it.

Variate values derived from industrial situations differ for many reasons. Machine operators develop different levels of competence, the materials used are not all of exactly the same quality, machines tend to develop their own characteristics—a fact to which owners of mechanical grass-mowers will readily testify—and, on a more human plane, a good spirited workman will produce better work than one whose morale is poor. These factors which induce variation are common to the manufacture under factory conditions of many products.

The less rigidly disciplined world of nature produces the same kind of variation. If a botanist bent upon a field study measures the length of all plantain leaves within the confines of his 'field-frame' he will of course discover that they are different. With the length of leaf as variate, he can draw up a distribution table. The botanist will be able to put forward reasons for the variation: local concentration of suitable fertiliser, the types of adjacent plants, the aspect, and possibly a few bricks embedded in the top soil beneath one corner of the site!

In any situation the data for which yields a bell-shaped or other form of distribution, the measure of variation is its variance. Normally one accepts that the numerical data is drawn from a homogeneous population, and often one knows that this is so. A factory process whose function is to measure the electrical resistance of an assembled component of a radio set would reject those whose resistances fell outside the specified limits. In doing so it discards those which, at some stage, have been manufactured or assembled by a process that, even momentarily, did not conform to the specification set for the population. The solder used for one joint might contain a fragment of impurity, or the emotional chatter of two females might have distracted one of them from her job. The factory process which measures the resistance of the assembled component is thus making the *accepted* population of components homogeneous, and its variance smaller.

There are however some populations which are probably heterogeneous. A dairy farming research institute might well conduct experiments on the quality and quantity of milk yield from herds of different breeds, Jersey, Friesian, Ayrshire and so on. The volume of milk produced and percentage of fat content would be charted for each beast for each milking, and the findings eventually summarised.

ELEMENTARY THEORY

Let us imagine that the following Table 12.1 gives the quantity of milk yield per cow over a certain defined period of time. The information is listed by breed and in units which are immaterial for our purpose. (The breeds are identified by letters so that no

TABLE 12.1

	Breed			
	A	B	C	D
	4	5	8	7
	2	4	5	6
	3	3	6	6
	3	4	5	9
Breed totals	12	16	24	28
Breed means	3	4	6	7

Grand total $= 80$. Grand mean $= \frac{80}{16} = 5$.

agriculturalist—or any cow for that matter—can be offended by an unintended inaccuracy in the simple hypothetical data.)

Now, to doubt the homogeneity of the total population is really to pose the question 'Is there a significant difference between the mean milk yield for the different breeds?' If we were concerned with only two breeds, the significance of the difference could be tested using the methods of Chapter 8. When more than two means are to be tested, however, a different approach must be adopted.

The problem is tackled by a technique aptly known as the *analysis*

N

of variance. In the context of the milk yield problem, the figures for each breed can initially be treated as a sample of the population. If we then set up the 'null hypothesis' that the figures for each breed taken separately are consonant with the population mean, it would follow that an estimate of the population variance based upon the variations arising *between* the breeds would be consistent with an estimate of the population variance based upon variations arising *within* the breeds, since the same influences would be responsible for such variations as manifested themselves.

Fortunately there is a test ready to hand to determine whether these two statistics are estimates of the same variance. It is the variance ratio or '*F* ratio test', which was introduced in Chapter 8. We must however remember, in using this test, to use the appropriate number of degrees of freedom when calculating the several variances. Now let us follow these concepts through the information in the table.

PRACTICAL EXAMPLE

First, consider the total sum of squares and number of degrees of freedom for the population. (Throughout the rest of this chapter the phrase 'sum of squares' will refer to the sum of squares *measured from the appropriate mean.*)

Remembering that the population mean is 5, the total sum of squares is obtained as follows.

	Breed		
A	B	C	D
$(4-5)^2 = 1$	$(5-5)^2 = 0$	$(8-5)^2 = 9$	$(7-5)^2 = 4$
$(2-5)^2 = 9$	$(4-5)^2 = 1$	$(5-5)^2 = 0$	$(6-5)^2 = 1$
$(3-5)^2 = 4$	$(3-5)^2 = 4$	$(6-5)^2 = 1$	$(6-5)^2 = 1$
$(3-5)^2 = 4$	$(4-5)^2 = 1$	$(5-5)^2 = 0$	$(9-5)^2 = 16$
Totals 18	6	10	22

Grand total = 56
Number of degrees of freedom = $16-1 = 15$

Secondly, consider the sum of squares and number of degrees of freedom for the estimate of variance based on *between* breed variations.

To do this we must eliminate all *within* breed variations by replacing each item by its breed mean:

Breed			
A	B	C	D
$(3–5)^2 = 4$	$(4–5)^2 = 1$	$(6–5)^2 = 1$	$(7–5)^2 = 4$
$(3–5)^2 = 4$	$(4–5)^2 = 1$	$(6–5)^2 = 1$	$(7–5)^2 = 4$
$(3–5)^2 = 4$	$(4–5)^2 = 1$	$(6–5)^2 = 1$	$(7–5)^2 = 4$
$(3–5)^2 = 4$	$(4–5)^2 = 1$	$(6–5)^2 = 1$	$(7–5)^2 = 4$
Totals 16	4	4	16

Grand Total $= 40$
Number of degrees of freedom $= 4 - 1 = 3$.

(There are four breed means, and knowledge of three determines the fourth.)

The estimate of variance for variations *between* breeds is therefore $\frac{40}{3} = 13\frac{1}{3}$.

Finally consider the sum of squares and number of degrees of freedom for the estimate of variance based on *within* breed variations. In this instance we have to eliminate the breed mean from each item.

Breed			
A	B	C	D
$(4–3)^2 = 1$	$(5–4)^2 = 1$	$(8–6)^2 = 4$	$(7–7)^2 = 0$
$(2–3)^2 = 1$	$(4–4)^2 = 0$	$(5–6)^2 = 1$	$(6–7)^2 = 1$
$(3–3)^2 = 0$	$(3–4)^2 = 1$	$(6–6)^2 = 0$	$(6–7)^2 = 1$
$(3–3)^2 = 0$	$(4–4)^2 = 0$	$(5–6)^2 = 1$	$(9–7)^2 = 4$
Totals 2	2	6	6

Grand total $= 16$

Now there are three degrees of freedom for each breed, and there are four breeds, so there are 12 degrees of freedom altogether. The estimate of variance for variation within breeds is therefore

$$\frac{16}{12} = 1\frac{1}{3}.$$

We can now summarise our findings:

Source of variation	Sums of squares	Degrees of freedom	Estimate of variance
Between breeds	40	3	$13\frac{1}{3}$
Within breeds	16	12	$1\frac{1}{3}$
Total	56	15	

The 'between breeds' and 'within breeds' sums of squares add up to the 'total' sums of squares although the three were calculated independently of one another. The same applies to the degrees of freedom. This shows that we have accounted for all the variation in the original data. Using the F-ratio Test on the above results:

$$F = \frac{\text{Larger estimate of variance}}{\text{Smaller estimate of variance}} = \frac{13\frac{1}{3}}{1\frac{1}{3}} = 10.$$

There are three degrees of freedom for the larger estimate of variance and twelve for the smaller estimate of variance. From the F-ratio Table (Table 8.4) we see that for 3 and 12 degrees of freedom at the 1% significance level, the tabulated value of F is 5·95, so that our observed value of 10 is significant at the 1% level.

Expressing this in the terms that began our enquiry, we are now satisfied, on the evidence presented, that the milk yields vary significantly with the breed of animal. The next step would probably be to collect further information with more animals and conduct a series of 'significance of means' tests between the breeds. The analysis of variance has pointed the way.

DEVICES FOR SIMPLIFYING THE CALCULATIONS

This example was a contrived one, to illustrate the basic ideas of the analysis of variance technique. Figures obtained from a real life situation would be less obligingly simple and the burden of computation would be considerably heavier. The burden can be eased however by the adoption of an alternative method of computation which simplifies the arithmetic while remaining algebraically equivalent to the original method. The mathematical basis for this alternative method is given in an appendix to this chapter, but its slavish application is demonstrated below.

Original Information

	Breed			
	A	B	C	D
	4	5	8	7
	2	4	5	6
	3	3	6	6
	3	4	5	9
Totals	12	16	24	28

Total of all items $T = 80$
Total number of items $N = 16$.

METHOD

1. Calculate a factor called the *correction factor*.

$$= \frac{T^2}{N} = \frac{6400}{16} = 400.$$

2. List the squares of the items:

	Breed			
	A	B	C	D
	16	25	64	49
	4	16	25	36
	9	9	36	36
	9	16	25	81
Totals	38	66	150	202
Grand total = 456				

3. Calculate the total sum of squares by subtracting the correction factor from the grand total of squares in (2) above:

$$456 - 400 = 56.$$

This agrees with our earlier calculation.

4. Calculate the *between* breeds total sum of squares by aggregating the squares of the breed *totals*, and dividing this total by the number of items in each breed: finally subtract the correction factor:

$$\tfrac{1}{4}(12^2 + 16^2 + 24^2 + 28^2) - 400$$
$$= \tfrac{1}{4}(144 + 256 + 576 + 784) - 400 = \tfrac{1}{4}(1760) - 400 = 40.$$

This agrees with our earlier calculation.

5. Calculate the *within* breeds total sum of squares. This is done by subtracting the result in (4) above from the result in (3) above:

$$56 - 40 = 16.$$

The number of degrees of freedom appropriate to each total sum of squares is derived as before, and the same table of results drawn up.

But the astute reader will still not be satisfied with this short-cut as an answer to his calculation difficulties. The new method will not be of much help with large figures for example. However, the variance of a set of numbers is unchanged if a constant is subtracted from or added to each of the constituent items. The variance of our original data, the milk yield for cows, would be unchanged if ten units were added to each of the collected values. (The reader can readily see that to add ten units to each item is merely to add ten units to the grand mean, so the resulting difference remains unchanged.)

A simple test of this rule in relation to the analysis of variance is to deduct four from each of the milk yields in the table and recalculate the separate variances.

Original Information Less 4

	Breed			
	A	B	C	D
	0	1	4	3
	−2	0	1	2
	−1	−1	2	2
	−1	0	1	5
Totals	−4	0	8	12

Total of all items $T = 16$
Total number of all items $N = 16$.

1. Correction factor $= \dfrac{T^2}{N} = \dfrac{16^2}{16} = 16.$

2. Squares of all items $=$

    $(4+1+1+1+1+16+1+4+1+9+4+4+25) = 72.$

3. Total sum of squares $= 72-16 = 56.$

4. *Between* breeds total sum of squares $=$

    $\frac{1}{4}(-4^2+0+8^2+12^2)-16 = \frac{1}{4}(224)-16 = 56-16 = 40.$

5. *Within* breeds total sum of squares $= 56-40 = 16.$
    The results correspond to those obtained previously.

There is yet a further device for simplifying the arithmetic involved in calculating the variance ratio F. The reader will recall that in Chapter 2 the work for finding the variance of a distribution was shortened by dividing all items by a constant, say, C. The variance was then measured in 'working units' which this division established —if x was the variable then the number of working units was $\dfrac{x}{C}$. If the variance calculated in this way was S^2, then the value of the variance in original units was C^2S^2. It therefore follows that if the estimate of the variance from a certain source in an analysis of variance is $C^2S_1^2$, and the estimate from another source is $C^2S_2^2$, then

$$F = \frac{C^2S_1^2}{C^2S_2^2} = \frac{S_1^2}{S_2^2} \text{ (assuming } S_1^2 > S_2^2\text{).}$$

The value C plays no part in determining the value of F. We can therefore say that

'In working through an analysis of variance, we may add to or subtract from, or multiply or divide, each item in the table by a constant amount without in any way affecting the value of variance ratio F.'

(This result is only a repetition of some work done in connection with the determination of the coefficient of correlation, which the reader may be turning over in his mind as he reads this.)

MORE THAN ONE SOURCE OF VARIABILITY

So far we have been dealing with a situation in which we were aware of only one main cause of variability in the data, namely the breed of cows. Very often problems are more complicated, in that the information presented to the statistician points to there being more than one probable source of variability. In education for example, we can see that student performance depends not only on the innate ability of each student, but also on the competence of the teachers. In trying to gauge the relative merits of different teaching systems it is always difficult to extract from the numerical results of examinations the variability arising from the differing quality of the candidates. One method of approach is to choose students who would seem, from past tests, to be of about equal ability and industry, and then to carry out an analysis of variance, calculating both 'between students' and 'between methods' sums of squares. In this way, any variation in the examination marks which is due to differences between students can be removed before considering the 'between methods' effect.

Consider the following problem.

A school of educational research believes that the 'programmed-instruction' method of teaching is superior to the 'class-lecture' method for a range of allied subjects A, B and C, which could be, e.g. algebra, trigonometry and geometry or, say, social, economic and political history. The body representing the teachers contests this, and suggests that if the teachers were to be relieved of all non-instructional duties and the ratio of teachers to students were to be improved, the 'class-lecture' method would be the better of the two.

As a way of converting the dispute to a numerical level, imagine that three students were selected by the teaching body as being of about equal ability in each of the three subjects and they were taught

Subject A by the 'Class-lecture' method,

Subject B by a mixture of both methods,

Subject C by the 'programmed-instruction' method.

There are difficulties inherent in such an approach. The teachers employed must also be of about equal ability, and the degree to which the methods should be mixed would have to be agreed between both parties. In any case the result of the exercise could not, of itself, be taken as decisive; it would be only one of a number of contributory tests.

The following example demonstrates the method of solving an analysis of variance problem where there are two known possible sources of variability. The authors in no way wish to participate in the argument, but only to make use of it.

Marks in Examinations

	Class-lecture (C/L)	Mixture (M)	Programmed instruction (P.I.)	Totals
Student 1	6	6	8	20
2	5	7	7	19
3	5	6	9	20
Totals	16	19	24	59

The first stage is to simplify the material by deducting (say) 6 from each item:

	(C/L)	(M)	(P.I.)	Totals
Student 1	0	0	+2	+2
2	−1	+1	+1	+1
3	−1	0	+3	+2
Totals	−2	+1	+6	+5

Total of all items = $T = +5$

Total number of items = $N = 9$

Correction factor $= \dfrac{T^2}{N} = \frac{25}{9}$.

1. Total sum of squares =

$$(0+1+1+0+1+0+4+1+9)-\tfrac{25}{9} = \tfrac{128}{9}.$$

Number of degrees of freedom = $9-1 = 8$.

2. *Between* students total sum of squares =

$$\tfrac{1}{3}(4+1+4) - \tfrac{25}{9} = \tfrac{2}{9}.$$

Number of degrees of freedom $= 3 - 1 = 2$.

3. *Between* methods total sum of squares =

$$\tfrac{1}{3}(4+1+36) - \tfrac{25}{9} = \tfrac{98}{9}.$$

Number of degrees of freedom $= 3 - 1 = 2$.

4. *Residual* total sum of squares is the difference between (1) and the sum of (2) and (3). We will have a look at the meaning of this after we have drawn up a table of results.

Source of variation	Sums of squares	Degrees of freedom	Estimate of variance
Between students	$\tfrac{2}{9}$	2	$\tfrac{1}{9}$
Between methods	$\tfrac{98}{9}$	2	$\tfrac{49}{9}$
Residual	$\tfrac{28}{9}$	4	$\tfrac{7}{9}$
Total	$\tfrac{128}{9}$	8	

Now the estimates of variance from the 'residual' and 'between students' sources show the residual to be the larger, so one would not trouble to consult the *F*-ratio Table as hitherto. We can therefore say, from the evidence produced in the table, there is no significant 'between students' effect, and the students can therefore be regarded as being of thesame calibre. However, the value of *F* as between the 'residual' and 'between methods' sources is again 7. Here the *F*-ratio Table value is 6·9, so the difference is *just* significant.

MEANING OF THE 'RESIDUAL'

We now have to make a more detailed study of the 'residual' source of variation. The form of the problem suggested that the variability that existed between the nine separate items was most probably due to differences between the students and between the teaching methods. Any other variability was due to random but unspecified influences. What does this mean in numerical terms? Let us look again at the original data.

	C/L	M	P.I.	Totals	Means
1	6	6	8	20	$\frac{20}{3}$
2	5	7	7	19	$\frac{19}{3}$
3	5	6	9	20	$\frac{20}{3}$
Totals	16	19	24	59	
Means	$\frac{16}{3}$	$\frac{19}{3}$	$\frac{24}{3}$	$\frac{59}{9}$	—

If each of the nine items were equal to the grand mean of $\frac{59}{9}$ there would be no variability and the variance would be zero.

We can eliminate the variability due to each student by calculating the difference between the grand mean and the mean for each student, and deducting these results from the appropriate members of the nine items:

Student 1 $\frac{20}{3} - \frac{59}{9} = +\frac{1}{9}$,

   2   $\frac{19}{3} - \frac{59}{9} = -\frac{2}{9}$,

   3   $\frac{20}{3} - \frac{59}{9} = +\frac{1}{9}$.

	C/L	M	P.I.
Student 1	$6-(+\frac{1}{9}) = \frac{53}{9}$	$6-(+\frac{1}{9}) = \frac{53}{9}$	$8-(+\frac{1}{9}) = \frac{71}{9}$,
2	$5-(-\frac{2}{9}) = \frac{47}{9}$	$7-(-\frac{2}{9}) = \frac{65}{9}$	$7-(-\frac{2}{9}) = \frac{65}{9}$,
3	$5-(+\frac{1}{9}) = \frac{44}{9}$	$6-(+\frac{1}{9}) = \frac{53}{9}$	$9-(+\frac{1}{9}) = \frac{80}{9}$.

Similarly we can eliminate the variability due to each teaching method by calculating the difference between the grand mean and the mean for each method and deducting these results from the appropriate members of the nine items in the adjusted table above:

C/L $\frac{16}{3} - \frac{59}{9} = -\frac{11}{9}$,

M $\frac{19}{3} - \frac{59}{9} = -\frac{2}{9}$,

P.I. $\frac{24}{3} - \frac{59}{9} = +\frac{13}{9}$,

	C/L	M	P.I.
Student 1	$\frac{53}{9}-(-\frac{11}{9}) = \frac{64}{9}$	$\frac{53}{9}-(-\frac{2}{9}) = \frac{55}{9}$	$\frac{71}{9}-(+\frac{13}{9}) = \frac{58}{9}$
2	$\frac{47}{9}-(-\frac{11}{9}) = \frac{58}{9}$	$\frac{65}{9}-(-\frac{2}{9}) = \frac{67}{9}$	$\frac{65}{9}-(+\frac{13}{9}) = \frac{52}{9}$
3	$\frac{44}{9}-(-\frac{11}{9}) = \frac{55}{9}$	$\frac{53}{9}-(-\frac{2}{9}) = \frac{55}{9}$	$\frac{80}{9}-(+\frac{13}{9}) = \frac{67}{9}$

Now let us calculate the total sum of squares for this table by using the standard method of aggregating the squares of the nine items and subtracting the correction factor. It can first be simplified by subtracting $\frac{58}{9}$ from each item:

	C/L	M	P.I.
Student 1	$+\frac{6}{9} = +\frac{2}{3}$	$-\frac{3}{9} = -\frac{1}{3}$	$0 = 0$
2	$0 = 0$	$+\frac{9}{9} = +1$	$-\frac{6}{9} = -\frac{2}{3}$
3	$-\frac{3}{9} = -\frac{1}{3}$	$-\frac{3}{9} = -\frac{1}{3}$	$+\frac{9}{9} = +1$
Totals	$+\frac{1}{3}$	$+\frac{1}{3}$	$+\frac{1}{3}$

Total of all items $= T = 1$.
Total number of items $= N = 9$
Correction factor $= -\dfrac{T^2}{N} = \frac{1}{9}$

Total sum of squares is $\frac{1}{9}(4+0+1+1+9+1+0+4+9) - \frac{1}{9} = \frac{28}{9}$.

This is the value of the residual sum of squares obtained by subtracting the 'between students' and 'between methods' sums of squares from the total sum of squares for the population; but we have now obtained it by direct measurement.

DESIGN OF EXPERIMENTS

So far we have presented analysis of variance as a technique for analysing data, without concerning ourselves with how the data was collected. In practice the statistician has a great deal to say about how data should be collected, and analysis of variance is in fact the main tool of a major branch of statistics known as the *Design of Experiments*. This subject is beyond the scope of this book but we can illustrate the type of approach.

We take as our example a hypothetical experiment to measure the effect on crop yield of three different fertilisers. We assume that we have a large plot of land on which to conduct our experiment. We divide this plot of land into nine sub-lots (nine because this particular design takes the number of treatments and squares this number; thus $9 = 3^2$).

The three experimental fertilisers are applied to the plot so that each fertiliser occurs once only in each row and column. The object of this arrangement is to neutralise the effect of any variations in fertility within the main plot.

Consider the following situation, using fertilisers A, B and C. The yield of bushels per acre is given in each square.

A	B	C
32	16	20
B	C	A
12	20	24
C	A	B
24	32	16

1. Divide each yield by 4:

A	B	C
8	4	5
B	C	A
3	5	6
C	A	B
6	8	4

2. Subtract 5 from each item:

A	B	C	Total
+3	−1	0	+2
B	C	A	
−2	0	+1	−1
C	A	B	
+1	+3	−1	+3

Total	+2	+2	0

	A	B	C
Treatment totals	+7	−4	+1

Total of all items = T =   +4

Total number of items = N = 9

Correction factor = $\dfrac{T^2}{N}$ = $\tfrac{16}{9}$.

1. Total sum of squares
$$= (9+4+1+1+0+9+0+1+1)-\tfrac{16}{9} = \tfrac{218}{9}.$$
Degrees of freedom $= 9-1 = 8$.

2. Between rows total sum of squares
$$= \tfrac{1}{3}(4+1+9)-\tfrac{16}{9} = \tfrac{26}{9}.$$
Degrees of freedom $= 3-1 = 2$.

3. Between columns total sum of squares
$$= \tfrac{1}{3}(4+4+0)-\tfrac{16}{9} = \tfrac{8}{9}.$$
Degrees of freedom $= 3-1 = 2$.

4. Between treatments total sum of squares
$$= \tfrac{1}{3}(49+16+1)-\tfrac{16}{9} = \tfrac{182}{9}.$$
Degrees of freedom $= 3-1 = 2$.

5. Residual total sum of squares
$$= \tfrac{1}{9}(218-26-8-182) = \tfrac{2}{9}.$$
Degrees of freedom $= (8-2-2-2) = 2$.

The table for these results is

Source of variation	Sums of squares	Degrees of freedom	Estimate of variance
Between rows	$\tfrac{26}{9}$	2	$\tfrac{13}{9}$
Between columns	$\tfrac{8}{9}$	2	$\tfrac{4}{9}$
Between treatments	$\tfrac{182}{9}$	2	$\tfrac{91}{9}$
Residual	$\tfrac{2}{9}$	2	$\tfrac{1}{9}$
Total	$\tfrac{218}{9}$	8	—

Using the estimate of variance for the residual as the denominator, and entering the F-ratio Table at the appropriate degrees of freedom:

	5% level	1% level
For between rows $F = 13$	19	99
For between columns $F = 4$	19	99
For between treatments $F = 91$	19	99

The result of the trials suggests that there is a significant difference between the yields derived from the different fertilisers, further trials can be conducted to confirm the findings.

The concepts of analysis of variance and the design of experiments can be developed far beyond the work of this chapter, but the reader should attempt nothing more until he has fully mastered the idea on which it is based and the examples set out for him.

The foregoing example was an agricultural one but the principles are perfectly general and find applications in industrial, medical and commercial experiments. Examples include:

Testing the effect of lubricating oils with different viscosities on fuel consumption in road transport.

Testing alternative processes for the production of chemicals.

Testing the effects of drugs and vaccines.

Testing different marketing strategies.

APPENDIX: ANALYSIS OF VARIANCE FOR ONE CRITERION OF CLASSIFICATION

Consider a random sample of N values of a variable x. Let these N values be subdivided into m classes according to some criterion of classification. Let the jth member of the ith class be written x_{ij} and let there be n_i members to the ith class. So that

$$\sum_{i=1}^{m} n_i = N.$$

The arrangement of the N values can be represented as:

Class 1 x_{11} x_{12} $x_{13}...x_{1j}...x_{1n_1}$

Class 2 x_{21} x_{22} $x_{23}...x_{2j}...x_{2n_2}$

Class 3 x_{31} x_{32} $x_{33}...x_{3j}...x_{3n_3}$

$\cdot$ $\cdot$ $\cdot$ $\cdot$ $\cdot$ $\cdot$ $\cdot$ $\cdot$ $\cdot$ $\cdot$ $\cdot$ $\cdot$

Class i x_{i1} x_{i2} $x_{i3}...x_{ij}...x_{in_i}$

$\cdot$ $\cdot$ $\cdot$ $\cdot$ $\cdot$ $\cdot$ $\cdot$ $\cdot$ $\cdot$ $\cdot$ $\cdot$ $\cdot$

Class m x_{m1} x_{m2} $x_{m3}...x_{mj}...x_{mn_m}$

When an analysis of variance experiment is being designed, it is usually arranged that there will be the same number of members in each class, but it is not essential.

Let the arithmetic mean of the ith class be $\bar{x}_i$.

Let the general mean of all values be $\bar{x}..$

Then

$$\sum_{j=1}^{n_i} (x_{ij} - \bar{x}_i.) = 0 \tag{1}$$

So that:

$$\sum_{j=1}^{n_i} (x_{ij} - \bar{x}..)^2$$

$$= \sum_{j=1}^{n_i} (x_{ij} - \bar{x}_i. + \bar{x}_i. - \bar{x}..)^2$$

$$= \sum_{j=1}^{n_i} (x_{ij} - \bar{x}_i.)^2 + 2(\bar{x}_i. - \bar{x}..) \sum_{j=1}^{n_i} (x_{ij} - \bar{x}_i.) + n_i(\bar{x}_i. - \bar{x}..)^2$$

$$= \sum_{j=1}^{n_i} (x_{ij} - \bar{x}_i.)^2 + n_i(\bar{x}_i. - \bar{x}..)^2$$

because of the result in equation (1).

Therefore:

$$\sum_{i=1}^{m} \sum_{j=1}^{n_i} (x_{ij} - \bar{x}..)^2 = \sum_{i=1}^{m} \sum_{j=1}^{n_i} (x_{ij} - \bar{x}_i.)^2 + \sum_{i=1}^{m} n_i(\bar{x}_i. - \bar{x}..)^2. \tag{2}$$

Now, consider equation (2).

The term on the left-hand side is the total sum of the squares of the deviations of each of the values from the general mean. It is therefore a measure of the total variation.

The second of the terms on the right-hand side of the equation is the variation which would have arisen if every value within each class had been equal to the present class mean. It is therefore the variation which would have occurred had there been no 'within class' variation. Hence it is the *variation between classes*, since it is summed over all classes.

The first of the terms on the right-hand side of the equation is the variation that exists *within the classes*. It is the residual variation, after the variation between classes has been extracted.

The result can be written briefly as follows:

Total variation = variation between classes + variation within classes.

If the classification criterion has no effect on the values of the variable, the values comprising each class can be considered as random samples of the population. The null hypothesis can therefore be set up that there is no variation between classes. If this is to stand, there must be no significant difference between the estimate of variance derived from the variation between classes and that derived from the variation within classes assuming that the population is Normal. The significance of the difference between the estimates can be assessed through the F-ratio test, using the 'within classes' estimate as the yardstick, since the variation of the values within the classes is not affected by any differences which may exist between the classes.

The following table can now be drawn up:

Analysis of Variance for one Criterion of Classification

Source of variation	Sum of squares	Degrees of freedom	Estimate of variance
Between classes	$\sum_{i=1}^{m} n_i(\bar{x}_i. - \bar{x}_{..})^2$	$m-1$	$\sum_{i=1}^{m} n_i(\bar{x}_i. - \bar{x}_{..})^2 \div m-1$.
Within classes	$\sum_{i=1}^{m} \sum_{j=1}^{n_i} (x_{ij} - \bar{x}_i.)^2$	$N-m$	$\sum_{i=1}^{m} \sum_{j=1}^{n_i} (x_{ij} - \bar{x}_i.)^2 \div N-m$.
Total	$\sum_{i=1}^{m} \sum_{j=1}^{n_i} (x_{ij} - \bar{x}_{..})^2$	$N-1$	—

SHORT-CUT METHOD

The 'short-cut' method worked through in the text is derived as follows:

Let $T = \sum_{i=1}^{m} \sum_{j=1}^{n_i} x_{ij}$ and let $T_i = \sum_{j=1}^{n_i} x_{ij}$

Then

$$\sum_{i=1}^{m} \sum_{j=1}^{n_i} (x_{ij} - \bar{x}_{..})^2 = \sum_{i=1}^{m} \sum_{j=1}^{n_i} x_{ij}^2 - N\bar{x}_{..}^2 = \sum_{i=1}^{m} \sum_{j=1}^{n_i} \bar{x}_{ij}^2 - \frac{T^2}{N}$$

since

$$N\bar{x}_{..}^2 = N \frac{\sum_{i=1}^{m} \sum_{j=1}^{n_i} x_{ij}}{N} \cdot \frac{\sum_{i=1}^{m} \sum_{i=1}^{n_i} x_{ij}}{N} = \frac{T^2}{N}.$$

Also

$$\sum_{i=1}^{m} \sum_{j=1}^{n_i} (x_{ij} - \bar{x}_i)^2 = \sum_{i=1}^{m} \left\{ \sum_{j=1}^{n_i} (x_{ij} - \bar{x}_i)^2 \right\}$$

$$= \sum_{i=1}^{m} \left(\sum_{j=1}^{n_i} x_{ij}^2 - \frac{T_i^2}{n_i} \right) = \sum_{i=1}^{m} \sum_{j=1}^{n_i} x_{ij}^2 - \sum_{i=1}^{m} \left(\frac{T_i^2}{n_i} \right).$$

Also

$$\sum_{i=1}^{m} n_i (\bar{x}_i - \bar{x})^2 = \sum_{i=1}^{m} \left(\frac{T_i^2}{n_i} \right) - \frac{T^2}{N}.$$

The three constituents of the analysis of variance equation are thus expressed in terms of x_{ij}, T, T_i, N and n_i.

Prove your prowess [1]

Because of the amount of numerical example in this chapter, only two problems are set for the reader. The first incorporates all the principles which have been set out in the chapter. The second illustrates a very useful application of Analysis of Variance to regression problems.

12.1. The Latin square printed below depicts the results obtained from an experiment similar to that resolved in the chapter. Three different varieties of seed A, B and C, are sown, in random fashion so that no variety occurs more than once in any row or column, over nine adjacent plots. The figures indicate units of the yields.

Conduct an analysis of variance, and calculate if there is a significant difference of yield as between varieties, rows or columns.

A	B	C
20	15	12
C	A	B
11	18	15
B	C	A
17	11	17

12.2.* For the data of Example 9.5, test the significance of the regression coefficients using Analysis of Variance.

[1] An asterisk against the exercise number indicates that the solution is given in detail.

REFERENCES

MORONEY, M. J. 1951. *Facts from Figures.* Harmondsworth, Penguin Books.

CROXTON, F. E. 1959. *Elementary Statistics with Applications in Medicine and the Biological Sciences.* New York, Dover.

SOLUTIONS

Chapter 1

1.1. (*a*).

Year	Solution (millions)	Year	Solution (millions)
1950	980·0	1960	1040·0
1951	1000·0	1961	1020·0 ⎫ Rounding to the
1952	990·0	1962	960·0 ⎭ nearest even digit
1953	990·0	1963	940·0
1954	1020·0	1964	930·0
		1965	870·0
1955	990·0		
1956	1030·0		
1957	1100·0		
1958	1090·0		
1959	1070·0		

1.1. (*b*). The reader can do this.

1.2.

Commodity	Rounded figures	Percentage error
Spirits	11 362 000	+0·003
Beer	1 143 000	+0·009
Fruit juice and table waters	223 000	−0·2
Cocoa preparations		
with sugar	548 000	+0·09
without sugar	265 000	+0·09
All other items	284 000	+0·15
Total		

1.3. Rearrange in following order (MEN):

	Frequency
20 to 29: 20. 25. 27. 25. 23. 20. 26. 28. 27. 21. 24. 25. 24.	13
30 to 39: 35. 31. 37. 31. 30. 34. 30. 39. 33. 35. 37.	11
40 to 39: 48. 41. 43. 47. 44. 46. 42. 45. 42. 42. 45. 40. 40. 44	14
50 to 59: 52. 57.	2
Total	40

Points to remember

1. The values of the constituent members of each class must be fairly equally distributed over the class.

2. The mean of the members of each class must approximate to the class midpoint.

The following frequency table then suggests itself. The totals of the variate values and their mean for each class are shown: these satisfy (as far as is possible with such few numbers) point 2 above. Point 1 is satisfied by inspection.

MEN

Age	Frequency	Total variate value	Mean variate value	Midpoint
20–24	6	132	22·0	22
25–29	7	183	26·1	27
30–34	6	189	31·5	32
35–39	5	183	36·6	37
40–44	9	378	42·0	42
45–49	5	231	46·2	47
50–59	2	109	54·5	54·5

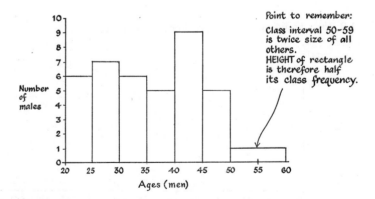

Point to remember:
Class interval 50–59 is twice size of all others.
HEIGHT of rectangle is therefore half its class frequency.

The reader can complete the questions for women's ages.

1.5

Angle on 'pie' chart (2) × 3·6

185·0	
40·6	
30·2	
26·4	
19·8	
58·0	
360·0	

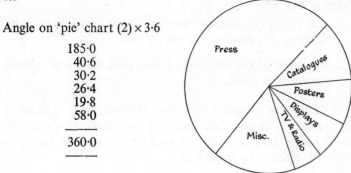

1.9. (a)

Amount of commission (£)	Cumulative frequency table
2·05–2·50	4
2·50–3·00	11
3·00–3·50	22
3·50–4·00	37
4·00–4·50	62
4·50–5·00	104
5·00–5·50	157
5·50–6·00	195
6·00–6·50	219
6·50–7·00	231
7·00–7·50	236
7·50–8·00	237

1.9. (b).   Ogive or cumulative frequency curve.

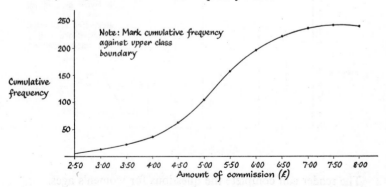

1.11

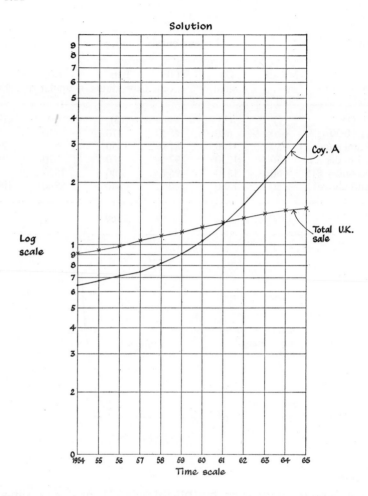

Notes.

1. Coy. A can be seen from its graph *shape* to be increasing its sales at an increasing rate.

2. 'Total U.K. sales' is represented by a graph line which is nearly straight. It is therefore increasing at a *constant* rate.

3. Scale is unimportant: all that matters is that the curves should be close together so they may be compared.

1.14   Lorenz curve

Expenditure (1)	No. of savers (2)	(2) Cumulative (3)	(3) Per cent (4)	Total expenditure (5)	(5) Cumulative (6)	(6) Per cent (7)
Under £0·50	620	620	45·5	210	210	8·6
£0·50 and under £1	480	1100	81·0	350	560	23·8
£1 and under £5	150	1250	92·0	360	920	39·0
£5 and under £15	60	1310	96·5	600	1520	65·0
£15 and above	50	1360	100	840	2360	100
	1360			2360		

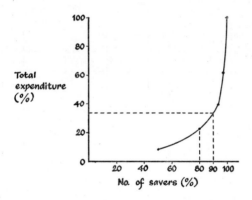

Notes.

1. 80% of all the savers contributed only 22% of all expenditure on the units.

2. The top 10% of all the savers contributed 64% of all expenditure on the units, i.e. the majority of savers are small savers.

1.16.   'Z' Chart

| | 1965 | | 1966 | | |
	(1)	(2)	(1)	(2)	(3)
January	27	27	30	30	541
February	38	65	35	65	538
March	40	105	49	114	547
April	42	147	49	163	554
May	55	202	50	213	549
June	50	252	52	265	551
July	65	317	50	315	536
August	55	372	40	355	521
September	35	407	32	387	518
October	42	449	70	457	546
November	51	500	61	518	556
December	38	538	42	560	560

Column (2) is Column (1) aggregated.
Column (3) is the sum of the twelve monthly totals (i.e. moving
annual total)

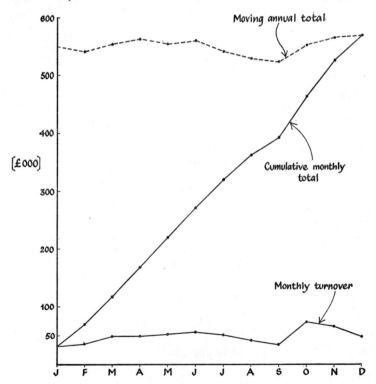

Chapter 2

2.1

Amount spent £ (1)	Number of salesmen f (2)	Class midpoint x (3)	Col. (3) less 11 d (4)	fd (5)	fd² (6)
<2	5	1	−10	−50	500
2<4	6	3	−8	−48	384
4<6	8	5	−6	−48	288
6<8	14	7	−4	−56	224
8<10	21	9	−2	−42	84
10<12	12	11	0	−244	0
12<14	9	13	+2	+18	36
14<16	9	15	+4	+36	144
16<18	6	17	+6	+36	216
18 and over	2	20	+9	+18	162
	92			+108 −244	2038
				−136	

Let $\bar{x}_a = 11.$

Then

$$\bar{x} = \bar{x}_a + \left(\frac{\Sigma fd}{n}\right)$$

$$= 11 + \left(-\frac{136}{92}\right) = 11 - 1\cdot5$$

i.e. $\bar{x} = £9\cdot50.$

$$s = \sqrt{\left\{\frac{\Sigma fd^2}{n} - \left(\frac{\Sigma fd}{n}\right)^2\right\}}$$

$$= \sqrt{\left\{\frac{2038}{92} - \left(-\frac{136}{92}\right)^2\right\}}$$

$$= \sqrt{(22\cdot2 - 2\cdot3)} = \sqrt{19\cdot9}$$

i.e. $s = £4\cdot40$ approx.

Note. The final class is open-ended, but its limit is unlikely to be in excess of 22. Therefore, assume this value for the upper class boundary, even though it deprives us of the facility of a working unit that an assumed value of 24 would yield. (Viz. class midpoint for '18<24' would be 21, for which d would be $21-11 = 10$.)

2.2.   (*a*) £21 350; (*b*) £6400.

2.4. (*a*) 3·95, (*b*) 0·46, (*c*) 0·116.

2.5.

Height inches (1)	Number of men (2)	Column (2) cumulative (3)
60<63	4	4
63<66	14	18
66<69	59	77
69<72	33	110
72<75	8	118
75<78	2	120
	120	

Median is value of 60th term. This is:

$$66+\left(\frac{42}{59}\times 3\right) = 66+2\cdot14 = 68\cdot14 \text{ inches.}$$

Lower quartile is value of 30th term. This is:

$$66+\left(\frac{12}{59}\times 3\right) = 66+0\cdot61 = 66\cdot61 \text{ inches.}$$

Upper quartile is value of 90th term. This is:

$$69+\left(\frac{13}{33}\times 3\right) = 69+1\cdot18 = 70\cdot18 \text{ inches.}$$

Mode is obtained from formula

$$L+\left(\frac{D_1}{D_1+D_2}\right)c$$

where L = Lower class boundary of modal class

   D_1 = Difference in frequency for modal class and class below it

   D_2 = Difference in frequency for modal class and class above it

   c = class interval size.

Therefore mode is

$$66 + \left(\frac{45}{45+26}\right) 3 = 66 + \left(\frac{45}{71}\right) 3 = 66 + 1\cdot9$$

$$= 67\cdot9.$$

2.6. (a) 9·68, (b) 9·75, (c) 10·07; 9·21, (d) 8·74; 9·1; 9·32.

2.7. 11·1(x); 7·7(y).

2.8. (a) 62·15, (b) 21·5.

2.9. Let v = annual percentage fall. Then:

end 1956 population was 546 000

end 1957 population was 546 000 − (546 000)v = 546 000(1 − v)

end 1958 population was 546 000(1 − v)^2

end 1959 population was 546 000(1 − v)^3 = 512 000

$$\therefore \qquad (1-v)^3 = \frac{512\ 000}{546\ 000},$$

$$v = 1 - \sqrt[3]{\frac{512}{546}}$$

$$= 1 - 0\cdot98 = 0\cdot02 \text{ or } 2\%,$$

i.e. there is an annual average loss of 2%.

2.10. 45%.

2.11. 39p per lb.

2.12. Plod; 55·5, 62·5, 59·5.

2.13. Mean of combination of all distributions $\bar{x}$ is

$$\tfrac{1}{600}[(100 \times 4\cdot1) + (200 \times 4\cdot0) + (100 \times 4\cdot2) + (200 \times 4\cdot1)].$$

$$\therefore \quad \bar{x} = \tfrac{1}{600}[410 + 800 + 420 + 820] = \tfrac{1}{600}[2450]$$

$$= 4\cdot08 = 4\cdot1 \text{ (to the nearest tenth).}$$

Let $\sigma^2 \equiv$ variance of combination of all distributions

$$\therefore 600\sigma^2 = 100(0\cdot1^2 + 0^2) + 200(0\cdot2^2 + 0\cdot1^2) + 100(0\cdot1^2 + 0\cdot1^2)$$

$$+ 200(0\cdot2^2 + 0^2)$$

(using formula

$$no^2 = n_1(\sigma_1^2 + d_1^2) + n_2(\sigma_2^2 + d_2^2) + n_3(\sigma_3^2 + d_3^2) + n_4(\sigma_4^2 + d_4^2))$$
$$600\,\sigma^2 = 1 + 10 + 2 + 8 = 21.$$
$$\sigma^2 = \tfrac{21}{600} = 0\cdot035.$$
$$\sigma = \sqrt{0\cdot035} = 0\cdot19.$$

Therefore the mean length of all 600 items is 4·1 in and the standard deviation of length is 0·19 in.

Chapter 3

3.1. (a) The 1964 weights, both actual and expressed as parts per thousand, the price relatives and the final index numbers are shown in the table below:

| | 1964 Weights | | | Price relatives | | |
	Actual	Parts per 000	1964	1965	1966	1967
Sheets	135 256	354	100	88·6	90·9	88·6
Plates	112 791	295	100	78·0	75·6	78·0
Heavy sections	65 312	171	100	103·1	103·1	103·1
Bright bars	17 160	45	100	97·0	90·9	84·8
Wire rods	51 765	135	100	88·6	80·0	82·9
Total	382 284	1000	Index: 100	88·3	87·0	87·0

(b) The 1967 weights and the quantity relatives are shown below:

| | 1967 Weights | | | Quantity relatives | | |
	Actual	Parts per 000	1964	1965	1966	1967
Sheets	106 977	368	112·1	103·2	103·5	100
Plates	75 104	259	117·2	110·1	106·6	100
Heavy sections	55 572	191	121·2	127·8	103·1	100
Bright bars	11 816	41	123·2	120·6	109·5	100
Wire rods	41 093	141	104·4	105·6	96·6	100
Total	290 562	1000	$\frac{1}{\text{Index}}$: 114·5	110·7	103·5	100

The values labelled '$\dfrac{1}{\text{Index}}$' are converted to the Paasche index numbers by taking their reciprocals (and multiplying by 100). We have

1964	1965	1966	1967
87·3	90·3	96·6	100

(Note that the Quantity Relatives are more properly expressed as ratios rather than percentages.)

In order to obtain the required index numbers, based on 1964 = 100. we must change the base of the set of values we have just calculated. We do this by multiplying through by the factor $\dfrac{100}{87·3}$, to obtain:

1964	1965	1966	1967
100	103·4	110·6	114·5

(c) The 1965 weights, the price relatives and the final index numbers are shown below:

	1965 Weights		Price relatives			
	Actual	Parts per 000	1964	1965	1966	1967
Sheets	110 370	338	112·8	100·0	102·6	100
Plates	82 688	253	128·1	100·0	96·9	100
Heavy sections	71 016	217	97·0	100·0	100·0	100
Bright bars	16 288	50	117·9	114·3	107·1	100
Wire rods	46 407	142	120·7	106·9	96·6	100
Total	326 769	1000	Index: 114·6	101·7	100·0	100

3.2. (a)

1964	1965	1966	1967
88·3	100	101·6	101·6

(b)

1964	1965	1966	1967
100	96·4	90·6	87·4

(c)

1964	1965	1966	1967
100	85·5	78·7	76·0

Note. The decline in value from 1964 to 1967 can be explained in terms of the price and quantity index numbers. From 3.1. (a) and 3.2. (b) the Laspeyre values for price and quantity indexes in 1967 are 87·0 and 87·4 respectively. The product of these values gives the value index i.e. 87·0 × 87·4 = 76·0. Thus the decline in the value of steel consumption was brought about by a joint decline in prices and quantities, of about the same magnitude.

3.4. Omitting heavy sections necessitates recalculating the remaining four weights so that they sum to 1000. These weights are then applied to the four sets of price relations (derived from the solution to 3.1. (a)) to obtain the required index numbers. We have:

	1964 Weights			Price relatives			
	Actual	Parts per 000		1964	1965	1966	1967
Sheets	135 256	427		100	88·6	90·9	88·6
Plates	112 791	356		100	78·0	75·6	78·0
Bright bars	17 160	54		100	97·0	90·9	84·8
Wire rods	51 765	163		100	88·6	80·0	82·9
Total	316 972	1000	Index:	100	85·3	83·7	83·7

3.5. (a) All items index = 117·0.

(b) All items except 'Services' = 116·8.

(c) Assuming that the question does mean a rise of 10% and not 10 percentage points, the 'services' index becomes 131·0 and the revised 'all items' index becomes 117·6, a rise of 0·6 percentage points, or 0·5%.

3.6. 129·8.

3.7. 1953   1956
     106·5   140·6

3.8. (a) 1966   1967
          139·9   144·0

(b) 1960   1961   1962   1963   1964
    91·4   92·0   96·0   95·1   96·5

3.9. We require the weighting pattern for each of the years 1964–1966 and the price relatives for 1964/65, 1965/66, 1966/67.

$I_{64,65}$		$I_{65,66}$		$I_{66,67}$	
1964 weights	Price relatives	1965 weights	Price relatives	1966 weights	Price relatives
354	88·6	338	102·6	378	97·5
295	78·0	253	96·9	258	103·2
171	103·1	217	100·0	191	100·0
45	97·0	50	93·8	46	93·3
135	88·6	142	90·3	127	103·6
1000	0·8833	1000	0·9841	1000	1·0003

$$I_{64,65} \times I_{65,66} \times I_{66,67} = 0·8833 \times 0·9841 \times 1·0003 = 0·8696$$

therefore $\qquad I_{64,67} = 87·0.$

3.10. $I_{64,67} = 87·0.$

3.11.

1956	1957	1958	1959	1960	1961	1962	1963	1964
103	104	105	106	108	109	109	110	112

3.12. The crude rate is calculated by simply expressing the total number of deaths as a rate per 1000 of the total population. Therefore

$$\text{crude rate} = \frac{1450}{110\ 000} \times 1000$$

$$= 13·2.$$

The Standardised Rate is obtained by first calculating the rate for each age-group separately (the *Age Specific Rates*) and then applying these rates to the standard population. The resulting standardised number of deaths is then expressed as a rate per thousand of the total standardised population.

Age group	Age specific death-rate	Standard population	Standardised deaths
0–9	16·67	221	3 684·07
10–24	3·40	298	1 013·20
25–44	6·19	285	1 764·15
45–64	20·82	149	3 102·18
65 and over	83·00	47	3 901·00
Total		1000	13 464·60

$$\text{Standardised rate} = \frac{13\ 464 \cdot 60}{1\ 000\ 000} \times 1000$$

$$= 13 \cdot 46.$$

Note. The column headed 'Standardised deaths' is not strictly necessary since the sum of the products of the two previous columns could be accumulated in the calculating machine.

3.13. *A* 0·50
    C 1·43
    B 1·52
    D 3·97
    F 4·28
    E 6·74

3.14. North: 15·2%; South: 14·7%.

The difference between the crude rates of return is due to the different product mix of the two factories. In fact neither met all three target rates of return on individual products, although North was closer to them than South. When profit margins are specified for individual products, as in the example, an overall target rate of return must assume a particular product-mix. Deviations from target can then be broken down into two main components: deviations from assumed product-mix and deviations from target rates of return on individual products.

Chapter 6

6.1. (a) $\frac{1}{216}$
    (b) $\frac{1}{36}$
    (c) $\frac{25}{216}$
    (d) $\frac{35}{36}$

6.2. 1 Head with 1 coin: $\frac{1}{2}$
    2 Heads with 2 coins: $\frac{1}{4}$
    3 Heads with 3 coins: $\frac{1}{8}$

    n Heads with n coins: $\dfrac{1}{2^n}$

6.3. 2520; 24

O

6.4. $\dfrac{\dbinom{8}{2}\dbinom{8}{2}}{\dbinom{52}{4}} = \frac{1}{345}$ approx.

6.5. $\frac{2}{13}$

6.6. $\frac{2}{3}$

6.7. 0·56

6.8. £31·87$\frac{1}{2}$

6.9. (a) $\dbinom{200}{10} 0\!\cdot\!05^{10}\, 0\!\cdot\!95^{190}$

    (b) $0\!\cdot\!95^{200}$

6.10. 0·39

Chapter 7

7.1(a). See 'Area under Normal curve' on facing page.

$\bar{x} = 4015 - 10(\frac{24}{100}) = 4015 - 2\!\cdot\!4 = 4012\!\cdot\!6$ or 4013 to nearest whole number.

$\sigma^2 = 10^2[\frac{140}{100} - (\frac{24}{100})^2] = 10^2[1\!\cdot\!4 - 0\!\cdot\!0576] = 134$; therefore $\sigma = 11\!\cdot\!5$.

Note that the total frequency for the *theoretical* distribution is 99; this arises because there is an accumulative rounding error. The sensible thing to do is to increase 33 to 34, thus bringing the total to 100.

(b) Proportion of population of packages with a weight of less than 4 ounces: area under curve for this proportion is

$$0\!\cdot\!5 - 0\!\cdot\!3708 = 0\!\cdot\!1292 = 13\%.$$

The student can fit a Normal frequency curve by the 'ordinate' method and compare results.

7.2. (a) $\dbinom{4}{1}\left(\dfrac{7}{8}\right)^1\left(\dfrac{1}{8}\right)^3 = 4 \cdot \frac{7}{8} \cdot \frac{1}{512} = \frac{7}{1024}$.

    (b) $\dbinom{4}{3}\left(\dfrac{7}{8}\right)^3\left(\dfrac{1}{8}\right)^1 = 4 \cdot \frac{343}{512} \cdot \frac{1}{8} = \frac{343}{1024}$.

7.1. (a)

('Area under Normal curve' method)

Weight (oz. × 1000) (1)	Midpoint (2)	Col. (2) less 4015÷10 (3)	Observed frequency (4)	Col. (3) × col. (4) (5)	Col. (3) × col. (5) (6)	True boundaries (7)	z (8)	Area under normal curve (table 7.6) (9)	Theoretical frequency ÷100 (10)	Theoretical frequency col. (10)×100 (11)
3980<3990	3985	-3	3	-9	27	3980	-2.87	0.4979	0.0207	2
3990<4000	3995	-2	6	-12	24	3990	-2.00	0.4772	0.1064	11
4000<4010	4005	-1	37	-37	37	4000	-1.13	0.3708	0.2682	27
4010<4020	4015	0	29	-58	0	4010	-0.26	0.1026	0.3317	33
4020<4030	4025	+1	16	+16	16	4020	+0.61	0.2291	0.2015	20
4030<4040	4035	+2	9	+18	36	4030	+1.48	0.4306	0.0600	6
			100	+34	140	4040	+2.35	0.4906		99
				-58						
				-24						

(c) 'At most 2' means the cumulative probability that 0, 1 and 2 will graduate.

$$= \binom{4}{0}\left(\frac{7}{8}\right)^0\left(\frac{1}{8}\right)^4 + \binom{4}{1}\left(\frac{7}{8}\right)^1\left(\frac{1}{8}\right)^3 + \binom{4}{2}\left(\frac{7}{8}\right)^2\left(\frac{1}{8}\right)^2$$

$$= \frac{1}{8^4} + 4 \cdot \frac{7}{8^4} + 6 \cdot \frac{49}{8^4}$$

$$= \frac{1}{8^4}[1 + 28 + 294] = \tfrac{323}{4096}.$$

(d) $\binom{4}{4}\left(\frac{7}{8}\right)^4\left(\frac{1}{8}\right)^0 = \tfrac{2401}{4096}.$

As a Check: Probability of [at most $2 + 3 + 4$] = Total probability

i.e. $\tfrac{323}{4096} + \tfrac{1372}{4096} + \tfrac{2401}{4096} = \tfrac{4096}{4096} = 1.$

7.3. (a) Mean $\mu = np = 1000 \times 0.001 = 1.$
Probability that 2 will suffer a reaction is

$$\frac{e^{-1} \cdot 1^2}{2!} = \frac{0.367\ 88}{2} = 0.183\ 94.$$

(b) Probability that at most 3 will suffer a reaction is

$$\frac{e^{-1} \cdot 1^0}{0!} + \frac{e^{-1} \cdot 1^1}{1!} + \frac{e^{-1} \cdot 1^2}{2!} + \frac{e^{-1} \cdot 1^3}{3!}$$

$$= e^{-1}(1 + 1 + \tfrac{1}{2} + \tfrac{1}{6}) = 0.367\ 88 \times 2.67 = 0.98.$$

(c) Probability that 4 will suffer a reaction is

$$\frac{e^{-1} \cdot 1^4}{4!} = 0.367\ 88 \cdot \tfrac{1}{24} = 0.0154.$$

7.5. (a) 174, (b) 293, (c) 174.

7.6. (a) 0.9802, (b) 0.02, (c) 0.0002.

7.7. (a) 0.057, (b) 0.061.

Chapter 8

8.1. Using the Normal approximation, we set 95% confidence limits to our observed mean as follows:

$$6 \cdot 25 \pm 1 \cdot 96 \cdot \frac{0 \cdot 05}{5},$$

i.e. $6 \cdot 25 \pm 0 \cdot 02$ approx.

Therefore we conclude that the machine has been set too high.

8.2. (a) $8 \cdot 67 \pm 1 \cdot 07$ oz.
(b) $8 \cdot 67 \pm 0 \cdot 89$ oz.

8.3. (a) $9 \cdot 56 \pm 1 \cdot 39$ oz.
(b) $9 \cdot 56 \pm 1 \cdot 16$ oz.

8.4. 15% approximately. *N.B.* From Table 8.1 it is only possible to conclude that the percentage lies between 10 and 20.

8.5. (a) The percentage of all garages selling below list price is given by:

$$\tfrac{170}{500} \times 100 = 34\%.$$

(b) From (a) we have $p = 0 \cdot 34$. Assuming that the sample of 500 garages represents less than 10% of the population of garages we can use the Normal approximation to estimate the 95% confidence limits. These are:

$$0 \cdot 34 \pm 1 \cdot 96 \sqrt{\left\{ \frac{(0 \cdot 34)(0 \cdot 66)}{500} \right\}}$$

i.e. $0 \cdot 34 \pm 0 \cdot 04.$

(c) The expression $1 \cdot 96 \sqrt{\left\{ \dfrac{(0 \cdot 34)(0 \cdot 66)}{n} \right\}}$ where n is sample size must equal $0 \cdot 02$. Hence

$$\frac{(0 \cdot 34)(0 \cdot 66)}{n} \cdot 1 \cdot 96^2 = 0 \cdot 0004$$

therefore

$$n = \frac{(3 \cdot 8416)(0 \cdot 2244)}{0 \cdot 0004}$$

$$= 2155.$$

8.6. (a) $0 \cdot 23 \pm 0 \cdot 03$
(b) $0 \cdot 63 \pm 0 \cdot 03$

8.7. $157,500 \pm 7,500$

8.8. $325 \pm 33{\cdot}4$ lb.

8.9. (a) $4{\cdot}25 > s > 2{\cdot}74$.
      (b) $4{\cdot}24 > s > 2{\cdot}67$.

8.10. Yes. $z = 2{\cdot}67$.

8.11. $z = 0{\cdot}60$. Not sig. diff.

8.12. $z = 11{\cdot}66$***

8.13. $z = 2{\cdot}44$, which is significant at the 5% level for both a one-tail and a two-tail test.

8.14. Yes. $z = 2{\cdot}01$*

8.15. $z = 14{\cdot}4$***

8.16. $t = 2{\cdot}23$*

8.17. $\chi^2 = 110{\cdot}77$***.

8.18. $\chi^2 = 7{\cdot}97$* i.e. we reject, at the 5% level, the hypothesis that the data come from a Normal distribution.

Chapter 9

9.1. $Y' = 5{\cdot}4 + 0{\cdot}59X$.

9.2. $0{\cdot}59 \pm 0{\cdot}26$.

9.3. $32 \pm 11{\cdot}2$ minutes.

9.4. $X' = -2{\cdot}1 \pm 1{\cdot}30Y$.

9.5. Let $Y =$ Crude Steel Production, $X_1 =$ Vehicles and $X_2 =$ Construction. For ease of computation we require sums of squares, products, etc. measured about their means i.e. the crude sums must be corrected before applying the formulae.

Crude Sums of Squares, etc.

$\Sigma Y = 230{\cdot}6$, $\Sigma X_1 = 1177$, $\Sigma X_2 = 1236$

$\bar{Y} = 23{\cdot}06$, $\bar{X}_1 = 117{\cdot}7$, $\bar{X}_2 = 123{\cdot}6$

$\Sigma Y^2 = 5374{\cdot}74$, $\Sigma X_1^2 = 139\ 409$, $\Sigma X_2^2 = 154\ 854$

$\Sigma X_1 Y = 27\ 338{\cdot}1$, $\Sigma X_2 Y = 28\ 759{\cdot}7$, $\Sigma X_1 X_2 = 146\ 611$

Corrected Sums of Squares, etc.

$$\Sigma y^2 = 57\cdot10, \quad \Sigma x_1^2 = 876, \quad \Sigma x_2^2 = 2084$$

$$\Sigma \bar{x}_1 y = 196\cdot5, \quad \Sigma x_2 y = 257\cdot5, \quad \Sigma x_1 x_2 = 1134$$

Then

$$b_1 = \frac{\Sigma x_1 y \Sigma x_2^2 - \Sigma x_2 y \Sigma x_1 x_2}{\Sigma x_1^2 \Sigma x_2^2 - (\Sigma x_1 x_2)^2}$$

therefore

$$b_1 = \frac{(196\cdot5)(2084) - (257\cdot5)(1134)}{(876)(2084) - (1134)^2}$$

$$= 0\cdot22$$

also

$$b_2 = 0\cdot01.$$

The regression line, still measured about the mean, is therefore:

$$y = b_1 x_1 + b_2 x_2$$

or, in original units:

$$Y' - \bar{Y} = b_1(X_1 - \bar{X}_1) + b_2(X_2 - \bar{X}_2)$$

therefore

$$Y' = \bar{Y} - b_1\bar{X}_1 - b_2\bar{X}_2 + b_1 X_1 + b_2 X_2$$

or

$$Y' = -4\cdot07 + 0\cdot22 X_1 + 0\cdot01 X_2.$$

Confidence Limits

We need s_{b_1} and s_{b_2}. As in the case of a single regressor variable we proceed by first computing the standard error of the estimate of Y from X. We have:

$$s_{Y \cdot X_1 X_2} = \sqrt{\left\{ \frac{\Sigma(Y - \bar{Y})^2 - b_1 \Sigma(X_1 - \bar{X})(Y - \bar{Y}) - b_2 \Sigma(X_2 - \bar{X})(Y - \bar{Y})}{n - 3} \right\}}$$

(compare this expression with that for $s_{Y \cdot X}$ on page 260.)

Since we already have the sums of squares, etc. measured about their means we can substitute directly into the expression above. Then

$$s_{Y \cdot X_1 X_2} = \sqrt{\left\{ \frac{57\cdot10 - (0\cdot22)(196\cdot5) - (0\cdot01)(257\cdot5)}{n - 3} \right\}}$$

$$= \sqrt{\frac{1\cdot6143}{7}}$$

$$= 0\cdot48$$

$$s_{b_1} = s_{Y.x_1x_2}\sqrt{\frac{\Sigma x_2^2}{\Sigma x_1^2 \Sigma x_2^2 - (\Sigma x_1 x_2)^2}}$$

$$= 0.48\sqrt{\frac{2084}{539\ 628}}$$

$$= (0.48)(0.06)$$

$$= 0.03.$$

$$s_{b_2} = s_{Y.x_1x_2}\sqrt{\frac{\Sigma x_1^2}{\Sigma x_1^2 \Sigma x_2^2 - (\Sigma x_1 x_2)^2}}$$

$$= 0.48\sqrt{\frac{876}{539\ 628}}$$

$$= (0.48)(0.05)$$

$$= 0.02.$$

$$t_{0.025} = 2.36 \text{ for 7 df.}$$

Therefore 95% confidence limits are given by

$$b_1 \pm 2.36 s_{b_1} \text{ and } b_2 \pm 2.36 s_{b_2}$$

i.e. 0.22 ± 0.07

and $0.01 \pm 0.05.$

Hence b_1 is significant but b_2 is not.

[We would now recalculate the regression equation, omitting X_2. In general it is not necessary to go right back to the original data although with only two regressor variables it is convenient to do so. With more than two regressor variables the computing routine which produces the coefficients and their confidence limits also provides a relatively quick method of adding or deleting one or more regressor variables.]

9.6. We have:

$$Y' = -4.07 + 0.22X_1 + 0.01X_2$$

i.e. $Y' = -4.07 + (0.22)(130) + (0.01)(150)$

therefore $Y' = 26.0$ million tons.

Confidence limits for this estimate are obtained from the standard error of the estimate Y'. We have:

$$s_{Y'} = s_{Y.x_1x_2} \sqrt{\left\{ 1 + \frac{1}{n} + \frac{\Sigma x_2^2(X_1 - \bar{X}_1)^2}{\Sigma x_1^2 \Sigma x_2^2 - (\Sigma x_1 x_2)^2} + \frac{\Sigma x_1^2(X_2 - \bar{X}_2)^2}{\Sigma x_1^2 \Sigma x_2^2 - (\Sigma x_1 x_2)^2} \right\}}$$

$$= (0{\cdot}48)(1{\cdot}68)$$

$$= 0{\cdot}81.$$

95% confidence limits are therefore given by

$$Y' \pm (2{\cdot}36)(0{\cdot}81)$$

i.e. $26{\cdot}0 \pm 1{\cdot}91.$

N.B. It will be obvious to the reader that not only are the calculations tedious but a large number of decimal places must be kept in order to arrive at accurate answers. The results given above are in general based on a larger number of decimal places than shown.

9.8. $\dfrac{406{\cdot}0}{526{\cdot}1} = 0{\cdot}7716$, hence $r = 0{\cdot}88$.

9.9. $b \cdot b' = (0{\cdot}59)(1{\cdot}30) = 0{\cdot}77$, hence $r = 0{\cdot}88$.

$\dfrac{1{\cdot}96}{\sqrt{(n-1)}} = 0{\cdot}65$, $\dfrac{2{\cdot}58}{\sqrt{(n-1)}} = 0{\cdot}86$, hence r is significant at the 1% level.

9.10. $0{\cdot}80$.

9.11. $\rho = 0{\cdot}16$. The data do not support the clerks' claim.

Chapter 10

10.1. Jan.: $m_t = F_1(t) = F_3(t) = F_6(t) = 18{\cdot}44$

Feb.: $m_t = F_1(t) = F_3(t) = F_6(t) = 19{\cdot}03$

Mar.: $m_t = F_1(t) = F_3(t) = F_6(t) = 18{\cdot}48$

10.2.

Jan.: $m_t = 18{\cdot}52$; $F_1(t) = 18{\cdot}67$; $F_3(t) = 18{\cdot}97$; $F_6(t) = 19{\cdot}42$

Feb.: $m_t = 19{\cdot}22$; $F_1(t) = 19{\cdot}42$; $F_3(t) = 19{\cdot}82$; $F_6(t) = 20{\cdot}42$

Mar.: $m_t = 18{\cdot}79$; $F_1(t) = 19{\cdot}05$; $F_3(t) = 19{\cdot}57$; $F_6(t) = 20{\cdot}35$

10.3.

Jan.: $m_t = 18.89$; $F_1(t) = 22.52$; $F_3(t) = 20.38$; $F_6(t) = 12.50$

Feb.: $m_t = 18.85$; $F_1(t) = 23.50$; $F_3(t) = 22.02$; $F_6(t) = 18.38$

Mar.: $m_t = 17.40$; $F_1(t) = 18.26$; $F_3(t) = 17.97$; $F_6(t) = 18.14$

Chapter 12

12.1. The yield from treatments is significantly different at the 5% level.

12.2. From Example 9.5 we already have the necessary sums of squares and cross-products. We form the following table:

	df	S.S.	M.S.	F
Regression on X_1 and X_2	2	45·81	22·90	14·22**
Regression on X_1 only	1	44·08	...	...
Increment from the addition of X_2	1	1·73	1·73	1·07
Residual	7	11·29	1·61	...
Total	9	57·10		

Ignore for the moment the second and third rows of the above table.

We have the Total S.S. and the Regression S.S. from Example 9.5. We enter these in the table with 9 and 2 df respectively. The Residual S.S. is obtained by difference as are the df. We compute F as the ratio of the Regression M.S. to the Residual M.S., obtaining a value which is significant at the 1% level.

We now test whether the inclusion of a second regressor variable has a significant effect on the regression sum of squares. In order to do this we first have to establish which of the two regressor variables has most effect, taken on its own. We do this by computing the regression sums of squares for X_1 and X_2 separately, i.e. as if we had fitted two simple regression lines.

We have:

$$\text{Regression S.S. for } X_1 = \frac{(\Sigma x_1 y)^2}{\Sigma x_1^2} = \frac{(196.5)^2}{876} = 44.08.$$

$$\text{Regression S.S. for } X_2 = \frac{(\Sigma x_2 y)^2}{\Sigma x_2^2} = \frac{(257 \cdot 5)^2}{2084} = 31 \cdot 82.$$

X_1 therefore accounts for the larger proportion of the total S.S. and the figure $44 \cdot 08$ is entered in the second row of the table above. The increment due to the addition of X_2 is obtained by subtracting $44 \cdot 08$ from the joint regression sum of squares, $45 \cdot 81$. The difference is entered in the third row of the table. Each of the new entries has 1 df, i.e. they show the joint regression df.

We now test the incremental mean square from the addition of X_2 against the residual mean square. This ratio is not significant, hence X_2 adds nothing to the simple regression on X_1. We reached the same conclusion in Example 9.5 by testing the significance of the regression coefficients and finding b_2 to be not significant.

INDEX